Conway Lloyd Morgan

Animal Biology

An Elementary Text-book

Conway Lloyd Morgan

Animal Biology
An Elementary Text-book

ISBN/EAN: 9783337216993

Printed in Europe, USA, Canada, Australia, Japan

Cover: Foto ©Andreas Hilbeck / pixelio.de

More available books at **www.hansebooks.com**

ANIMAL BIOLOGY

ANIMAL BIOLOGY

AN ELEMENTARY TEXT-BOOK

BY

C. LLOYD MORGAN

PROFESSOR OF ANIMAL BIOLOGY AND GEOLOGY IN UNIVERSITY COLLEGE
BRISTOL, AND LECTURER ON COMPARATIVE ANATOMY IN THE
BRISTOL MEDICAL SCHOOL

WITH ILLUSTRATIONS

RIVINGTONS

WATERLOO PLACE, LONDON

MDCCCLXXXIX

PREFACE TO SECOND EDITION.

THE call for a second edition of this volume has enabled me to revise the text, to substitute in several cases improved woodcuts, and to add a brief Classification of the Types, and a Glossary. C. LL. M.

UNIVERSITY COLLEGE, BRISTOL,
April 1889.

PREFACE

IN preparing this volume I have endeavoured to meet the requirements of those who intend to present themselves for the London Intermediate and Preliminary Scientific, for the Oxford and Cambridge Local, and for other examinations of like range based upon the type system.

Special attention has been paid to Embryology.

The larger portion of the work has been devoted to the Vertebrate types, because I have had in view, in the first place, the requirements of those who intend to follow a medical career, and who will, I believe, after working through such a course as is developed in this volume, be able to make a better start, and thereafter more rapid progress, in their study of human anatomy and physiology than would be otherwise possible; and, in the second place, because there are many who are led to take a deep interest in human anatomy and physiology, but who are of necessity able to make their study *practical* only through the dissection of the lower Vertebrates.

The illustrations have been engraved by Mr. George E. Lodge, after outline sketches of my own, chiefly from dissections or preparations made in the Biological Laboratory of the University College, Bristol. The dissections and preparations have in many cases been made in accordance with instructions given, in text or figures, by previous authors—especially by Professor Howes, in his excellent *Atlas of Biology*, by Professors Foster and Balfour in their *Elementary Embryology*, and by Professor T. J. Parker in his *Zootomy*. In the few cases in which my indebtedness is more direct I have placed the name of the author in brackets. With hardly any exceptions they are to be regarded, not as pictures, but as outline sketches, to serve as guides to practical work, and to be considered only in relation to the actual objects displayed by dissection. In a few cases they are to be looked upon as purely diagrammatic, to aid in the comprehension of the text. In my opinion, the more widely different such pure diagrams are from pictures the better.

My sincere thanks are due, and are here most gratefully tendered, to Professors W. K. Parker, F.R.S.; G. B. Howes, F.Z.S.; William Ramsay, Ph. D.; and Mr. G. Munro Smith, for valuable aid and advice.

April 1887.   C. Ll. M.

CONTENTS

PART II.

SOME INVERTEBRATE TYPES.

INTRODUCTORY.

CHAPTER I.

INTRODUCTORY.

WE live in a world that is teeming with life. The air, the surface of the land, the waters of ocean, river, and pond, swarm with living organisms, each more or less perfectly adapted to the conditions of its existence. Many problems arise with regard to this world of living things. What is their form and structure? How do they move and breathe and reproduce their kind? How, and on what do they feed; and how does the food minister to their growth or their activity? How are they distributed over the earth's surface? What is the method and manner of their origin? These, and other questions of like nature, arise in connection with the world of life. And the science that deals with these problems is the *Science of Biology*.

This science branches into two main departments, in accordance with the division of living things into Animals and Plants. The two branches start, indeed, from a common stem, for there are certain characteristics common to plant life and animal life, and the lowest plants are scarcely to be distinguished from the lowest animals. But the structure and functions of the vast majority of animals differ so markedly from the structure and functions of the vast majority of plants, and the problems of animal life differ so materially from the problems of plant life, that the *Science of Animal Biology*, which deals with the former, is justly entitled to a distinct position as a separate branch of study. It is with this branch that this volume deals.

The essence of science is organisation and exactness. Most

people have some knowledge of the animal world in which they live and of which they form a part. But this general information, useful and interesting as it is, lacks that organisation and exactness which is a distinguishing mark of science. To rise from general information to science, we must use the methods of science, which are : (1) observation and experiment; (2) inference and hypothesis ; (3) verification. The beginner will, however, wisely rest content with repeating the observations and experiments (not merely reading about them but *repeating them*), and conscientiously verifying the inferences, of his masters in science. This he must do if he is to learn science as science, and not as history. To learn about science is valuable. But to be taught science itself through the direct teachings of Nature is far more valuable. The student of science must learn his facts at first hand, and must regard books as guides to that object. Observation, experiment, and verification are to be regarded as primary duties by every student of science as such. They are therefore incumbent on the student of Animal Biology.

What is the nature of those living things on the scientific study of which we are now to enter? We are more or less acquainted with a considerable number of very different kinds of living animals, such as dogs, butterflies, worms, sea-anemones, star-fish, jelly-fish, and so forth. There are also simpler and more minute forms of life, with which, however, the student is presumably at present unfamiliar. Taking such animals as these, therefore, what is there about them to distinguish them on the one hand from not-living things, and on the other hand from plants? Let us endeavour to organise and make exact the general knowledge on these questions which we already possess.

A marked characteristic of life is *growth*. But we speak also of the growth of not-living things, of clouds, of the river in flood, of crystals from solution, of tapioca grains on boiling. What then distinguishes living growth? We cannot, perhaps, describe it better than by saying, (1) that it is an organic growth, that is a growth of the various organs of the living animal in due proportion; (2) that it is a growth, not merely by the addition of new material, but by the incorporation of that new material into the very substance of the old; and (3) that the material incorporated during growth differs from the material absorbed from without, which has thus undergone a chemical transformation within the animal. The growth of the organism is dependent upon the continued absorption of new material from without, and its transformation into the substance of the body.

But, after a while, the growth of the living animal ceases. It is, as we say, fully grown. Why, then, continue the process of *intus-susception*, as it is called, that is, the incorporation of new material from without? Because, if it be not continued, the animal wastes away and dies. And thus a new fact comes to light, that of constant waste, which must be made good by constant repair. So that we may say that a living animal is a centre of continual waste and repair, of nicely-balanced constructive and destructive processes.

Only so long as the constructive processes outbalance the destructive processes does growth continue. During the greater part of a healthy man's life, for example, the two processes, waste and repair, are in equilibrium. In old age waste slowly but surely gains the mastery; and, at death, it sets in unchecked by repair.

So far, then, a living animal is a centre of continual waste and repair, which may or may not be accompanied by growth. Let us now look at this growth a little more closely. There is something more than growth in the passage of the infant into the man. There is *development* as well. But take a more marked case. In spring and early summer there is plenty of frog-spawn in the ponds. A number of blackish specks of the size of mustard seeds are imbedded in a jelly-like mass. They are frogs' eggs. They seem unorganised. But watch them, and the organisation will gradually appear. The egg will be hatched, and give rise to a little fish-like creature (see Fig. 2, p. 3). This will gradually grow into a tadpole, with a powerful swimming tail. Legs will appear. The tail will shrink in size and be drawn into the body. The tadpole will have developed into a frog. So, too, in the case of the chick. We know that the egg we eat for breakfast, if it had been placed for three weeks or so under a hen, would have developed into a little chicken. But not only the frog and the chick, but the dog, the worm, the butterfly, the star-fish, are one and all developed from an egg which is at first just a little speck of living matter. So that we may say that living animals, during their growth, pass from a comparatively simple condition to a comparatively complex condition by a process of change which we call development.

Now it is clear that, since we have no knowledge of dead matter springing into existence as living matter, life on the earth would soon cease if there were not something more than growth, development, decay, and death. Since death is the heritage of living things, we have the necessity for *reproduction*. This process is essentially the detachment of a part of the parent organism, which part itself, in turn, develops, reproduces, decays, and dies. In the higher animals reproduction becomes possible when growth and development are ceasing. The excess of repair over waste is seen, not in growth, but in the periodic detachment of a portion of the organism to continue its kind.

Another fact must now be introduced since it is one that is eminently characteristic of living things. In any animal the line along which the series of changes (growth, development, etc.) takes place is not indeterminate, but is determined by inheritance. Every mammal, *e.g.*, begins life as a minute speck of an egg. There is often nothing about the egg to tell us to what particular animal it will give rise. And yet this is already quite determined by inheritance. So that we may say, summing up so far, that a living animal is a centre of continual waste and repair ; it undergoes a series of successive developmental changes constituting its life-history, the special nature of which is determined by inheritance ; it reproduces its kind by the detachment of a portion of its own substance.

And what is that substance ? The essential constituent of a living animal is *protein*, composed of carbon, hydrogen, oxygen, and nitrogen, with a little phosphorus and sulphur. This, with much water, forms the chief constituent of *protoplasm*. At present it is only known as the product of life. The protoplasm, moreover, is not distributed throughout the body in a continuous uninterrupted mass, but is disposed in separate individual particles called *cells*. If you scrape gently the inside of your lower lip with a pen-knife, you will remove some of the cells of that part of the body. Mounted in a little saliva, and examined under the microscope,[1] they will be seen (Fig. 24, ii.) to be flattened plates of irregular shape, sometimes curled up at the edges, and containing a rounded spot at the centre, called the *nucleus*. Of such nucleated cells (or their products), differing in form and appearance in different parts, the whole body is compounded.

Is there anything further to be added? It may be suggested that animals move about and are guided by feelings. Now, with regard to the feelings, it will be well to leave them on one side. Each of us knows a great deal about his own feelings, and very little of the feelings of his neighbours. Our knowledge of the feelings of animals is only arrived at by a complex process of inference; and with regard to the lowest animals we are completely ignorant whether they have feelings or not. Leaving the feelings on one side, therefore, we may notice that moving about is only a particular and conspicuous manifestation of that general activity or vital energy which is characteristic of the animal organism. A large proportion of this activity arises from the fact that the organism is eminently sensitive—using this term not as necessarily implying feeling, but in the same sense as a photographer would use the word when speaking of a sensitive plate. The animal is sensitive in its prompt and ready response and reaction to the stimulus of surrounding conditions, just as a sensitive plate promptly reacts under the

---

[1] The student must provide himself with a good working compound microscope. The instrument I am in the habit of recommending to my students in the University College, Bristol, is a stand on the Hartnach model, with a No. 2 or 3 Zeiss ocular, and Zeiss' objectives A and E, or a "1-inch" and "⅛th-inch" of English make. The former is spoken of in this book as *low power*, the latter as *high power*. Such an instrument may be obtained from any first-class instrument maker. The student is warned against purchasing low-priced second-hand instruments by inferior makers.

stimulus of light. But while a large proportion of the activity of the organism is due to its sensitiveness, a certain proportion, which increases as we rise in the scale of organisation, is spontaneous, for the animal contains the springs of action within itself. It will be well, therefore, for us to replace the particular statement, that animals move about, by the more general, comprehensive, and exact statement, that they exhibit certain activities, prompted from within or called forth by surrounding conditions. So that we may now finally state the characteristics of living animals as follows :—A living animal consists of an aggregate of protoplasmic cells, together with certain cellproducts; it is a centre of waste and repair; it undergoes a series of developmental changes constituting its life-history, the special nature of which is determined by inheritance; it exhibits certain activities by which it maintains its relation to surrounding conditions; it reproduces its kind by the detachment of a portion of its own substance.

This definition or description applies to all the higher animals; but we shall learn hereafter that there are many of the lower animals that are not cell-aggregates, but are each of them constituted by a single cell. These unicellular animals are called *protozoa* or *cytozoa*, while the multicellular organisms, in which, as we shall see, different cells have different modes of activity, are called *metazoa* or *histozoa*.

Let us now consider the essential points in which one of these higher animals—the animals with which we are ordinarily acquainted—differs from one of the higher plants—the plants we see around us. But first we may note the points of resemblance. They run almost through the whole description given above. Both animal and plant are cellular and protoplasmic; both grow by intus-susception; both are centres of waste and repair; of both there is a developmental life-history; in both is reproduction similar in principle. The activities of the animal, however, differ widely from those of the plant. But the main difference is in the nature of the food, and the manner of its intus-susception. Plants can build up protoplasmic matter out of such inorganic materials as contain the requisite elements.

Animals cannot do this. They require ready-made protoplasm in the form of vegetable or animal matter. Plants raise the inorganic into the organic; animals then take up the process and often carry it forward to more complex products. But no animal can raise the inorganic into the sphere of the organic. That is the function of the plant. Plants alone can manufacture protoplasm. Animals require the manufactured article.

But the manufactured article is not directly incorporated as such into the substance of the consumer. The protoplasm of sheep cannot become the protoplasm of man without being largely taken to pieces, chemically, and then put together again. Hence the necessity for digestive and assimilative organs to enable the animal to do this. Even the protoplasm of the mother's milk must be digested before it can be assimilated by the young. In plants anything of the kind is exceptional and subsidiary. They take the elements of protoplasm from the air that bathes their leaves, and from the water that bathes their roots. Hence the branching and spreading form of that part which is exposed to the air, and the far-reaching ramifications of that part which is implanted in the earth. We may sum up this distinction by saying that animals differ from plants in that they require protoplasmic food-stuff which must undergo a more or less complex process of digestion, within or without their bodies, before it can be assimilated.

Out of this main distinction there flows a secondary distinction of some importance. Plants manufacture organic tissues out of such inorganic raw materials as contain the requisite elements. One of these raw materials is the carbonic acid gas of the air, from which the green plant abstracts the carbon, returning the oxygen to the air as a by-product for which it has no use. But carbonic acid gas is one of the chief products of that continued waste which is entailed by the ceaseless activity of the animal. And to enable that waste—which, as we have seen, is essential to animal life—to continue, the animal requires a more or less liberal supply of oxygen. So that animals and the higher green plants perform opposite and complementary functions in the economy of nature. By the plant carbonic acid gas is decomposed, and oxygen set free into the air. In the animal the carbonic acid gas is recomposed and breathed forth into the air.

The higher animals thus differ from the higher plants (1) in that they recompose the carbonic acid gas which the plants decompose; (2) in the nature and amount of their activities; and especially (3) in requiring protoplasmic food-stuff, which must undergo a process of digestion before it can be assimilated.

Having now gained some idea of the nature of those living things which it will be our task to study, let us proceed to consider what is the nature of the problems that are likely to be suggested by the study of animal organisation and animal life. And that we may not fall into generalities beyond the reach of the student, let us consider the special case of such an animal as the common frog. In the first place, the fully-grown frog has a tolerably constant external form and appearance, by which it may always be recognised and distinguished from other organisms, such as the toad, or the edible frog of the Continent. Dissection after death further shows that it has a definite and tolerably constant internal structure. A number of organs such as heart, liver, stomach, kidneys, are found within the body, and these have a constant form and constant relations to the nerves and blood-vessels which ramify throughout the body. And minute examination with the microscope further shows that there is a definite and constant minute structure. The organs are seen to be built up of cells or cell-products aggregated in special ways into what are called tissues. So that we have to consider: (1) the general form and structure of the organism; (2) the special form and structure of the organs; and (3) the minute form and structure of the tissues.

. But the frog does not stand alone among organisms. Hence it is necessary to compare its general structure, the structure of its organs, and the structure of its tissues, with the general, organic, and minute structure of other organisms, with the object of ascertaining what are the points of resemblance and the points of difference. By this means the range and importance of the structural problems is enormously increased.

Now all those problems which deal with **form** and **structure** are called morphological problems ; and *Morphology*, which deals with these problems, forms a well-marked department of Animal Biology. From what has just been **said**, moreover, it follows that this department comprises three divisions :

1. The morphology of organisms (zoology).
2. The morphology of organs (anatomy).
3. The morphology of tissues (histology).

But the form and structure of the frog does not remain the same in all periods of its existence. Beginning life as an egg, it is hatched as a tadpole, and only reaches the full stature of froghood after several metamorphoses. There is, therefore, a *Development* of the organism as a whole, of the organs within the organism, and of the tissues of which the organs are composed. Hence there arises a second and most important set of morphological problems of development, which may be divided into

1. The development of organisms $\begin{cases} a. \text{ of the individual (ontogeny).} \\ b. \text{ of the race (phylogeny).} \end{cases}$
2. The development of organs (organology) $\Big\}$ embryology.
3. The development of tissues (histogenesis) $\Big\}$

So far we have only considered the problems that arise out of a study of form or structure. We have regarded the frog merely as a piece of mechanism. But every structure has its special office. Every organ in the frog's body has its particular work to do, its *function* as it is called. The muscles contract or shorten, and so produce motion, the heart beats and ministers to the circulation of the blood, the lungs are for respiration, the eye for seeing, the glands for secreting, and so forth. In these various organs, moreover, special tissues perform the special part of the work, and within these tissues special cells are developed. Finally, all these functions are to be made subservient to the general good of an organism which is set in the midst of more or less complex surrounding conditions. And then the frog himself, as one organism among many, has to perform his special function in the economy of Nature. Hence arises a new set of problems connected with functions and the chemical or *metabolic* changes which accompany the exercise of function. These fall within the department of *Physiology*. And just as there is a morphology of organisms, of organs, and of tissues, so too is there—

1. The physiology of organisms.
2. The physiology of organs.
3. The physiology of tissues and tissue cells.

Other biological problems fall under the head of *Distribution*. Frogs do not occur all the world over. They have a definite distribution in space. Nor do they occur in all geological strata. They have a definite distribution in time. Hence the problems connected with the distribution of animals fall naturally under two heads:

1. The distribution in space (chorology).
2. The distribution in time (chronology).

Finally, there is a fourth department of Animal Biology, which deals with the *causes* of those *facts* which are studied in the other three departments. This is of recent growth. To the questions, What are the structures and functions of tissues, organs, and organisms? What is the series of developmental changes undergone? What is the distribution in space and time? is added a further question, How has all this been brought about? And thus to the three departments of Biology which deal with the facts of structure, the facts of function, and the facts of distribution, is added a fourth, called *Ætiology*, which deals with the causes to which these facts are due. And this fourth department is co-extensive with the other three, which may be grouped together under the head of Descriptive Biology. Descriptive Biology thus deals with facts, and Ætiology with causes. And since no fact can be said to be understood until we have some knowledge of its manner of causation, it is clear that Ætiology is necessary to and supplementary to the otherwise incomplete science of Descriptive Biology. Descriptive Biology without Ætiology lacks life: Ætiology not founded on

Descriptive Biology is baseless speculation. Both must grow together and minister to each other's wants.

It only remains, in this introductory chapter, to indicate the aim and scope of this volume.

The number of animals—each a centre of so many biological problems—is well-nigh countless. So is the number of houses in England. And yet, on entering the house of an acquaintance whom we have never before visited, we know at once what we shall find. The hall and passages, the dining-room, drawing-room, library; kitchen and offices below; bedrooms above—we already more than half know them all. And why ? Because we already know the type of house that belongs to a certain grade of society. If we visit a lord or a labourer, our expectations are different. We know that some men have to live in hovels, in which one room has to suffice for all the needs of the family. We know that at the other end of the scale of social life there is a separate room for every function of that more complex life. So is it also with animals. There is a lowly type of animal, where a single cell constitutes the whole house, and all the functions of life have to be performed in and through that single cell. There are others composed of many cells, in which differentiation of structure and specialisation of function have been carried far. But a special type of structure is characteristic of each grade of animal life, just as a special type of house characterises each grade of social life. And if we know the type which marks that special grade, we more than half know what will be the state of matters with regard to differentiation of structure and specialisation of function in any individual case. If we know, for example, the typical structure of a mammal, a bird, and an amphibian, we know pretty well what to expect in any other mammal, bird, and amphibian, and are some way on our road towards a knowledge of the structure of a fish or a reptile. And so with regard to other grades of animal life.

It is the object of this volume to assist the student in acquiring such a knowledge of some of the more important facts of the morphology and physiology of certain typical animals as may form a base-line sufficiently accurate and extensive to enable him, by further study and research, to carry on his survey of the animal kingdom. Space will not permit of the study of the facts of distribution. And with regard to Ætiology, the aim will be rather to pave the way for a study of causes by an accurate presentation of facts than to deal at any length and **more** than incidentally with the theory of Evolution or the Doctrine of Descent.

PART I.

VERTEBRATE ANATOMY AND PHYSIOLOGY

AS EXEMPLIFIED BY

THE FROG, THE PIGEON, AND THE RABBIT,

with occasional Reference to other Types.

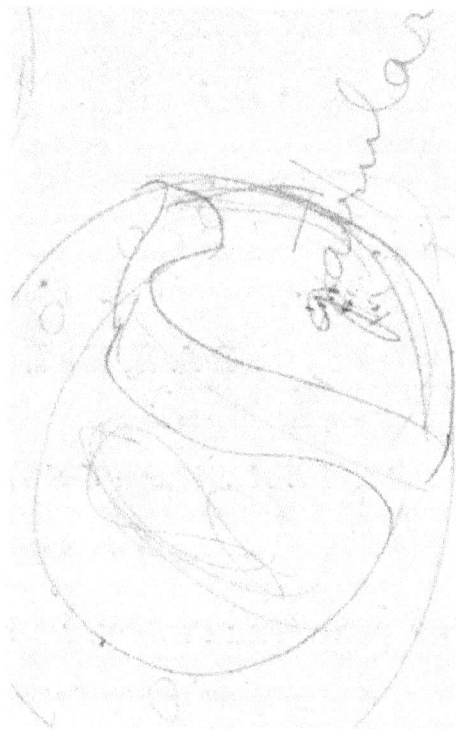

CHAPTER II.

NATURAL HISTORY AND EXTERNAL CHARACTERS.

1. **The Frog.**—The common frog, *Rana temporaria*, is tolerably abundant in summer in damp places and by the side of ponds. In winter it is not readily to be found, for at that season of the year frogs hibernate, often in groups, buried in the mud and under water. The warmth of spring rouses them from their torpor, and they then congregate and pair with much sound of croaking. The female lays a number of eggs, about one-tenth of an inch in diameter, each of which is surrounded by a thin layer of albumen. And as they are laid the male pours upon them a fertilising fluid. If a little of this fluid, or the water into which it is shed, be examined under the microscope (high power), a great number of minute active bodies of delicate tapering form will be seen. They are the *spermatozoa*, the essential elements in fertilisation. After the eggs have been fertilised by the entrance into each egg of a spermatozoon, the albuminous coating of the yolk swells to many times its original thickness by the absorption of water; so that the frog-spawn then has the appearance of a white gelatinous mass, made up of largish jelly-like spheres, in the midst of each of which is the dark ovum, which is seen, on closer inspection, to have a darker and a lighter hemisphere.

In a few hours after fertilisation a groove forms on the darker hemisphere, gradually extends round the egg, and, becoming deeper and deeper until it reaches the centre, cleaves the ovum into two hemispheres. Let us note clearly what has taken place here. The ovum, to begin with, is a single cell, and within it is a specialised portion which would seem to be of special import-

1

ance in cell life, the *nucleus.* The first thing that happens in development is the division of the original nucleated cell into two nucleated cells, called *blastomeres.*

Once begun the cleavage rapidly continues. The two cells become four, the four become eight, the eight become sixteen (Fig. 1, ii. iii. iv.), and so on, until by combined vertical and horizontal cleavage the primitive nucleated cell of the ovum has become split up into a great number of much smaller blastomeres, each of which is, however, a nucleated cell. The cells in the darker hemisphere are smaller than the cells in the lighter hemisphere; but after a while the darker portion begins to encroach upon the lighter, and this goes on until the whole surface has become dark, except a little patch, which finally becomes only a small depression, called the *blastopore* (Fig. 1, vi. *bl.*).

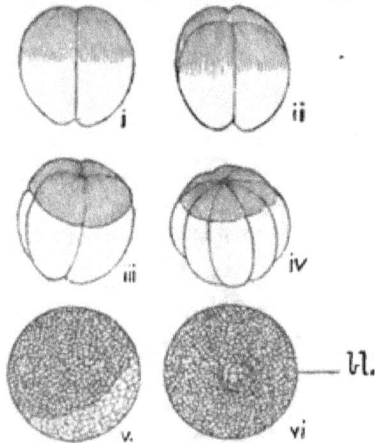

FIG. 1.—CLEAVAGE IN FROG'S OVUM.

i.-iv. Stages with 2, 4, 8, and 16 blastomeres. v. A later stage when the smaller blastomeres cover half the ovum. vi. A still later stage when the smaller blastomeres have enveloped the whole ovum, except at one spot, the blastopore, *bl.*

The first indication of the future frog is a broad shallow groove, the edges of which soon rise up to form ridges or folds. It is known as the neural groove, and the ridges as neural folds (Fig. 2, i. *ng.* and *nf.*). As development proceeds the folds rise up further, and bending over meet along the middle line, so as to convert the neural groove into a neural canal. The neural tube thus formed will give rise to the brain and spinal cord of the future frog.

The body now begins to elongate, becoming at first oval, but soon showing unmistakeable signs of head and tail. The mouth is indicated by a faint depression, behind and on either side of which are two well-marked suckers (*s.*), tending to run into each other posteriorly in the middle line. In front of the mouth depression (*stomodæum*) is a fold of skin (ii. *fr.*), at the upper

angles of which are the nasal pits (*na.*).   The position of
eye (*opt.*) and ear (*au.*) are marked out.   At the sides of the head
are four bar-like elevations, indicating the position of so-called
visceral arches.   The first (*md.*) is called the mandibular arch ;

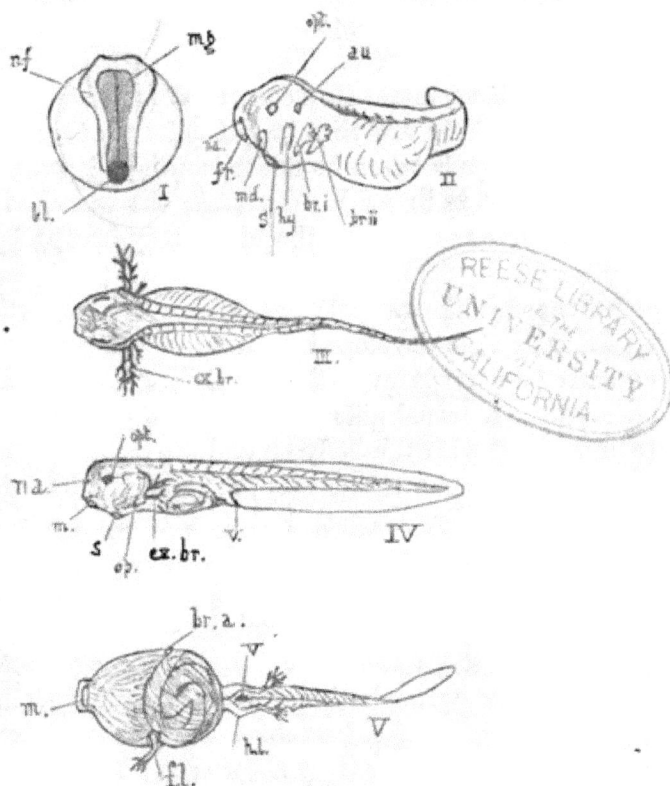

FIG. 2.—METAMORPHOSIS OF FROG.

i.-v. Stages in development of tadpole. *au.* Auditory organ. *bl.* Blastopore.
br. i.-ii. First and second branchial arches. *br. a.* Branchial aperture.
ex. br. External branchiæ. *f. l.* Fore-limb. *fr.* Frontal process. *h. l.* Hind-
limb. *hy.* Hyoidean arch. *m.* Mouth. *n. f.* Neural fold. *n. g.* Neural groove.
md. Mandibular arch. *na.* Nasal pit. *op.* Operculum. *opt.* Optic pit. *s.*
Suckers. *v.* Vent.

the second (*hy.*) the hyoidean ; the other two are branchial
arches, and bear the rudiments of external gills, or branchiæ.
Posteriorly, beneath the root of the tail, is a depression for the
vent (*proctodæum, r.*).

The embryo is now getting ready to leave the egg within which it lies, with its tail somewhat curved to one side (ii.). At last it becomes more active, breaks through the jelly-like mass and is hatched. It soon attaches itself to plants, or to the outside of the remaining gelatinous material, by means of the pair of suckers (*s.*) near the mouth, and somewhat later enters on a free-swimming existence. There are now (iii. *ex. br.*) three pairs of external branchiæ. In front of the first, and behind the first and second, there are slits or gill-clefts passing inwards to the throat. Water is taken in at the mouth and passes out through the clefts just as in a fish. Through the branched filaments of the gills the circulation of the blood may be watched under the microscope.

As the fish-like tadpole enlarges, a membranous fold (*operculum*, iv., *op.*) is developed in front of the gills, and gradually extends backwards over them. Before long it has completely covered the external gills on the right side, and closely adheres to and unites with the body behind the clefts. Thus the exit of water on the right side is stopped. On the left side, however, the fold does not altogether unite with the body, but leaves a branchial aperture (v., *br. a.*) which remains until near the end of tadpole life. Beneath the opercular membrane on each side there is a branchial chamber, and the two chambers communicate below. But the water which passes through both chambers, makes its exit through the single branchial aperture. During the formation of the branchial chambers the external gills *atrophy*, that is, disappear by absorption; but to take their place internal gills are developed on the inner sides of the branchial clefts.

Meanwhile the tadpole has become an active and vigorous feeder, mostly vegetarian, but by no means despising animal food. The mouth (v., *m.*) lies at the bottom of a somewhat protrusible cup, with a circular lip covered over with horny asperities, while within the cup the mouth is armed with two great horny crescentic jaws, with which it crops the water weeds, or browses on the flesh of some dead comrade.

The first sign of limbs is the appearance of two little rudiments of legs near the root of the tail. They are at first enclosed in the skin of the tail, and acquire a considerable size before any sign of fore-legs appears, for the fore-limbs lie hidden beneath the operculum, so as to be invisible without dissection. In Fig. 2, v., the right foreleg (*f. l.*) has been pulled out through a hole cut in the opercular membrane; the left remains invisible.

Thus by acquiring legs the organism passes from a fish-like to a truly amphibian condition. It is still, however, a tailed amphibian, like the newt or triton of our ponds. But as the legs increase in size, the tail shortens and begins to atrophy; and at last, by a final *ecdysis*, or throwing off of skin, the opercular membrane is got rid of, and the fore-legs are set free; the horny jaws are lost, and the mouth loses its suctorial form; the eyes, hitherto covered by skin, become freely exposed; the gills atrophy, and the gill-slits close; and the little frog breathes entirely (he has for some time breathed partially) by means of lungs. The short and stumpy rudiment of a tail gradually disappears, and the long series of changes is complete.

The student should be careful to verify the facts for himself. The frog-spawn is readily obtainable in the spring; and the tadpoles are easily reared in an aquarium.

Such a series of changes as is undergone by the frog is called *metamorphosis*, which essentially consists in the reduction and atrophy of provisional embryonic organs, and the appearance of adult organs in their place, the series of changes taking place during the free life of the organism. It is well to restrict the word metamorphosis to this use. If the changes take place before birth or before hatching, the word *transformation* should be used for the prenatal embryonic changes.

The time occupied in the hatching and metamorphosis of the frog varies with the temperature. At about 15° C. the eggs are hatched in ten days, and the metamorphosis is complete in about seventy-three days; but at about 10° C. the eggs are not hatched for twenty-one days, and metamorphosis is not complete until about the 235th day.

Even when metamorphosis is complete the frog still continues to develop,—that is, there is a change in relative dimensions of parts. There is, for example, a well-marked hump in the back of a frog. If now we take measurements from this point to the tip of the snout and the end of the back in an old frog, we shall find that the length of the body anterior to this point is not quite twice the length behind it; whereas in a minute frog, just emerged from its tadpole state, the length in front is more than three times as great as that behind. So that when metamorphosis is over there is still development.

Let us now note some of the more important points in the external characters of the fully-grown frog.

The body is oval, without neck or tail. The head is large, and within it can be felt a smooth bony mass, the *skull*. There is a hump in the back, anterior to which the separate *vertebræ* of the back-bone can be felt. Behind this hump there is a smooth rod of bone, the *urostyle*. The skull, vertebræ, and urostyle constitute the *axial skeleton*. In the region of the chest the breast-bone (*sternum*) can be felt; but there are no ribs.

There are four legs, within which the bony supports may be felt. The fore-legs are small, and are jointed to the *shoulder-girdle*: the hind-legs large and powerful, and attached to the *hip-girdle*. The bones of the limbs and the supporting arches or girdles constitute the *appendicular skeleton*. In the fore-limb are the following parts: *brachium*, or arm; *antibrachium*, or fore-arm; and *manus*, or hand. The corresponding parts of the hind-limb are: *femur*, or thigh; *crus*, or shank; and *pes*, or foot. These corresponding parts are said to be *homologous*. The manus has four digits, and is not webbed. The pes has five long digits, and is webbed. Thickened pads (*callosities*) are developed beneath the joints of all the digits, there being an extra one just at the base of the innermost digit of the pes, where there is apparently the rudiment of a sixth toe. The part of the manus answering to our wrist is short and small; but the part of the pes answering to our ankle is very much elongated—so much so as to give the appearance of an extra division of the hind limb between the shank and the foot.

The normal position of the limbs of the frog (Fig. 3, A) should

be compared with the primitive vertebrate position which is
well seen in the newt as it lies in a
sprawling attitude at the surface of
the water (B). In the fore-limb the
chief change is the bending forwards
of the antibrachium and inwards of the
manus, so that the little finger is anterior. In the hind-limb the femur
is swung forwards, and the whole
limb bent upon itself like an Z; so
that the great toe, which in the
newt is anterior in position, is in the frog, interior.

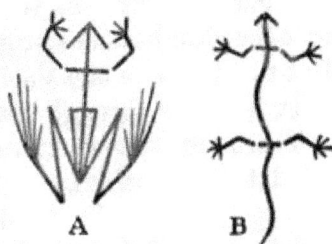

FIG. 3.—POSITION OF LIMBS IN
FROG (A) AND NEWT (B).

The integument of the frog is smooth and moist, and is
devoid of scales, feathers, hairs, or any form of *exoskeleton.* On
the dorsal aspect (the back) the colour is yellowish, or reddish
brown, with dark brown or greenish spots; on the ventral
aspect (the belly) it is pale yellow, with fewer spots. But
the colour varies a good deal in different frogs; and it also
varies in the same individual. In bright light the colour becomes brighter: in the dark it becomes duller. If a frog be
kept for some time in a dark cupboard, and then brought out
into a bright light, its skin will be found to be dull and pale;
but soon it will become much brighter, and more diversified
with spots.

Examination of a small transparent piece of the skin under
the microscope (low power) shows that the colour is due to a
great number of minute specks of pigmented material, called
pigment cells, or *chromatophores* (Fig. 24, xii., p. 66). These are
of various colours, white, light-yellow, orange, red, brown, and
black; but the black will probably be the most conspicuous.
Some of them will be rounded or oval; others star-shaped, or
arranged like a piece of network, the fibres of which are of unequal and inconstant thickness. In some places one colour, and
in others another colour, predominates.

It has been ascertained that the cells change their form in
accordance with the brightness of the light that falls upon the
eye of the frog. Darkness stimulates the cells to activity, and

causes them to contract, so that the skin becomes paler, duller, and more uniform in tint. Light causes their relaxation, so that the skin becomes brighter and more diversified with spots. But the darkness and the light do not seem to act directly on the cells, but only through their effect on the eye of the living frog.

There are four apertures in the frog's body, two median and two lateral or paired. In front is the large mouth with a broad long gape extending back to behind the eyes. On opening widely the mouth, two slit-like apertures are seen at the back, one ventral, clean cut, and longitudinal, the *glottis*, leading to the lungs : the other dorsal, irregular, and transverse, the opening of the *œsophagus*, or gullet. At the hinder end of the body is the vent or *cloacal opening*, whence issue the fæces, the excretion of the kidneys, and the products of the genital organs.

The lateral apertures are the external nares, one on either side of the snout. They may be seen to open and shut as the frog breathes by the alternate rise and fall of the floor of the throat. A bristle passed into one of the nares passes downwards and slightly backwards, and emerges in the mouth, by the posterior nares, tolerably far forwards in the roof. The nares are the only lateral apertures. There are no external ear openings. The membrane of the drum of the ear is close to the surface and only covered over with skin. It is readily visible in the midst of the characteristic triangular brown patch above and behind the angle of the mouth. If a hole be pricked in this *tympanic membrane*, and a bristle passed in, it will emerge at the side of the mouth, through one of the two large *Eustachian recesses.*

The *tongue* is large and white. It is fixed by its anterior end, the free posterior end being bifurcated. It is darted or slung out with great rapidity. The prey (insect, worm, or slug, for the adult frog feeds exclusively on animals) adheres to its sticky surface, and is drawn back into the mouth. Its escape is prevented by the *teeth*, which may be readily felt in the upper jaw and in the roof of the mouth (*vomerine teeth*). There are no teeth in the lower jaw.

The *eyes* are large and prominent, and can be withdrawn into sockets, and thus beneath the upper lids. The delicate and

filmy lower lids close up over the eye when **thus withdrawn.**
Round the dark oval *pupil* is a bright yellow *iris.*

Between the eyes is a minute *brow-spot* (see p. 218).

2. The Codfish.—The cod (*Gadus morrhua*) lives in the temperate and Arctic regions of the ocean.   It is permanently aquatic, and is hatched from ova which **float** at the surface **of** the sea.   Its form is admirably adapted for motion through **the** water, **the** head passing into the trunk directly, without any **neck, and** the trunk gradually tapering to the tail.   Motion is effected by the fins, of which there are ten—six median, and two pairs lateral.   There are three median dorsal fins (*d.* i., *d.* ii., *d.* iii.), (in the perch two, in the herring one) along the back, and two median anal fins (*a.* i., *a.* ii.), (in the perch and the herring one) on the ventral surface behind the anus.   The powerful caudal fin (*c.*) forms the main **part of the tail.**   The paired fins are **the** pectoral (*pl.*) on either side, just behind the head, and the pelvic (*pc.*) or ventral fins, somewhat **lower down and** further **forward.**   These four paired **fins answer to the** four legs of the frog.   The pectoral fins are supported on a shoulder-girdle, and the pelvic on bones which probably represent the hip-girdle.   Note that the pelvic fins, representing the hind-legs, are carried so **far** forward as to be anterior

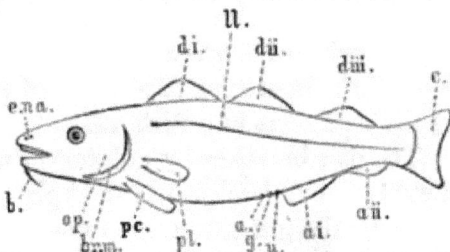

FIG. 4.—COD FISH : EXTERNAL CHARACTERS.

a. Anal aperture. *a.* i. *a.* ii. Anal fins. *b.* Barbule. *br. m.* Branchiostegal membrane. *c.* Caudal fin. *d.* i. ii. iii. Dorsal fins. *e. na.* External nares. *g.* Genital aperture. *l. l.* Lateral line. *op.* Operculum. *pc.* Pelvic fin. *pl.* **Pectoral fin.** *u.* Urinary aperture.

to the pectorals, representing the fore-legs.   This is not so in all fishes—*e.g.* the salmon or the herring.   The fins are supported on bony fin-rays, which are, at the ends, soft and flexible.   In the perch the rays of the first dorsal **fin are** stronger, and are produced upwards into sharp spikes.

There are ten apertures, four **median and three** pair lateral. The first median aperture is the **mouth.**   This passes back into the gullet; but at the sides of the pharynx there are five gill-clefts opening into the branchial chambers. The tongue is small, and cannot be protruded. Both upper and lower jaws are armed with teeth ; and, in addition, there are teeth in the roof of the mouth (*vomerine teeth*), and in the pharynx (*superior and inferior pharyngeal teeth*). There are no posterior nares opening into the mouth. The second, third, and fourth median openings are, in order, the anal (*a.*), genital (*g.*), and urinary (*u.*).

Of the lateral apertures two pairs are nasal, situated close together, and near the anterior end of the snout (*e. na.*). The other lateral apertures are the openings of the gill-chambers. Each lies behind a flap-like gill-cover, which is bony anteriorly (*operculum, op.*), and softer posteriorly (*branchiostegal membrane, br. m.*), containing bony rays—the *branchiostegal rays*. On raising these gill-covers there will be seen four complete *gills*, each supported on a bony branchial arch, and composed of a number of free deep-red *branchial filaments*. There is a fifth more rudimentary branchial arch which bears no gill. Attached to all the branchial arches are *gill-rakers*, horny filaments which bound the margins of the five clefts, and act as strainers. On the inner side of the opercular flap is a red patch *(pseudobranchia)*, which is the rudiment of a fifth gill.

The fish breathes by gulping in water through the mouth and forcing it out backwards through the clefts, over the gills, and beneath the gill-cover. To prevent the water passing out again through the mouth there are two flaps of skin, one on each jaw, which, as the water passes inwards, fold down against the jaw, but as the water attempts to pass outwards are raised and, coming into contact, bar the passage. Their action may be well seen in the pike in an aquarium.

The body is invested with an *exoskeleton*, consisting of overlapping *scales*, over which there is spread a thin layer of slimy skin (epidermis) containing pigment cells. Each scale consists of a thin oval plate, which under the microscope (low power) is seen to be built up of concentric rings. The free border is smooth and even (*cycloid* scale), and is not, as in the perch, produced into a number of comb-like processes (*ctenoid* scale). Along a definite line down each side of the fish, called the lateral line (*l. l.*), the scales are peculiarly modified, so as probably to minister in some way to special sensation. The *barbule* (*b.*) beneath the chin is probably an organ of touch. The eyes are large, and have no definite eyelids. There is no external aperture of the ear, nor is there, as in the frog, any visible tympanic membrane.

3. The Rabbit.—The wild rabbit (*Lepus cuniculus*) frequents furzy sandy heaths, taking shelter when disturbed in deep burrows, which it digs in the sand. In wet soils, instead of digging burrows, it forms "runs" or galleries in the matted vegetation. Its food is green vegetable matter, especially the young shoots of the furze.

The rabbit begins to breed at the age of six months, and has several litters of from three to nine young in the year. At this rate a single pair of rabbits might, at the end of about five years, look round with pardonable pride on a colony of something like a million descendants. The mother forms a special chamber in which the young are born and suckled. At birth they differ considerably from the adult. The head is much larger, the ears comparatively short, the tail a respectable length, and the fore and hind limbs of about equal size. They have a much more average mammalian appearance than the adult; specialization setting in as they grow up. They are suckled for a fortnight or so, and are adult in five or six months. During that time the rate of growth of the trunk is greater than that of the head; that of the hind-legs much greater than that of the fore-legs. The ears grow very fast, and the tail hardly at all: the former tend to droop, while the latter acquires its characteristic upward curve. In all this there is something more than growth—there is development. But there is no metamorphosis.

In the adult rabbit there is a distinct head, pretty well marked off into a *facial region* in front, and a *cranial region* behind. The neck is short, but distinct. The body is stout and slightly elongated, and divided into a *thoracic region* anteriorly, and an *abdominal region* posteriorly. The sides of the thoracic region are guarded with *ribs*, which meet the *sternum* or breast-bone below. The walls of the abdominal region are soft.

The whole skin is invested with an exoskeleton consisting of hairs, greyish brown on the back and limbs, white on the belly and under the tail. On the general surface of the body the hairs are of two kinds; larger and longer contour hairs (*pili*), and shorter and softer fur (*lana* or *lanugo*). In the seal at the Zoological Gardens we see the hair; but in the dressed sealskin the hair is removed, and we see only the prepared and dyed fur. Special hairs are developed on the face: the *vibrissæ* or *mystaces* on the upperl ip; the *supra-orbital hairs*, answering to our eyebrows; the *eyelashes*, and the *malar vibrissæ* or *cheek whiskers*. These vibrissæ may be regarded as long and delicate sense-levers. The hair is reflected into the mouth, so as to line the inner side of the cheeks, and is continued over the under surface of both fore and hind feet.

If a few hairs be mounted in glycerine, and examined under the microscope (low and high power), each will be seen to have three layers: (1) a delicate external layer or *cuticle*, composed of slightly overlapping scales, best seen near the tip of the hair; (2) beneath this a longitudinally fibrous layer, the *cortex*; and (3) a central layer of irregular structure, the medulla (Fig. 30, vi. vii.).

There are nine external apertures, three median and three pairs lateral. The anterior median aperture is the mouth, bounded by upper and lower lips, the former having a groove which passes upwards to the external nares. In front of the mouth are the large gnawing or *incisor* teeth; further back and separated from these by a long space or *diastema* are the grinding teeth (*premolars* and *molars*). In a very young rabbit there are three incisors on each side of the mid-line in the upper jaw, one larger and two smaller, one of the latter being just behind the larger tooth. In the lower jaw there is only one incisor on each side. There are three grinding teeth on each side in the upper jaw, and two in the lower. In the adult rabbit there are only two incisors in the upper jaw, the little ones to the side

being lost and not replaced, and one incisor on each side in the
lower jaw. The grinding teeth of the young have been shed
and replaced by permanent teeth, the *premolars*, and in addition
to these, three more grinding teeth on each side, and in each jaw,
are formed behind the premolars. They are the *molars*. The
arrangement of teeth in the young is called the *milk dentition*,
which consists of eighteen teeth ; that in the adult is called the
permanent dentition, which comprises twenty-eight teeth.

The teeth are implanted in sockets, are devoid of fangs, and
grow continuously during life, constant waste of substance being
constantly made good. Each is composed of three substances of
different hardness : *cement* (softest and bony), *dentine* (harder),
and *enamel* (hardest). The cement forms a thin incrusting
external layer. The enamel encases the dentine except where
it has been infolded. In the large incisors it forms, for
example, a casing to the dentine ; but the layer of enamel is
much thicker along the anterior face, especially in the mid-line,
where it is, so to speak, tucked in along a median groove. On
the posterior face it is very thin or absent. In consequence of
this arrangement of hard and soft substance, the anterior edge
of the tooth wears away less rapidly than the posterior, so that,
by the friction against each other of the upper and lower
incisors, a chisel-like cutting edge is maintained. In the little
incisors the front edge wears away slightly more rapidly than
the hinder edge, so that the larger and smaller incisors together
form a sort of notch into which the incisors of the lower jaw fit.
The premolars and molars vary from each other somewhat in
form and arrangement of their substance ; but the four median
grinding teeth have, from constant use, crowns which present two
furrows running across the axis of the jaw. These are separated
by a central ridge. Enamel forms the outer layer of the tooth
(coated with a little cement), and is infolded along the central
ridge, which owes its existence to this harder infolded layer.
Dentine occupies the troughs of the furrows.

Two pairs of passages communicate from the mouth to the
nasal chambers, (1) the large *posterior nares*, which lie far back
behind the roof of the mouth, (2) the minute *naso-palatine*

passages, which open from the organ of Jacobson into the mouth a little behind the incisors. There are also small *Eustachian tubes* leading to the cavity of the drum of the ear. At the back the mouth communicates (1) with the lungs by the *glottis*, (2) with the stomach by the *œsophagus*. The tongue is fixed behind and free anteriorly. At its sides are oval wrinkled patches (*papillæ foliatæ*), probably gustatory.

The other median apertures are the *anus*, and the *urinogenital* opening, of which the latter is anterior, and differs according to sex. Of the six lateral apertures, the most anterior pair are the *external nares*; the second pair the *ear openings*, guarded by the large external ears, and passing down to the tympanic membrane, which is not exposed at the surface as in the frog; and the posterior pair, *ducts of the perineal glands*, situated on hairless spaces on either side of the anus. These glands pour forth an odorous secretion. In addition to these lateral apertures, there are in the female those on the teats of the mammary glands.

The fore-legs are considerably shorter than the hind-legs. Each is jointed to the shoulder girdle, and has a brachium, an antibrachium, and a manus with five digits. The hind limb has a femur, crus, and pes with four digits. Thus the great toe is suppressed in the rabbit, and the thumb in the frog. The main hairy cushion under the sole of each foot is the *pulvinus*, the smaller cushion under each digit the *pulvulus.* The ankle is not elongated as in the frog, and there is a backwardly projecting heel. All the digits bear strong claws, or are *unguiculate.* The rabbit walks on the toes of its manus, but allows the whole plantar surface of the pes to touch the ground. Some animals, like the cat, walk on the toes of both fore and hind feet, and are called *digitigrade.* Others, like the bear, walk on the whole surface of both manus and pes, and are called *plantigrade.*

In their normal position the limbs of the rabbit lie parallel with the body; but whereas the femur runs forward from the hip girdle, the brachium runs backward from the shoulder-girdle. So that we must regard the fore-limb as having been folded backward from the typical position (Fig. 3, B.), while the hind-limb has been folded forward.   Both have then been more or less bent at the joints.   But if you fold your own arm backwards from the typical position into the position indicated for the rabbit (Fig. 5), and then bend it at the joints without further change, you will find that the hand is palm upwards — you must turn the hand over to apply the palm to the surface, a motion that is rendered possible by the partial rotation of the outer end of the fore-arm.

FIG. 5.—POSITION OF LIMBS IN RABBIT.

This further change of position has been effected in the rabbit so as to allow the palmar surface of the manus to rest on the ground.

4. The Pigeon.—The Rock Dove (*Columba livia*), of which our tame pigeons are domesticated varieties, builds, in cliffs and ruined towers, an untidy nest of sticks and leaves.   Here she lays two white eggs, upon which she sits for sixteen days, imparting to the embryo within them the warmth of her own body.   The young, when they emerge from the egg, are provided with patches of yellow down, like (but much scantier than) that which covers the newly-hatched chick.   Unlike the chick, which, so soon as it is hatched, is bright-eyed, active, and can feed itself, the little doves are at first quite helpless, with closed eyelids, and must be tended by their parents, who feed them with a creamy fluid secreted by the crop.   At the end of three weeks, however, they are fledged, and, after a few days' education by their parents, are able to fly forth and fend for themselves in the world.   Those birds which, like the dove, have to be nursed for a while by their parents are called *altrices*; those which, like the fowl or the duck, are able at once to run or swim and feed themselves, are known as *præcoces*.   In both development accompanies growth; but it is more marked in the altrices.

In the head of the pigeon the facial portion is produced into a horny beak with *upper* and *lower mandibles.*   At the base of the upper mandible is a swollen and featherless patch of skin, the *cere*. The cranial division of the head is well rounded. In the feathered pigeon the neck passes gracefully into the body, but in the plucked bird the distinction between head, neck, and trunk is obvious. The body tapers backwards, the thoracic region being well developed, guarded by ribs, and provided with a large ventrally-keeled sternum. The feathered tail is fan-shaped; but the plucked tail is an insignificant upturned protuberance. There are two pairs of limbs; the anterior converted into wings, the posterior into cursorial legs.

There is a well-developed exoskeleton consisting of feathers. The beauty of form is given mainly by the contour feathers (*pennæ*), but if these be plucked there still remain the more delicate plumose *filoplumes*. There are strong quill-feathers in the wings (*remiges*) and in the tail (*rectrices*). The hues of the living bird are largely due to the metallic tints of the feathers.

Although the external appearance of the pigeon would lead one to suppose that the feathers are developed uniformly over the whole body, yet closer inspection shows that they are arranged in more or less definite feather tracts (*pterylæ*), separated by featherless spaces (*apteria*). This may be better seen in a young blackbird or sparrow. There is a *spinal tract* along the mid-line of the back, broadening or bifurcating posteriorly. On the ventral surface are two parallel bands constituting the *ventral tract*, and separated by a median inferior space. The spinal and ventral tracts are separated by lateral spaces. There are also special feather tracts on the wings and legs. In the plucked bird, which loses by this process the characteristic grace and symmetry of outline, the feather tracts may be traced by the scars left by the removal of the feathers.

A quill feather consists essentially of a proximal part (that is, a part nearer the body), the quill, and a distal part (further from the body), the feather or vane. The quill (*calamus*) is cylindrical and hollow; at its proximal end is a hole (*inferior umbilicus*) into which a little fleshy *feather-papilla* is inserted; at the distal end, where the quill joins the shaft of the vane, there is an oblique aperture (the *superior umbilicus*). The vane (*vexillum*) has a central shaft (*rachis*) continuous with the quill, but differing from it in being quadrate in section, grooved on its under side, and filled with light white pith. On either side of the shaft are the *barbs*, attached to it much in the same way as the teeth are set on a comb, and, like them, flattened in a direction at right angles to the axis. The barbs will be found to adhere together, so that they cannot be separated without the application of some gentle force, when they suddenly tear asunder. When the continuity of the feather has thus been broken between any two barbs, simple pressure of the two barbs together will not readily mend it; but if the lower or proximal part of the broken vane be raised and hitched over the upper part, the barbs will once more adhere together and the broken vane will be mended.

The cause of this will be evident if we cut off a little piece of the vane, and, after soaking it for a few minutes in alcohol, examine it under the microscope (low power). Each barb will be seen to give rise to smaller barbs, or *barbules*, arranged on either side of it. The distal barbules (those nearer the feather tip) carry two or three hooks apiece; the proximal barbules (those nearer the quill) are simple and without hooks. When the vane is perfect, the hooklets on the distal barbules hook over the proximal barbules; and when we mend the broken vane, in the way above described, we hitch the invisible hooks over a series of invisible bars.

In the filoplumes the vane is rudimentary, and the barbules are provided with no hooks.

There are seven apertures, three median and two pairs lateral. The most anterior median aperture is the mouth, guarded by horny mandibles, and provided with no teeth. The posterior nares open by a common longitudinal slit in the roof of the mouth, bounded by palatal folds. At the posterior end of this slit are the openings of the Eustachian tubes. The tongue is narrow, pointed, and horny. The oval glottis and the wide gullet may be seen with difficulty at the back of the buccal cavity. At the posterior end of the body there is a common aperture (*cloaca*) for the exit of the fæces and the urino-genital products. The third median aperture is at the summit of a little papilla above the tail. It is the orifice of the duct of the uropygial oil-gland. The lateral apertures are the anterior nares, just in front of the cere, and the ear-openings at the side of the head. The fore-limb is divided into brachium, antibrachium, and manus; but the manus is not divided into separate digits, such of the fingers as exist being united into a continuous mass, while the rudimentary pollex (thumb) forms an insignificant projection. The quill feathers attached to the manus are called *primaries*; those attached to the antibrachium *secondaries*. The second and third primaries are longer than the first, and form the tip of the wing. The pollex bears a little group of feathers called the bastard wing. The quills of both primaries and secondaries are overlapped above and below by wing coverts. On the dorsal surface of the brachium is the *humeral* feather tract, and on that of the antibrachium the *alar* tract.

Thus the fore-limb is especially developed for flight. It is a *homologous* organ with that of the rabbit, but not an *analogous* organ. *Homologous organs are those that are built upon the same plan; analogous organs those that perform the same function.* The hind-limbs of the rabbit and the pigeon are both homologous and analogous, but the fore-limbs are homologous but not analogous; while the wings of a bird and of a butterfly are analogous, for they perform the same function, but not homologous, for they have no community of plan.

The hind-limb reminds us of that of the frog, in that it has four divisions. The full meaning of these divisions will only become clear when we study the osteology of the hind-limb. For the present we may regard them as femur, crus, and pes; the latter having an undivided portion, and a divided portion with four digits. The hallux, or great toe, is directed backwards, the other three toes forwards; the fifth is absent. There is a well-developed femoral tract of feathers on the femur, and a less developed crural tract on the crus. The pes throughout is devoid of feathers,[1] but has instead well-marked red scales.

The position of the fore-limb during flight is nearly that which

---

[1] Not so in some artificially modified pigeons.

we have seen to be primitive in the newt. In its closed position
it is bent upon itself Z-fashion, the
brachium and the manus being directed
backwards, and the antibrachium forwards. (Fig. 6.) In the hind limb,
the folding forwards which we noticed
in the rabbit has taken place. But the
undivided portion of the pes is so
thoroughly raised off the ground, that
it appears to belong rather to the leg
than to the foot, and, indeed, as we

FIG. 6.—POSITION OF LIMBS
OF PIGEON.

shall see hereafter, belongs partly to the one and partly to the
other.

General Conclusions.—Thus in these vertebrate types there
is much fundamental resemblance overlain and partially masked
by many well-marked differences. In the possession of a skull
and vertebrated back-bone; in the mode of attachment of the
limbs, and their general structure in frog, pigeon, and rabbit;
in the position of mouth, nares, eyes, and ears, there is resemblance. In the possession of a common cloaca, the frog and the
pigeon differ from the rabbit, as this in turn does from the fish,
in having a common urino-genital aperture. In the absence of
exoskeleton, the frog differs from the other types. The scales
of the fish are *dermal* structures formed in the deeper layer of
the skin. Feathers and hairs are *epidermal* structures formed in
the superficial layer of the skin. In their modes of life each
differs from the other. The cod-fish is aquatic and marine; the
frog amphibious near fresh water; the rabbit is terrestrial; the
pigeon fitted for aerial life. Hence the special modification of
the limbs and the form of the body. The cod breathes oxygen
dissolved in sea-water, the basic salts of which absorb the carbonic acid expired: the tadpole breathes the oxygen dissolved
in fresh water: the frog, pigeon, and rabbit breathe the oxygen
of the air, but the frog requires less than one-tenth of that
required by the pigeon. In their food they differ widely. The
cod feeds on shell-fish and other marine animals; the tadpole

is herbivorous; the frog insectivorous; the pigeon gramini-
vorous; and the rabbit herbivorous; and their mouths are
to some extent fitted for their special food. The cod has
powerful crushing jaws and pharyngeal teeth; the tadpole,
horny jaws for cropping the weeds; the frog, a protrusible
tongue for catching prey, and minute teeth for holding the prey
when caught; the rabbit, special gnawing and grinding teeth,
in relation to its mode of feeding; but the beak of the pigeon
cannot be said to be specially modified in relation to its peculiar
diet, though its stomach is specially modified, being converted
into a gizzard, with thick callous walls, in which the grain is
triturated by the aid of small stones swallowed for that purpose.
Finally, in the temperature of the body there is a marked difference. The temperature of the cod and of the frog varies with
the temperature of the water or the air by which they are bathed.
Hence the hibernation of the frog in winter, and its renewed
activity in the warmth of spring. The temperature of the
rabbit's body is constant at about 98° Fah., its clothing of hair
enabling it to maintain this temperature even in winter. The
temperature of the pigeon's body is 104° or 105° Fah., its thick
clothing of feathers, in which is entangled much air—an
excellent non-conductor of heat—aiding it in maintaining this
temperature.

The student is advised to lose no opportunity of comparing
the structures of different animals, and ascertaining for himself
how far they resemble—in the external characters we have now
considered, and in the internal structure, on the study of which
we have next to enter—the cod, the frog, the pigeon, and the
rabbit; and how far they differ from these types.

CHAPTER III.

GENERAL ANATOMY.

1. The Frog.—The skin of the frog is very loosely attached to the underlying parts of the body, being, indeed, separated therefrom by a system of *subcutaneous lymph spaces* (Fig. 8, *Sc. l, s.*). If, therefore, a frog which has been killed with chloroform be pinned out under water in a dissecting dish,[1] it is an easy matter to slit open the skin along the whole mid ventral line without injuring the body-wall beneath. Note in doing so the partitions between separate lymph spaces. By reflecting the flaps of skin on either side, the whole of the ventral body-wall may thus be displayed. It will be seen to be white, and largely composed of fibrous bands of muscle running in various directions. In the throat they run transversely; in the median line of the trunk they are longitudinal; on either side of this line they run downwards (*i.e.* ventrally) and backwards; in the pectoral region they converge towards the shoulders. In the mid-line of the pectoral region is the elongated sternum, on either side of which may be felt the bony bars of the shoulder girdle. On each reflected flap of skin is a large vein (*great cutaneous*), the main trunk of which passes under the arm-pit. It carries blood from the skin towards the heart. Through the abdominal muscles there is seen a median vein (*anterior abdominal*). At the posterior end of the body may be felt the hip girdle.

[1] A useful vessel is made of tinned iron about 8 inches square, with sides slightly sloping outwards, and 2 inches high, bound with thick iron wire to give firmness. A thin piece of cork nailed on to a thin sheet of lead should fit loosely into the bottom.

With the point of a scalpel a small incision may now be made in the muscular body-walls just in front of the hip-girdle, and sufficiently to the left (the frog's left) of the middle line to avoid injuring the median vein.   A small hole will thus be made into the *body-cavity* or *cœlom*, which contains the chief *viscera*.   By inserting the small forceps into this hole the body-walls may be raised, and an incision may be carried forward just to the left of the median vein as far as the bony bar of the shoulder-girdle. A similar incision may be made to the right of the median vein, and with a little care it may be dissected away from the strip of the body-wall to which it is attached.   On raising the sternum the heart will be seen lying beneath it in a membranous bag. Avoiding injury to this organ, the left incision may now be carried forward by dividing with strong scissors the bony bars of the shoulder-girdle to the left of the sternum.   Those to the right may be similarly divided, and the sternum carefully removed. By the removal of the ventral wall of its membranous bag the heart may be displayed.   After trimming away the sides of the body-walls the dissection will now resemble that given in outline in Fig. 7.   Fig. 8 represents a transverse section of the body of a frog which had lain in a solution of 1 per cent. chromic acid for three or four days till the bones were softened.   The student should make other such sections through the eyes, through the tympanic membrane, and through the shoulder-girdle.

We are now in a position to take a preliminary view of the following organs, or systems of organs :—

1. The Alimentary System.
2. The Respiratory System.
3. The Heart and Circulatory System.
4. The Urino-genital System.
5. The Nervous System.
6. The Skeletal System.
7. The Muscular System.
8. The Integumentary System.

1. *The Alimentary System.*—The *alimentary canal* is a tube running right through the body from the mouth to the vent, and

attached to it are certain *glands* which pour their **secretion into** the canal. The **parts of** the alimentary canal **are** *mouth, œso-*

phagus, stomach (St.), small intestine (Sm. In.), and *large intestine (L. I.).* The glands attached are the *liver* and the *pancreas (Pn.).* The cavity in which the canal and its glands lie is called the *pleuro-peritoneal cavity* or *cœlom.*

The stomach *(St.)* is elongated, and lies well over to the **left** (the frog's left) side. **It is much** hidden by **the large brown** liver. **But if we lift the** lobes of the liver we shall see it pass anteriorly **into** the œsophagus. Posteriorly it passes **at the valvular** *pylorus* into the small **intes-** tine *(Sm. In.),* which is coiled. This, in turn, passes somewhat suddenly into the large intestine *(L. I.),* and this into the cloaca, open- ing **outwards** by the vent. **The parts of the canal are**

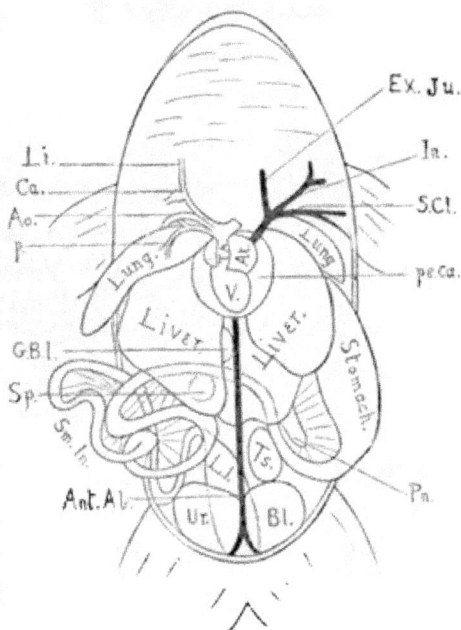

FIG. 7.—VISCERA OF MALE FROG.

In the female the large ovaries and the coiled white oviducts occupy much space.

Ant. ab. Anterior Abdominal vein. *Ao.* Aortic arch. *At.* Atrium. *Ca.* Carotid gland and artery. *Ex. Ju.* External jugular vein. *G. Bl.* Gall bladder. *In.* Innominate vein. *L. I.* Large intestine. *Li.* Lingual artery. *p.* Pulmonary artery. *pe. ca.* Pericardial cavity. *Pn.* Pancreas. *S. Cl.* Sub- clavian vein. *Sm. In.* Small intestine. *Sp.* Spleen. *T. A.* Truncus arteriosis. *Ts.* Testis. *Ur. Bl.* Urin- ary bladder. *V.* Ventricle.

connected together, and with **the dorsal wall of the** cavity, by a transparent membrane, the *mesentery.* The whole pleuro- peritoneal cavity is, moreover, lined with a glistening pig- mented membrane, the *peritoneum,* which forms the double layer of the mesentery, and so passes on to the viscera. Thus there is an outer *(parietal* or *somatic)* layer lining the walls of the cavity, and an inner *(visceral* or *splanchnic)* layer re- flected down the mesentery on to the alimentary organs. This is represented by the dark line in Fig. 8. It secretes a moistening *serous fluid.*

The stomach and succeeding (*duodenal*) portion of the small

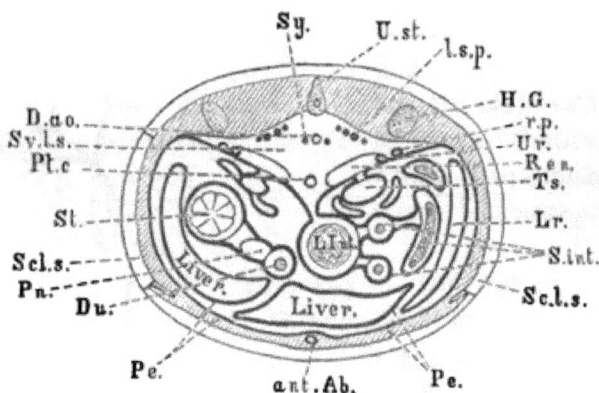

FIG. 8.—TRANSVERSE SECTION OF THE POSTERIOR PART OF THE BODY OF A
MALE FROG.

ant. Ab. Anterior abdominal vein. *D. ao.* Dorsal aorta. *Du.* Duodenum.
H. G. Hip-girdle. *l. s. p.* Lumbo-sacral plexus. *L. Int.* Large intestine.
Lr. Liver. *Pe.* Peritoneum. *Pn.* Pancreas. *Pt. c.* Postcaval vein. *r. p.*
Renal portal vein. *Ren.* Kidney. *S. int.* Small intestine. *Sc. l. s.* Subcutaneous lymph space. *Sv. l. s.* Subvertebral lymph space. *St.* Stomach.
Sy. Sympathetic nerve chain. *Ts.* Testis. *U. st.* Urostyle. *Ur.* Ureter.

intestine form a loop united by mesentery. In this loop lies a triangular yellow organ, the pancreas (*Pn*). Through the midst of it there passes a delicate tube opening into the small intestine. This is the *bile duct* leading the product of the liver from that organ, and from the gall-bladder (*G. Bl.*) attached to it, into the alimentary canal, and receiving also the pancreatic fluid secreted by the pancreas. Note how the large liver is attached to the anterior end of the cavity, and how it is divided into *lobes.*

2. *The Respiratory System.*—The organs of respiration of the adult frog are the *lungs.* They lie in the pleuro-peritoneal cavity on either side near its anterior end. The right lung is slightly displaced in the figure (7). The layer of peritoneum reflected on to the lungs is called the *pleura.* Air passes into them from the *glottis*, and they may be inflated from that aperture. The *nares* are the external apertures of the respiratory

system. When the frog breathes the anterior nares are opened, and the floor of the mouth is depressed. Air rushes in. Then the nares are closed and the floor of the mouth is raised. The air is forced into the lungs. When the nares again open the air rushes out of the lungs, which tend to collapse from their own elasticity.

3. *The Heart and Circulatory System.*—The *heart* lies in its own special *pericardial cavity* (7, *pe. ca.*), which projects backwards into the pleuro-peritoneal cavity. This cavity is lined with a glistening pigmented membrane, the *pericardium*, which has a parietal and a visceral layer, the one lining the cavity, the other reflected on to the heart. The heart itself has four parts, (1) *sinus venosus*, (2) *atrium*, (3) *ventricle*, (4) *truncus arteriosus*. The ventricle is the posterior fleshy part of the heart (*V.*). The truncus arteriosus (*T. A.*) passes forwards from the ventricle as a fleshy tube. The atrium is seen in Fig. 7 (*At.*) on either side of *T. A.* It is thin-walled. The sinus venosus is also thin-walled and dorsal in position, lying above the atrium. It may be seen by lifting up the ventricle. Into this division of the heart the blood is received from the various parts of the body, and is passed on to the atrium. Thence it flows into the ventricle, and thence into the truncus arteriosus to be distributed throughout the body. In a recently-killed frog the heart may still be beating. Its parts will be seen to contract in the order given.

After all the other points in the general anatomy of the frog have been made out, the heart may be removed and dissected under water. The atrium will be found to be divided by a septum into two chambers, a right and a left *auricle*. The ventricle has an undivided cavity (Fig. 9, iv. *v.*). The right auricle receives the blood from the general mass of the body through the sinus venosus (*s. v.*). The left auricle receives blood from the lungs by the *pulmonary vein* (*p. v.*). Blood from both auricles passes into the ventricle, and thence to the truncus arteriosus for distribution.

The truncus arteriosus soon divides into a right and a left branch, each of which gives origin to three vessels (Figs. 7 and 10). Of the three arteries into which it splits, the most anterior carries blood to the head (*carotid, ca.*), and to the tongue (*lingual, li.*); the mid-branch curves back and passes to the dorsal side of the heart to supply the body with blood (*systemic aorta—ao.* in 7, and *sy. ao.* in 10); the posterior branch passes

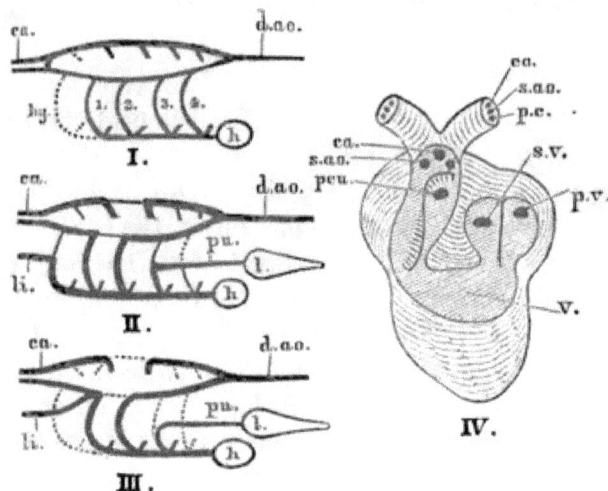

FIG. 9, iv.—HEART OF FROG.
(i., ii., and iii. may here be disregarded.)

to the lung (*pulmonary, p.*), giving off also a (*cutaneous*) branch (10, *cu.*) to the skin. The left branch of the two into which the truncus arteriosus splits divides in the same way. The two systemic aortic arches, right and left, curve round and pass to the back of the heart, and after giving off a branch to the fore-limb, ere long meet in the middle line beneath the back-bone, fusing together to form the *dorsal aorta* (Figs. 8 and 10, *d. ao.*),[1] which runs backwards in the large *subvertebral lymph sinus* (*Sv. l. s.*) shown in Fig. 8. Where the two aortic arches join to form the dorsal aorta (or from the left arch just before it joins its fellow), a large branch (*cœliaco-mesenteric*, 10, *c. m.*) passes off to supply the stomach and intestines with blood. Other branches are given off further back to the kidneys (*ren.*, Fig. 8). While posteriorly the aorta bifurcates to supply the two legs (Fig. 11).

All these blood-vessels which supply the organs with blood

[1] See note, p. 28.

direct from the heart are called *arteries*. When they reach the organ they supply, they branch again and again, and eventually break up into minute tubes called capillaries. Then these unite

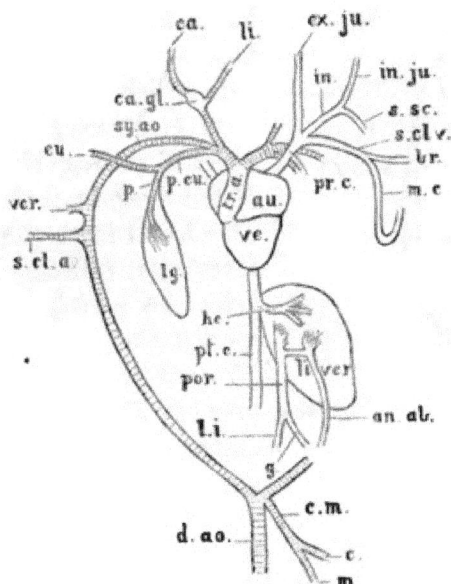

FIG. 10.—HEART AND SOME GREAT VESSELS OF FROG.

with each other, and by repeated union give rise to larger blood-vessels, which lead the blood away from the organ. These are called *veins*. So that we have the series, (1) artery, (2) capillary plexus, (3) vein. Where two blood-vessels unite to form a common trunk they should be called *factors*; and where a trunk splits into two vessels they should be called *branches*.

The blood carried to the lungs by the pulmonary artery is returned thence to the left auricle of the heart by the *pulmonary rein*. The blood supplied to the head and fore-limbs is returned to the sinus venosus by (*precaval*) veins (10, *pr. c.*), of which that on the left side is shown in Fig. 7, that on the right side being removed. This vein is seen to be made up of three factors (*external jugular, ex. ju.; innominate, in.; subclavian, s. cl.*). The blood from the hind-limbs may take one of two courses, either by the *anterior abdominal* vein (*ant. ab.*), in which case it passes to the liver, and there breaks up into a capillary plexus, or by two veins (*renal portal, r. p.*),[1] which will be seen passing to the outer faces of the kidneys, in which case it breaks up into capillary plexuses in these organs. From these organs it passes by means of small vessels into a large vein which lies between the kidneys (*post-caval*, 10 and 11, *pt. c.*),[1] and runs up through the base of the liver to the sinus venosus. The blood from the stomach and intestines collects into a large vein (the *portal*, 10, *por.*), which carries the blood to the liver.[1] There it breaks up into a capillary plexus. From the liver, the blood delivered to that organ by the anterior abdominal, and portal veins, and by the *hepatic artery* (a branch of the cœliaco-mesenteric),

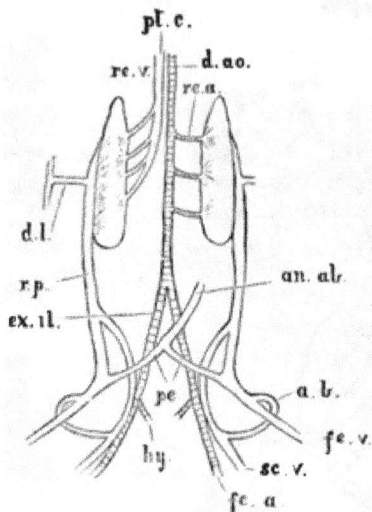

FIG. 11.—RENAL PORTAL AND PELVIC VESSELS : FROG.

is conveyed into the post-caval by the *hepatic veins* (10, *he*).

As it passes through the capillaries of the various organs of the body, the watery fluid of the blood oozes out and forms the *lymph*, which collects in irregular cavities called lymph spaces. It is pumped back into the blood-stream by four lymph-hearts (see p. 195).

4. *The Urino-genital System.*—The urinary organs are the kidneys (8, *ren.*), elongated red organs lying in the subvertebral lymph space ; attached and closely applied to them are the yellowish *adrenal* bodies. From the outer edges of the posterior ends of the kidneys arise their ducts (*ureters*, 8, *ur.*), which open into the cloaca by minute apertures in its dorsal wall. The ureters of the male have connected with them glandular bodies, the *vesiculæ seminales.* In the ventral wall of the cloaca is a

---

[1] When the portal vein has been made out, the stomach, intestines, and liver should be carefully removed, and the subvertebral lymph space opened out without injury to the kidneys, testes, etc. The dorsal aorta, post-caval, and renal portal will then be readily made out, as shown in Fig. 11.

median aperture leading into the large bi-lobed urinary bladder
(7, *Ur. Bl.*). Into it the urinary fluid, therefore, does not pass
directly from the ureters, but indirectly through the cloaca.

The genital organs of the male are the *testes* (7 and 8, *Ts.*).
They are closely connected with the kidneys by a mesenteric
membrane (the *mesorchium*), in which fine white lines may be
seen, which are the ducts by which the seminal fluid is conveyed
to the kidney, to pass thence by the ureter, which should therefore be termed the *urino-genital duct* in the male.

The genital organs of the female are, (1) the large *ovaries*
crowded in the breeding season with large ova, and suspended
by the *mesoarium*, and (2) the coiled *oviducts*. The oviducts are
anteriorly thick and glandular and swell up in water ; posteriorly
they are thin-walled, and open into the cloaca by slits in its
dorsal wall, just in front of the apertures of the ureters. The
oviducts do not open into the ovaries, but open at the very
anterior end of the pleuro-peritoneal cavity close to the base of
the lungs. The ripe ova are shed into the pleuro-peritoneal
cavity, and escape thence by the oviducts.

The yellow finger-like processes seen in both sexes are the
fat bodies (*corpora adiposa*).

5. *The Nervous System.* — In the subvertebral lymph space
there are parts of two nerve systems.

On either side of the dorsal aorta is a delicate dark thread,
from which minute threadlets pass off to much more conspicuous
white threads passing backwards beneath and on either side of
the urostyle. The delicate dark thread belongs to the *sympathetic
nerve system* (8, *Sy.*). The larger longitudinal white threads
belong to the *cerebro-spinal nerve system*. They form part of the
lumbo-sacral plexus (8, *l. s. p.*), from which the nerves of the hind-
leg take their origin. Further forward, two of the spinal nerves
form a *brachial plexus* running out to the arm. Other spinal
nerves will be seen between these two plexuses. All these
spinal nerves take their origin in the *myelon* or *spinal cord*, which
lies in a dorsal cavity within the arches of the vertebræ of the
backbone. In the neighbourhood of the head there are cranial
nerves, taking their origin in the brain, which lies within the
skull.

To see this central nervous system the frog should be turned over and pinned out back upwards. An incision should be made in the skin, and the muscles cleared away from the vertebral column. The dorsal cavity should then be opened out, beginning at the junction of the skull and vertebral column, and removing the roof of the skull and the upper portions of the vertebræ. The brain and spinal cord will thus be displayed. In

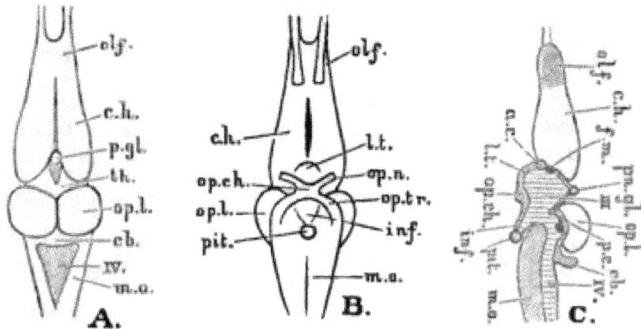

FIG. 12.—BRAIN OF FROG.

A. From above.   B. From below.   C. In longitudinal section.

the brain three parts will readily be seen: a *fore-brain*,[1] composed of two *cerebral hemispheres* (12, *c. h.*) lying side by side, and passing anteriorly into the *olfactory lobes* (*olf.*), which are joined in the mid-line; a *mid-brain*, composed largely of two large rounded *optic lobes* (*op. l.*); and a *hind-brain*, composed of a band-like *cerebellum* (*cb.*), and behind this the *medulla oblongata* (*m. o.*), the upper surface of which is marked by a triangular depression, covered over by a plexus of blood-vessels. Dissection will readily show the cerebral hemispheres and optic lobes to be hollow.

The *spinal cord* or *myelon* is continuous with the medulla oblongata; it presents two enlargements where the nerves (1) for the arms, and (2) for the legs, are given off. It ends in a fine point, the *filum terminale*.

6. *The Skeletal System.*—Somewhat has been said of this

---

[1] The primitive fore-brain and its outgrowths may be regarded as *fore-brain.* See Chapter XI.

system incidentally in the last chapter. Further study we will reserve for Chapter VIII.

7. *The Muscular System.*—Of this system little can here be said. Suffice it to remark that most of the muscles have their *origin* in some bone or other part of the body which is relatively fixed, and their *insertion* in some bone or other part of the body which has to be moved. There are some muscles, however, which encircle certain parts of the body, generally tubular, and by their contraction serve to lessen the diameter of the tube or completely close it.

8. *The Integumentary System.*—Nothing need here be added to that which was said in the last chapter.

In conclusion, attention may be drawn to the following points for the sake of future comparison :—

1. The lungs are contained in the pleuro-peritoneal cavity or cœlom, together with the liver, alimentary canal, generative organs, spleen (7, *Sp.*), etc.

2. The heart has one ventricle and two auricles; a sinus venosus, and a truncus arteriosus.

3. There are two systemic aortic arches.

4. Blood is distributed to the kidney by renal portal veins from the posterior end of the body, as well as by renal arteries from the aorta.

5. There is a large anterior abdominal vein.

6. The kidneys lie in the large subvertebral lymph sinus.

7. There is a common duct for the testis and the renal organ in the male.

8. The apertures of the urino-genital duct of the male, and of the ureters and genital ducts in the female, are on the dorsal side of the cloaca.

9. The urinary bladder opens into the ventral side of the cloaca.

10. There is a common cloacal aperture for the exit of the fæces from the alimentary canal and for the urino-genital products.

11. Respiration is by means of gills in the tadpole and lungs in the frog.

The Cod.[1]—To open up the body-cavity of the cod, the fish should be laid on its side, a ventral incision made from the hip-girdle (under the throat) to within half-an-inch of the anus, and the side wall removed from the *peritoneal cavity* or *cœlom* thus exposed. The cavity is lined with peritoneum, which is reflected on to the viscera. It extends backwards, somewhat behind the anus, and is bounded anteriorly by the *pericardio-peritoneal septum*, which separates it from the chamber which contains the heart. Along the dorsal wall of the cavity runs the large *air-bladder*, terminating anteriorly in a coiled *cæcum*, and not communicating, as in some fishes (*e.g.* trout, herring), with the gullet.

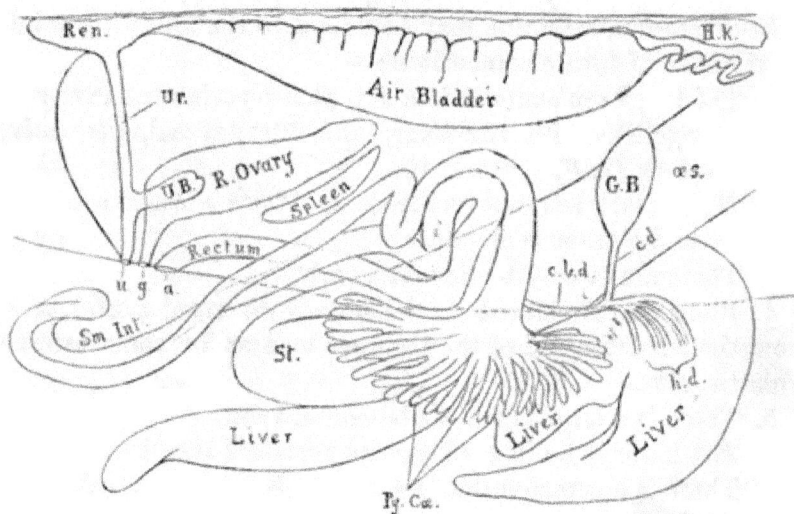

FIG. 13.—VISCERA OF COD-FISH.

a. Anal aperture. *c. b. d.* Common bile duct. *c. d.* Cystic duct. *g.* Genital aperture. *G. B.* Gall bladder. *h. d.* Hepatic ducts. *H. K.* Head kidney. *œs.* Œsophagus. *Py. Cœ.* Pyloric cæca. *Ren.* Kidney. *Sm. Int.* Small intestine. *St.* Stomach. *u.* Urinary aperture. *U. B.* Urinary bladder. *Ur.* Ureter.

1. *The Alimentary System.*—The gullet or œsophagus (*œs.*, Fig. 13) is of about the same diameter as the stomach (*St.*), but

[1] Instead of the cod, the closely-allied haddock or whiting may be dissected. Where this (or any other animal) is too large for dissection under water, it should be pinned out with blanket-pins (or awls) on to a dissecting board of convenient size.

has thinner walls. The stomach is bent upon itself, and where it passes into the intestine there are a number of *pyloric cæca* (*Py. Cœ.*). A little beyond these processes is the *common bile duct* (*c. b. d.*) formed by the confluence of *hepatic ducts* (*h. d.*) from the liver. The *cystic duct* (*c. d.*) leading to the *gall bladder* (*G. B.*), in which the bile is stored, is an offshoot from the common duct. The intestine is somewhat coiled, and passes almost without change of diameter to the *anus* (*a.*).

2. *The Respiratory System.*—The respiratory organs are the gills which we have seen in the last chapter. There are no lungs.

3. *The Heart and Circulatory System.*—To expose the heart the hip-girdle must be divided in the mid-line and the upper half (in the present position of the fish on its side) of both this girdle and the shoulder-girdle carefully removed. The pericardial cavity thus displayed is a small chamber lined with pericardium, which is reflected over the heart. The heart itself has a thin-walled dorsal sinus venosus (14 A, *s. v.*), an irregular thin-walled atrium (*au.*), not divided by any septum into two auricles, a fleshy ventricle (*v.*) with a single cavity, and a whitish *bulbus arteriosus* (*b. a.*).

The bulbus arteriosus passes forward into the median *ventral aorta* (14 B, *v. ao.*), which forthwith sends off branches, *afferent branchial arteries* (*af. br. a.*), right and left to the four gills on either side. In the capillaries of the gills the blood is aerated and collects into vessels (*efferent branchial arteries, ef. br. a.*), by which it is conveyed to vessels, one on each side, which run along the dorsal ends of the gill-bearing branchial arches (right and left *epibranchial arteries, ep. br. a.*). Traced forwards, these run into the (*carotid, car.*) arteries which supply the head and brain. They are united together by a transverse vessel beneath the back of the skull (*tr.*). Traced backwards, the epibranchials fuse together to form the dorsal aorta ; but before they unite each gives off an artery (*subclavian, s. a.*) to supply the pectoral fin, and that on the right side gives off in addition arteries (*cœliac, cœ. a.*, and *mesenteric, me. a.*) to supply

3

the stomach and intestines. The dorsal aorta passes backwards beneath the back-bone, and becomes posteriorly the *caudal artery.* The *circulus cephalicus* formed by the epibranchials and transverse vessel is shown very diagramatically in Fig. 14, which must be taken to indicate the principle of the arrangement of the vessels.

The sinus venosus receives the blood that has circulated through the body. At its posterior end it receives the two

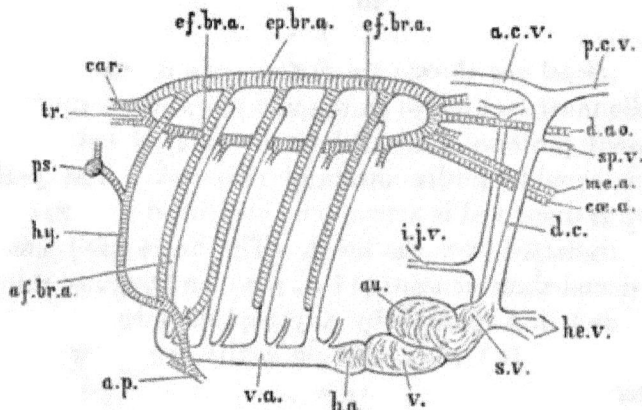

FIG. 14.—DIAGRAM OF HEART AND GREAT VESSELS: COD.

hepatic veins (*he. v.*) from the liver. Dorsal to these, there enter two large veins running downwards on either side of the gullet. They are the *ductus Cuvieri* (*d. c.*). Each results from the union at its dorsal end of two factors; an anterior factor (*anterior cardinal, a. c. v.*) bringing blood from the head, and a posterior factor (*posterior cardinal, p. c. v.*) bringing blood from the posterior regions. The right posterior cardinal is continuous through the substance of the kidney, but the left is aborted for some distance. Posteriorly the two cardinals unite in the caudal vein, which, with the caudal artery, runs in a canal (the *hæmal arch*) protected by bony (hæmal) processes of the vertebræ. The right ductus Cuvieri, slightly above the sinus venosus, receives a vein (*inferior jugular, i. j.*) from the lower parts of the head. The left ductus is seen in Fig. 14 to receive a factor (*spermatic*) from the male organs of generation.

The blood from the stomach and intestines is delivered to the liver by a *portal* system of veins. There it breaks up into a capillary plexus, and is collected again into the hepatic veins which pass straight to the heart. There is a connecting branch between the portal system and the caudal vein.

4. *The Urino-genital System.*—The renal organ (13, *ren.*) is an elongated body lying behind the dorsal wall of the air-bladder. Its anterior enlargement, the *head-kidney* (*h. k.*) lies above the coiled cæcum of the air-bladder; its posterior enlargement lies within the anterior portion of the hæmal canal; between the two there is on each side an irregular longitudinal band. From the posterior mass an unpaired ureter passes down to the urinary aperture (*u.*).

The *testes* of the male (soft roe) are elongated lobular bodies, uniting together in the median line for the posterior quarter of their length, and sending off to the genital aperture a common duct.

The *ovaries* of the female are somewhat conical pink bodies (hard roe), which also unite with each other posteriorly, and send off a common *oviduct* to the genital aperture (*g.*). The ovaries are hollow, and the ova are, when ripe, shed into their cavities and pass out by the oviduct without entering the peritoneal cavity.

5. *The Nervous System.*—The *brain* may be exposed by skinning the top of the head, removing the muscles, and then breaking away the roof of the skull. It lies somewhat loosely in the cranial cavity. In it the three divisions may readily be made out. The cerebral hemispheres of the fore-brain are relatively shorter than in the frog. In front of them the large optic nerves form a X, and above them lie the long peduncles of the olfactory lobes. The optic lobes of the mid-brain resemble those of the frog. But the cerebellum of the hind-brain is much larger than in the frog. The medulla oblongata passes almost insensibly into the spinal chord. At the base of the brain are two bean-shaped inferior lobes.

The side of the fish that has not been dissected away should be carefully skinned in order to see the distribution of two of the cranial nerves. This is shown in Fig. 15. The dark line is the fifth-nerve, of which one branch goes along the back and gives off nerve branches to the dorsal fins; one branch crosses the body to innervate the anal fins; and a third passes downwards behind the gill slits to supply the pectoral and pelvic fins. The dotted line is the tenth-nerve with two branches: an upper, which follows the lateral line for about two-thirds of its length and then thins out; a lower, which runs along the line of division between the dorsal and the ventral set of muscles, and along the posterior third of the lateral line.

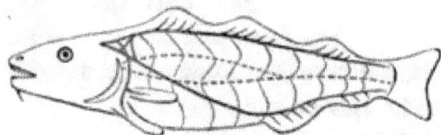

FIG. 15.—CRANIAL NERVES V AND X : COD.

6. *The Muscular System.*—The same Fig. (15) shows in the light zigzag lines, the divisions between the myotomes or segments of the body muscles. The myotomes are, however, far more numerous than here figured.

Of the skeletal and integumentary systems nothing need here be said.

In conclusion, attention may be drawn to the following points for the sake of comparison with other types :—

1. The heart has one auricle and one ventricle; a sinus venosus and a conus arteriosus.

2. The blood after aeration is at once distributed throughout the body, and is not first brought back to the heart.

3. The kidney receives blood from the caudal vein as well as from the renal arteries.

4. There are separate apertures for the ureter, genital duct, and alimentary canal.

5. There is a common ureter for the two kidneys, which also unite in the middle line.

6. There is a common genital duct for the two testes or ovaries, which also unite in the middle line.

7. There is no connection between the kidney and the testis.

8. The urinary bladder (13, *U. B.*) opens on the antero-ventral side of the ureter.

9. The posterior nares do not open into the mouth, nor are there any Eustachian tubes.

10. Respiration is throughout life by means of gills.

3. The Rabbit.—The skin of the rabbit is not separated from the body walls by large lymphatic spaces as in the frog. On the contrary, when we place a rabbit on its back and make a longitudinal incision from chin to pelvis, the skin has to be separated from the body walls with the aid of the flattened handle of the scalpel, and even then a thin sheet of (*cutaneous*) muscle, containing numerous minute blood-vessels, will probably adhere to it.

The abdominal cavity (cœlom) may now be opened out by the removal of its thin muscular walls. Its anterior boundary is a concave moveable partition (the *diaphragm*) separating it from an anterior thoracic cavity containing the lungs and the heart. This diaphragm behind the lungs is a very characteristic mammalian structure. The pink lungs may be seen through it. And if we prick it at one side, the lung of that side will be seen at once to collapse. The thoracic cavity may now be opened out, as in Fig. 16, by cutting through the ribs with bone forceps, and removing most of the sternum. A ventral bridge supporting the diaphragm should however be left.

1. *The Alimentary System.*—Four pairs of *salivary glands* pour their secretion into the mouth. These are—

a. The *Submaxillary*, a reddish gland lying between the two divisions (*rami*) of the lower jaw. The gland of either side meets its fellow in the mid-line.

b. The *Sublingual*, small and slightly anterior to the Submaxillary. Both these glands send ducts forward which open into the floor of the mouth.

c. The *Parotid*, seen by dissecting away the skin in front of and below the base of the external ear. Its duct passes forward into the mouth.

d. The *Infra-orbital*, seen by removing the eye-ball. It lies in the anterior and inferior division of the socket, its duct passing downwards to the mouth. Note at the same time two other glands in the eye-socket, (1) the *Harderian*, anterior and superior; and (2) the *lachrymal*, posterior and superior. They are not salivary.

From the mouth the *œsophagus* passes through the thoracic cavity, pierces the diaphragm, and at once enters the large curved thick-walled *stomach*. (The elongated deep red body near its broader end is the *spleen*.) Then follows the long-coiled *small intestine*, of which the first part (*duodenum, du.*) forms a great U-shaped loop, the two limbs of the U being closely connected by a mesenteric fold, and the ascending limb, that further from the stomach, being closely connected by a membranous fold, with the terminal portion (*rectum, rm.*) of the large intestine. Within the loop of the duodenum is a ramifying mass of yellowish-red lobules. This is the *pancreas*. Its duct opens into the duodenum near the lower end of the loop. Into the duodenum, shortly below the stomach, opens the *common bile duct*, formed by the union of several factors from the various lobes of the large *liver* and one from the *gall*

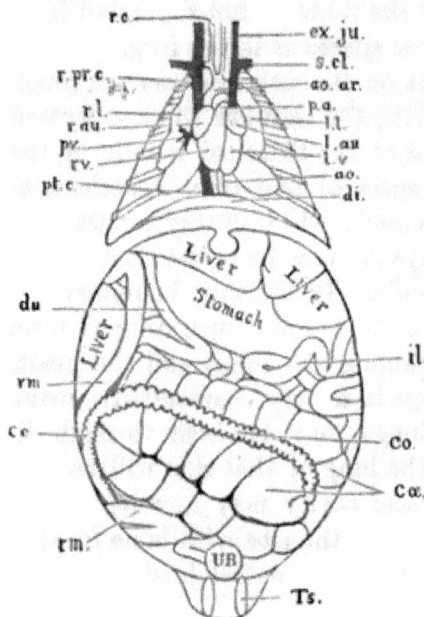

FIG. 16.—VISCERA OF RABBIT.

ao. Aorta. *ao. ar.* Aortic arch. *Cœ.* Cæcum. *Co.* Colon. *di.* Diaphragm. *du.* Duodenum. *ex. ju.* External jugular vein. *il.* Ileum. *l. au.* Left auricle. *l. l.* Left lung. *l. v.* Left ventricle. *p. a.* Pulmonary artery. *p. v.* Pulmonary vein. *pt. c.* Postcaval vein. *r. au.* Right auricle. *r. c.* Right common carotid artery. *r. l.* Right lung. *r. pr. c.* Right precaval vein. *r. v.* Right ventricle. *rm.* Rectum. *s. cl.* Subclavian vein. *Ts.* Testes. *U.B.* Urinary bladder.

bladder. **The rest of the** small intestines (*ileum, il.*)—that part not included in the duodenal loop—is suspended by a separate fold of mesentery. At its distal end there are several raised whitish oval patches (*Peyer's patches*).

The small intestine passes into an oval sac (*sacculus rotundus*), with walls resembling Peyer's patches. Just beyond this there is given off a very large dark thin-walled diverticulum (the *cæcum, Cæ.*), round which there passes a spiral constriction. It ends blindly in a finger-like process (*appendix vermiformis*), the walls of which are thicker, and again resemble Peyer's patches.

Just beyond the sacculus into which the small intestine passes, and beyond the cæcal appendage, is the commencement of the large intestine, which seems at first sight to be a continuation of the cæcum. Soon, however, it narrows and becomes puckered with lateral and median sacculations. Gradually the sacculations of this anterior portion (*colon, Co.*) of the large intestine become less marked, and it passes into the terminal smooth portion (*rectum, rm.*) of the alimentary canal, which contains pill-like balls of fæces. It passes from the right anterior corner of the abdominal cavity backwards through the pelvic arch to open externally by the *anus*, near which rectal glands are developed on either side of the rectum. The relative lengths of the various parts of the alimentary canal should be measured. The total length is some fifteen or sixteen times the length of the body. For the examination of its parts it must be turned over in the cœlom.

2. *The Respiratory System.*—The roof of the mouth, which forms also the floor of the nasal chambers, is transversely ridged. This is continued backwards as the soft *velum palati* above which the two nasal passages become confluent. These nasal passages may be regarded as the air-track, the mouth as the food-track. Note that the air-track is here dorsal. The food-track continues as the *œsophagus*, the air-track as the *trachea*, through the *glottis*, which is guarded by the tongue-like *epiglottis* in front and the *cornicula laryngi* behind. Note that here the air-track is ventral, so that in the *pharynx* the air-track and the food-track cross.

Behind the glottis is the *larynx* or organ of voice (the brown glandular mass on either side of and beneath which is the *thyroid gland*). The air-track continues as the *trachea*, strengthened with cartilaginous rings, strong ventrally, but incomplete dorsally. Posteriorly the trachea bifurcates to form the two *bronchi*, one for each lung, which break up within the spongy substance of the lungs into a multitude of bronchial tubes.

The mechanism of respiration differs from that in the frog. Each lung lies in a separate pleural cavity, lined with a delicate membrane, the *pleura*, a visceral layer of which is reflected on to the lung. This cavity the lung completely fills. The pressure of the air in the lung forces it to fill this cavity, which itself contains no air or other gas. Between the lungs lies the heart in its pericardial cavity. When the thorax is complete, therefore, it is completely filled by the distended lungs on either side, and the heart in the middle. Any alteration in the size of this thoracic box will cause a corresponding alteration of size in the distensible lungs. In two ways can the size of the thoracic box be altered, (1) By a slight change in the position of the ribs; (2) By a change in the position of the diaphragm. When the size is increased air is sucked in, when it is diminished air is forced out. In the frog we had a buccal force-pump. In the rabbit we have a costal and diaphragmatic suction pump.

3. *The Heart and Circulatory System.*—The pericardial cavity in which the heart lies is between the pleural cavities in which lie the lungs. The student should tear away a little of the parietal layer of the pleura of one side, and satisfy himself that the pericardium is a separate membrane. As he does so he may notice between the two a white thread, the phrenic nerve which passes to the diaphragm.

The heart itself is slightly conical, with its blunt apex lying somewhat to the left of the median line. Posteriorly are the ventricles (16 and 17, *l. v.*, *r. v.*), there being two distinct cavities, and not only one as in the frog and the fish. They are thick-walled and fleshy. Anteriorly are the two auricles (*r. au.*, *l. au.*), thin-walled, and distended with blood, and, especially if the rabbit is young, partially hidden by a soft fatty mass (the *thymus gland*), which should be removed. There is no truncus arteriosus or distinct sinus venosus, such as we saw in the frog.

A large thick-walled artery crosses over the anterior end of the heart to the left; it is the *pulmonary artery* (*p. a.*) carrying blood to the lungs. Another thick-walled artery arises rather to the right of this (but from the left ventricle), and after giving off branches (*common carotid*) which may be traced upwards beside the trachea to the head, and branches to the arms, of which that on the right side is given off from the right carotid (*r. c.*), passes back over (dorsal to) the heart. It is the *aortic arch* (*ao. ar.*), of which there is but one in the rabbit. It curves over to the left, crossing the left bronchus, and represents the left aortic arch of the frog. As it passes along the roof of the peritoneal cavity, the thick-walled pink *aorta* (*ao.*) gives off *renal* branches to the kidneys, three to the stomach and intestines, one in front of and two behind the renal arteries, and finally divides into two branches (18, *com. il. a.*) for the hind-limbs.

Close by this pink thick-walled empty aorta will be seen the larger thin-walled *postcaval vein* (*pt. c.*). This carries blood to the right auricle of the heart from the posterior parts of the body, and from the kidneys, to which there passes no renal portal vein. The postcaval shortly before reaching the heart receives the hepatic vein from the liver, which conveys from that organ not only the blood received from the hepatic artery, but the much larger quantity delivered to it from the stomach and intestines by the *portal vein*. The right auricle, which receives the post-caval vein from the posterior regions of the body, receives also *precaval* veins (*r. pr. c.*) from the head and anterior regions. Of these precavals the large (*external jugular, ex. ju.*) veins seen on each side of the trachea, the smaller (*internal jugular*) veins lying still closer to the trachea, and large (*subclavian, s. cl.*) veins from the arms are factors.

The left auricle receives, by *pulmonary* veins (*p. v.*), the blood which has passed through the capillaries of the lungs.

The distribution of the vessels will be seen by reference to Figs. 17 and 18, and will be more fully considered in Chapter X.

4. *The Urino-genital System.*—The kidneys, which have the characteristic mammalian form, are attached to the body-wall

of the dorsal part of the peritoneal cavity, the peritoneum
passing over their ventral surfaces.   The left is a good deal

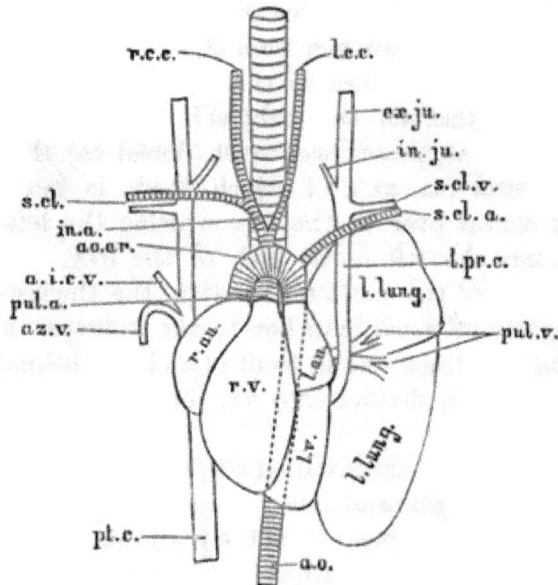

FIG. 17.—HEART AND GREAT VESSELS : RABBIT.

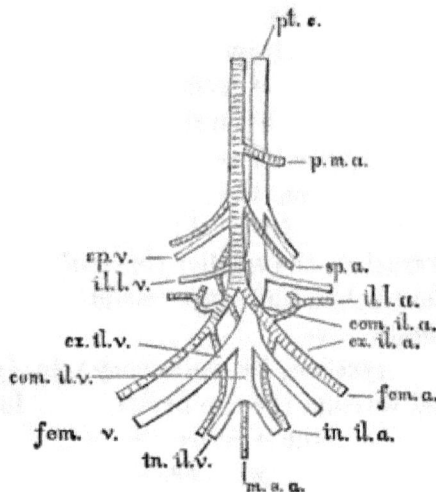

FIG. 18.—PELVIC VESSELS : RABBIT.

posterior to the right. Close to the right kidney, but somewhat anterior to the left, is on either side a yellowish oval

body (*adrenal*). From the inner edge (*hilus*) of each kidney a *ureter* passes back and enters the *bladder* (16, U. B.), which is an oval sac lying at the posterior end of the cœlom. It communicates with the **exterior in a manner** which differs considerably in male **and female.**

There is no connection between the *testis* of **the male and the** kidney. The testes (*Ts.*) of the rabbit undergo, during development, a remarkable change of position, by which they **pass** out of the cœlom, and come to lie in little pouches **of the** skin below the pelvic arch, one on either side **of the penis.** They are provided with special ducts (*vasa deferentia*), **which** will be seen to form a loop on either side over the ureters. **They** then join the urethra, which forms the canal of the penis.

The *ovaries* in the female are somewhat elongated oval **bodies,** upon which, in the adult, **are** a number of minute projections, the *Graafian follicles*, which contain the ova. Close to each ovary is the funnel-shaped **opening of the** *Fallopian tube*, into which the ova are received **when they are** shed. This is **somewhat** convoluted, but gradually expands to form the *uterus*, in which the ova, fecundated by spermatozoa within the Fallopian tube, undergo prenatal development. The two uteri unite in the middle line into the common *vagina*, which opens into the vestibule leading to the external aperture or *vulva*.

5. *The Nervous System.*—At the back of the peritoneal cavity may be seen nerves; several of these in the region of the arm form a *brachial plexus*, those in the thorax run out between the ribs, and several of those in the region of the hind-limb form a *lumbo-sacral plexus*. The *sympathetic* nerve chain may be seen in the thoracic region, lying on the heads of the ribs on either side of the vertebral column; while in the region of the neck a flat nerve, lying beneath the carotid artery, alongside a rounded cranial nerve, is a continuation of the sympathetic chain.

The *brain* must be displayed by breaking away the roof of the skull, and the spinal cord by removing the arches of the more anterior vertebræ. The brain is relatively very much larger, and considerably more complex in form than that of the frog, and it fits much **more** closely into its bony case. A tough membrane (*dura* *mater*) lines the skull and spinal canal, while a delicate membrane (*pia mater*) covers the brain and myelon. Between the two is a third (*arachnoid*) membrane. (These membranes also exist in

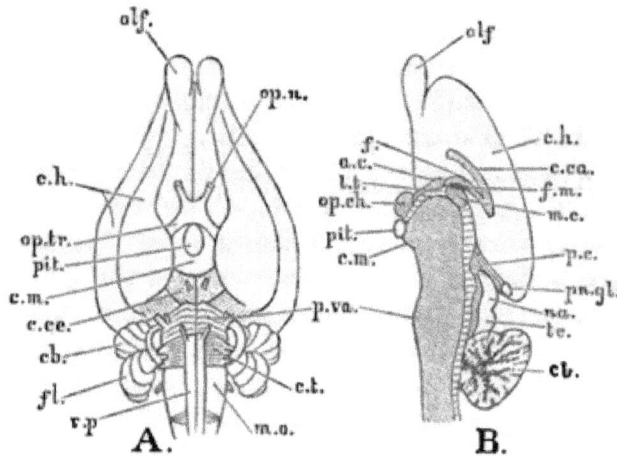

FIG. 19.—BRAIN OF RABBIT.

A. From below. B. In longitudinal section.

our other types.) The brain should be carefully removed, together with about an inch of the spinal cord. A number of nerves which proceed from the brain will have to be carefully severed. Before putting it into strong spirit for future study its chief parts may be noted as seen from the dorsal aspect. Posteriorly is a part which looks like an enlargement of the spinal cord (*medulla oblongata*, Fig. 19, *m. o.*). Over it hangs a convoluted ridged mass, with a median and two lateral portions (*cerebellum*, *cb.*). The medulla oblongata and cerebellum are parts of the *hind-brain*. In front of this is a smooth, somewhat pear-shaped portion, divided in the middle line. The two divisions constitute the *cerebral hemispheres* (*c. h.*). Anterior to these are two club-like prolongations, the *olfactory lobes* (*olf.*). The cerebral hemispheres and olfactory lobes are parts of the *fore-brain*. The mid-brain is hidden from view. But on lifting the posterior edges of the cerebral hemispheres, four rounded prominences (*optic lobes, na.* and *te.*), answering to the two we saw in the frog and the fish, will be seen. They are parts of the mid-brain.

Attention may now be drawn to some of the more salient points in the general anatomy of the rabbit :—

1. The lungs are contained in separate pleural sacs, and these, together with the pericardial sac, are contained within the thoracic cavity, which is well marked off from the peritoneal cavity. The cœlom is thus sub-divided.

2. There is a complete diaphragm behind the lungs.

3. The heart has two ventricles and two auricles ; there is no separate sinus venosus or truncus arteriosus.

4. There is one aortic arch, the left.

5. There is no renal portal vein distributing blood from the posterior regions to the kidney.

6. There are separate ducts for the testis and the kidney in the male.

7. There is no cloaca. The urino-genital aperture being distinct from the anus.

8. The urinary bladder receives the ureters and opens into a urino-genital canal.

4. The Pigeon.—After plucking the ventral aspect of the pigeon, tie a piece of fine string round the beak at the base of the cere, insert the blow-pipe into the mouth, and tighten the string. Then inflate until the crop and abdomen are moderately distended, and withdraw the blow-pipe, tightening and knotting the string so as to prevent the escape of air. Place the bird on its back, and make a median incision in the skin from beak to cloaca. Reflect the skin and make out the structures already visible in the neck. The soft thin-walled gullet or *œsophagus* is seen to expand into a large bi-lobed *crop*. Over it anteriorly lies the ringed *trachea*, which passes to the left, and so over to the dorsal side of the crop (Fig. 20, *Tr.*). Behind the crop, in its present position, lie the *jugular* veins gorged with blood, and connected with ramifying plexuses of blood-vessels in the skin of the neck. A stout nerve (the *vagus*) accompanies each jugular. Paired oval reddish lymphatic glands will also be visible in the neck.

Make an incision in the abdominal walls from the posterior

edge of the sternum to the cloaca. The contents of the peritoneal cavity or cœlom will be partially hidden by the *great omentum*, a sheet of membrane loaded with fat. By lifting it, the coils of the intestine and the membranous *air-sacs* will be displayed. Make transverse incisions along the posterior border of the sternum for a short distance on either side of the median line, being careful to injure as little as possible the air-sacs. It will be well at once to describe the position of the chief air-sacs. They are membranous bags with transparent walls, and are connected with the respiratory system by tubes which pass through the substance of the lungs.

The air-sacs are nine in number, eight lateral in four pairs, and one median, resulting from the confluence of a fifth pair. There is one pair *cervical*, or prebronchial, lying above the lobes of the crop. Then follows the median *interclavicular*, resulting from the union of a sub-bronchial pair, and sending out prolongations towards the arm-pits. It lies in the fork of the merrythought. The third and fourth pairs (*anterior* and *posterior thoracic*, or *intermediate*) are covered over by the sternum. The last pair (*abdominal* or *posterior*) are those seen on opening the cœlom. The ends of the posterior thoracic are just visible beneath (dorsal to) the sternum without further dissection.

The sternum must now be removed, an operation that requires some little care. The central keel or *carina* will be seen in the mid-line of the thorax. Having freed the crop from its anterior edge, cut into the muscles of the breast by a longitudinal incision on each side of the carina. This incision should be just deep enough to sever the great pectoral muscle without cutting the subclavian muscle that lies beneath it (on its dorsal side). With the handle of the scalpel these two muscles may be separated in the mid-region of the sternum, and the edges of the great pectoral may then be separated from their anterior attachment to the merrythought (*furcula*), and their posterior attachment to the hinder border of the sternum. In reflecting the pectorals, the large pectoral veins will be brought into view, probably gorged with blood. If they be cut the blood will gush orth.[1]　The subclavian muscles may next be cut near the carina, and reflected on either side, and the reflected flaps may be removed.　The body of the sternum will thus be exposed, and anterior to it the furcula (merrythought) and two largish bones (*coracoids*), also belonging to the pectoral arch.　With strong scissors the hinder half of the body of the sternum on either side of the carina may now be removed.　A median sheet of membrane, the *falciform ligament*, will be seen to extend forwards beneath (dorsal to) the carina, and the anterior and posterior air-sacs may perhaps be seen.　The rest of the body of the sternum, its central keel, the furcula, and the greater part of the coracoids should now be removed with the utmost care.　Any clotted blood must be washed away, and the dissection cleaned up by the removal of such muscle or connective tissue as hides the relations of the heart, blood-vessels, and digestive organs.　The student may now make out the relations of the viscera by turning them over and making such dissection of the parts as he may find necessary.

1. *The Alimentary System.*—In Fig. 20 the alimentary canal and its glands have been removed from the body and pinned out so as to display its parts.　There are no large salivary glands opening into the mouth.　The gullet expands into a large crop, the walls of which, in the nursing bird, secrete a white nutritive fluid.　Behind the crop the œsophagus is wide and thin-walled, and soon expands into the pink thicker-walled *proventriculus*, which lies dorsal to the heart and liver, and is marked dotted in Fig. 20 (*Prov.*) as if seen through the liver.　This passes into the rounded hard muscular *gizzard* (*g.*), the cavity of which is small, lined with horny matter, and generally contains small stones.　From the gizzard the *duodenum* (*Du.*) takes its origin at a point near that at which the proventricular aperture is situated.　The duodenum forms a U-shaped loop, within which lies the *pancreas* (*Pn.*), a compact reddish gland having three ducts, two of which enter the duodenum above *dd*, while the third enters near the *g* on the gizzard.　The two bile ducts are marked with dark

[1] They may be ligatured by tying two pieces of thread round each of them, one-third of an inch apart, and then cut between the two threads.　But the operation is a somewhat difficult one.

lines in the figure : **both start from the right lobe of the** liver, the shorter entering the duodenum near the gizzard, the other in the distal limb of the loop. The *liver* itself is a large reddish brown mass, with a **right** and left lobe, hollowed on its dorsal surface for the reception of the gizzard and duodenum. There is no gall bladder. Following on the duodenal loop is a **short** loop of *jejunum* beginning at *Je.* ; and on this, in turn, follows **the** *ileum*, of which the proximal part forms a coil (*il.*), **and** the distal part a single loop (*il'.*). The junction of the small and large intestines (or *rectum, rm.*) is marked by two small *cæca* (*cœ.*). The rectum opens into the *cloaca*, **at the** dorsal side of which is, in young birds, a glandular oval sac, the *Bursa Fabricii* (**B. F.**) An inch or so of the small intestine

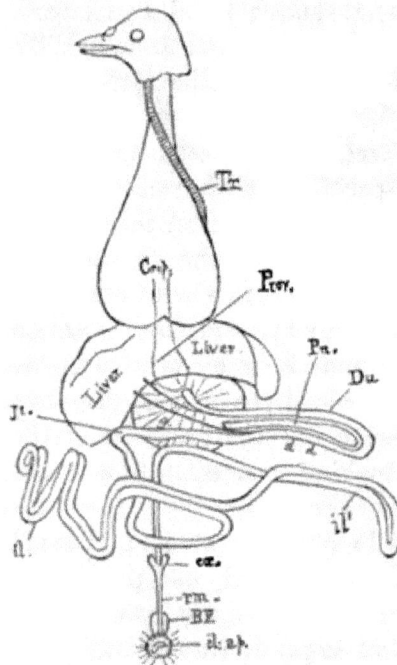

FIG. 20.—VISCERA OF PIGEON.

B. F. **Bursa Fabricii.** *cœ.* **Cæcum.** *cl. ap.* **Cloacal** aperture. *d d* Positions of two of **the** pancreatic ducts. *Du.* Duodenum. *g.* Gizzard. *il. il'.* Loops of ileum. *Je.* Jejunum. *Pn.* Pancreas. **Prov. Proventriculus.** *rm.* Rectum. *Tr.* **Trachea.**

should be cut out and slit open under water. It will probably contain a semi-fluid pulpy substance, the *chyme*. The inner walls will be velvety in appearance, from numerous minute processes termed *villi*.

2. *The Respiratory System.*—No definite epiglottis guards the glottis which opens into the *trachea*, at the anterior end of which is the *larynx*. There are, however, no vocal chords, and this organ is not in birds a functional organ of voice. The trachea is strengthened with bony rings, and at its lower end, just before it forks to send a *bronchus* to each lung, some of the lower tracheal rings are united together to form a kind of box, the *syrinx*, which is the organ of voice in birds.

The lungs are flattened and oval in shape, and are firmly fixed on each side of the vertebral column. Each is made up of a close plexus of air-tubes connected with the branches of the bronchi. There is a direct communication through the lungs with the air-sacs. The muscular fibres of the so-called diaphragm, which is totally different from that of the rabbit, arise from the ribs, outside the margins of the lungs, and from the vertebral column, and are attached to a strong aponeurosis (flattened shining masses of tendinous connective tissue) upon the ventral surface of the lungs.

The mechanism of respiration differs from that of either the frog or the rabbit. In the frog the lungs are inflated by a buccal force-pump; in the rabbit they are inflated by a thoracic suction-pump; in the pigeon they cannot be inflated at all. It is the air-sacs which are inflated, the air being sucked into them through the lungs, as the thoracic cavity is enlarged by movements of the ribs and sternum. Thus air passes backwards and forwards through the lungs with scarcely any inflation of the lungs themselves. The structure and mechanism of the lung-system are as perfect as that of the rabbit. But they are on a different plan.

3. *The Heart and Circulatory System.*—The *heart*, which lies in its pericardial cavity, beneath the anterior end of the sternum, is very large, and has four distinct chambers. The *right ventricle* (Fig. 21, *r. v.*) is smaller and thinner-walled than the left (*l. v.*). It gives off the pulmonary artery (*pul. a.*), which at once forks to send off a right and left branch to the corresponding lung. From the firmer larger *left ventricle* (*l. v.*) proceeds the **aortic arch** (ao.), and to the left of this two large (*innominate*) arteries (*in. a.*), which diverge like a **V**. Each sends off branches to the head (*carotid, c. c.*), to the wing (*brachial, br. a.*), and to the muscles of the breast (*pectoral, pc. a.*). The arch itself, which is smaller than the innominate arteries, passes back over the right bronchus, and represents the right arch of the frog, not the left as in the rabbit. At the back of the abdominal cavity it becomes the *dorsal aorta*, which gives off three branches to supply the stomach and intestines, two in front of (*cœliac, cœ. a.*, and *anterior mesenteric, a. m. a.*) and one behind the kidneys (*posterior*

4

mesenteric, p. m. a.). A large *sciatic* artery (*sc. a.*) passes through the kidney on each side, and gives off branches to that organ, which is also supplied with arterial blood by an anterior branch

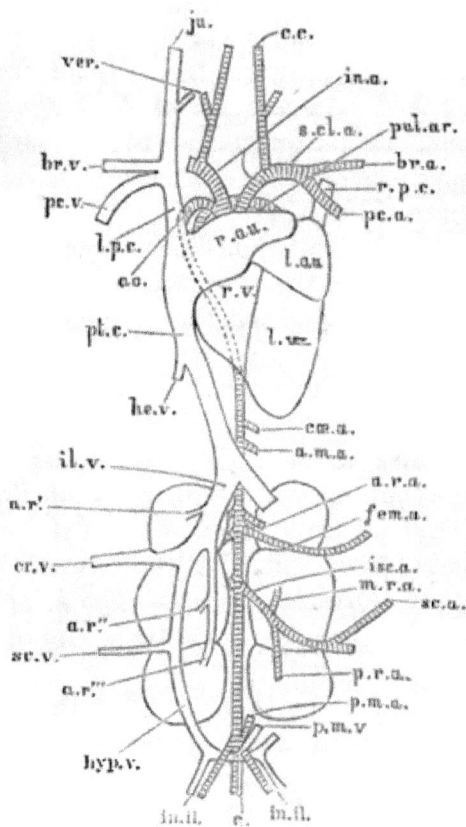

Fig. 21.—Circulation of Pigeon.

springing directly from the aorta. Posteriorly the aorta splits up into four branches, one median (*caudal, c.*); one median and ventral, the posterior mesenteric (*p. m. a.*); and one on each side (*hypogastric, hyp.*).

The blood from the stomach and intestines is delivered to the liver by a *portal* vein, and thence passes by *hepatic* veins into the postcaval (*pt. c.*). This postcaval results from the confluence of two large *iliac* factors (*il. v.*), which receive blood from the hind limbs by the large *crural* veins (*cr. v.*). Where the crural vein passes into the iliac, it is joined by a vessel which passes up through the substance of the kidney, receiving *en route* a *sciatic* factor (*sc. v.*). Traced backwards, this vessel (*hypogastric, hyp. v.*) emerges from the kidney, and may be traced to the roof of the pelvis. The hypogastric of each side, however, is connected with its fellow by a transverse vessel, and from this transverse connecting vessel a median vein (the *coccygeo-mesenteric* vein, *c. m. v.*) passes forwards, and forms a direct communication with the portal system.

The *precavals* (*l. pc., r. pc.*) are large trunks bringing blood from the head, wings, and breast, and are formed by the union of three main factors, the *jugulars* (*ju. v.*) from the head and neck, the *brachials* (*br. v.*) from the wings, and the *pectorals* (*pc. v.*) from the muscles of the breast. By severing the gullet and trachea a little below the head, and tracing the jugulars forward, they will be seen to curve round and unite in the middle line, receiving a number of small factors from the head.

4. *The Urino-genital System.*—The *kidneys* are dark-red elongated trilobed bodies closely attached to the dorsal body-walls. To their anterior ends are attached yellowish elongated *adrenals.* The ureters pass back from the point of junction of the anterior and middle lobes to the cloaca. There is no bladder.

The testes in the male have not undergone the peculiar change of position we noted in the rabbit, nor do they deliver their products to the kidneys as in the frog. From them run special ducts (*vasa deferentia*), which are convoluted and dilated posteriorly, and which open into the cloaca by apertures at the apices of conical papillæ.

In the adult female there is only one ovary, that of the left side. In it the mature, or nearly mature eggs, form large prominences. There is only one oviduct, the left, a convoluted thick-walled tube, with a funnel-shaped membranous opening, best seen by dissection under water. It opens posteriorly into the cloaca. A little process from the right side of the cloaca is the rudiment of the right oviduct.

The cloaca is divided into three chambers, an anterior, a median, and a posterior. The former receives the rectum ; the middle division receives on its dorsal side in the male the openings of the ureters, and the vasa deferentia, and in the female those of the ureters and the left oviduct. The posterior chamber has on its dorsal wall, in young birds, the aperture of the Bursa Fabricii.

5. *The Nervous System.*—In the thoracic region, white threads, the thoracic nerves, pass outwards between the ribs. The anterior thoracic and posterior cervical nerves form a brachial plexus, whence the wing is innervated. In the pelvic region there are two plexuses: one anterior, the lumbar; the other posterior, the sacral or sciatic. The gangliated sympathetic chain passes up on either side of the vertebral column, and in the neck runs within a canal embraced by bony processes of the vertebræ, and called the vertebrarterial canal.

The central nervous system must be displayed from the dorsal aspect. Great care must be taken in exposing the brain, which fits closely to the skull, the walls of which must be shaved or scraped away until the brain is exposed, and then carefully broken away piece by piece. The brain and an inch or two of the spinal cord may be removed and placed in strong spirit for future study We may note in passing the large smooth *cerebral hemispheres* (Fig. 22, *c. h.*), with the small anterior *olfactory lobes*

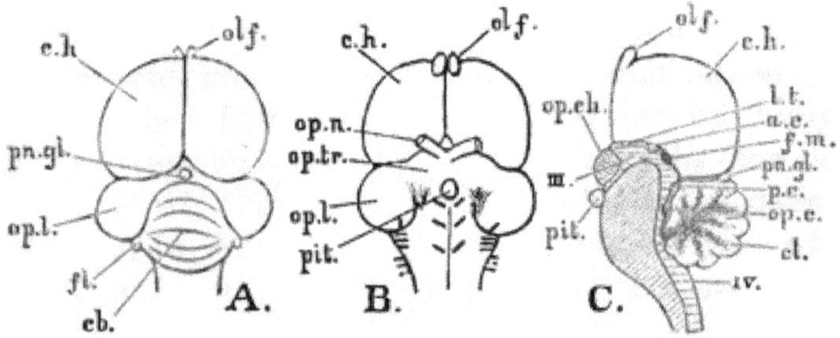

FIG. 22 —BRAIN OF PIGEON.

A. From above.   B. From below.   C. In longitudinal section.

(*olf.*) constituting the visible parts of the *fore-brain*; the ridged *cerebellum* (*cb.*), and underlying *medulla oblongata*, constituting the *hind-brain*; and between the two the rounded *optic lobes* (*op. l.*) of the *mid-brain*.

In the spinal cord there are two enlargements; a brachial enlargement between the shoulders, and a lumbar enlargement between the thighs. In the latter is a rhomboidal depression (the *sinus rhomboidalis*).

6. Attention may now be drawn to one or two salient points:

(1) The lungs are closely attached to the body-walls. They are not distensible, and only their ventral surface is covered by a pleural membrane. They communicate with distensible air-sacs.

(2) The heart is large, and has two ventricles and two auricles. There is no separate sinus venosus or truncus arteriosus.

(3) There is one aortic arch, the right.

(4) There is no definite renal-portal system, but the hypogastric veins pass through the kidney, and perhaps give off to that organ a few branches. Posteriorly the transverse branch connecting the hypogastrics is in connection with the portal.

(5) There are separate ducts for the kidneys and the testes in the male.

(6) There is a cloaca, in the dorsal walls of which are the apertures of the urino-genital ducts.

(7) The right ovary and right oviduct atrophy in the adult.

(8) There is neither urinary bladder nor gall bladder.

5. **General Anatomical Conclusions.**—Notwithstanding well-marked points of difference, our three air-breathing types, the mammal, the bird, and the amphibian, agree in possessing many fundamental points of resemblance.

1. All possess a skull and vertebrated back-bone.

2. The central nervous system is dorsal in position, and protected by the skull and arches of the vertebræ.

3. The brain is in each case divisible into fore-, mid-, and hind-brain, and the spinal cord is in each case similar in essential form and structure.

4. All have a visceral nervous system, the sympathetic.

5. As we shall see hereafter, there is a fundamental similarity in the distribution of the cranial nerves; and in all the heart, the lungs, and the stomach are largely innervated by a cranial nerve.

6. In all three the position of the heart is the same, just above the sternum.

7. Notwithstanding pronounced differences, chiefly due to suppression, the arrangement of the great vessels which enter and leave the heart is fundamentally the same.

8. Although they differ considerably in form and structure, and also to some extent in position, the relations of the lungs to the trachea, and of the trachea to the œsophagus are essentially the same.

9. In all. the alimentary canal is suspended in the body-cavity or cœlom, like a smaller tube within a greater ; but in the head and neck the greater tube is wanting.

10. In all the liver and pancreas have similar positions and relations.

11. In all the kidneys are similarly placed, and the ureters pass backward to open near the termination of the alimentary canal.

12. Although the testes of the rabbit have undergone a remarkable change of position, these organs are in all three types essentially similar structures ; but the connection of the testes of the frog with the kidneys, and the possession by that animal of a common urino-genital duct constitutes an important anatomical difference.

13. Notwithstanding the suppression of the right ovary and oviduct in the pigeon, the position and relations of these parts in our three types are fundamentally similar.

14. Many of these fundamental resemblances hold good for the cod also. The student should note for himself how far they extend.

CHAPTER IV.

GENERAL PHYSIOLOGY.

A LIVING organism is wont to be guided in its actions by impressions received from external sources. A simple impression, such as a sharp sound, a flash of light, a prick on the skin, is called a *stimulus*. If we apply such a stimulus as a light pinprick to the leg of a frog, the animal will spring away. Thus a very simple stimulus gives rise to a very complex action, one involving a great number of exceedingly delicate muscular adjustments. It is characteristic of the higher animals that very complex activities may be the outcome of very simple stimuli.

This is sometimes explained by saying that the stimulus affects the mind or consciousness of the animal, and then the mind or consciousness causes the activities which follow. But we know so little about the mind of frogs, that it will be well rather to confess our ignorance of what goes on within the organism than to accept this as an explanation.

If we kill a frog, rendered insensible by chloroform, by rapidly severing the muscles between the skull and vertebral column, so as to lay bare the central nervous cord, and then extirpating the brain by thrusting a large pin or stout wire within the skull, we shall find that it still responds to certain stimuli. If, for example, its side be touched with acetic acid, the hind-limb of that side will be drawn up and the foot passed over the spot so as to remove the source of irritation. Although the frog, as an organism, is quite dead, the tissues of its body are still living. We have no reason to suppose that the brainless frog possesses anything like consciousness, and yet complex activities follow a simple stimulus. Such a response is called a *reflex action.* We have reason for supposing that a wave of change started by the stimulus passes along a nerve-fibre to the spinal cord; that it there stimulates into activity certain groups of nervous cells within the cord; and that as a result of such stimulation, a wave of change passes down other nerve-fibres to certain muscles and causes them to contract in an orderly fashion. The nerve-fibre that carries the wave of change inward to the spinal cord is called an *afferent* fibre; the group of cells within the cord is called a *ganglionic* centre; the fibres that carry the waves of change outwards to the muscles are called *efferent* fibres. The series—afferent fibre, ganglion centre, efferent fibre—is called a *nervous arc.* If any part of the nervous arc be incomplete, reflex action does not take place. This much we may be said to know. How the incoming stimulus wave is converted into the outgoing wave, initiating orderly activities, we do not know. We are at present ignorant of the minute physiology of the process. But we believe that there is such a minute physiology.

The experiments of competent and qualified investigators have further taught us that if only the fore- and mid- brain be destroyed, the hind-brain remaining intact, the frog does not die. It breathes; it may be artificially fed; and it performs a number of simple reflexes, often with great vigour. And if only the fore-brain be destroyed, the hind- and mid- brain being left intact, the frog will eat if you feed it; it will jump on the stimulus of a needle-prick; it will swim if you put it in water; if you place it upon a board and then very gradually tilt the board, it will slowly climb and balance itself on the edge; if you put it on its back, it will turn over again and sit up; the male will croak and clasp your finger if, in the breeding season, you stimulate the pad on the manus. It is thus capable of a number of responsive actions—that is, actions in answer to definite stimuli. But all internal spring of action is gone. It is utterly incapable of initiating any action, even the simplest; and if left to itself, it will remain motionless in one position till it dies.

The inference from all which is, that hind-brain, mid-brain, and fore-brain have for their physiological function the co-ordination, and eventually the initiation, of successively more complex activities. But in what way they are able to do this— what is the exact nature and order of the changes which go on within them—these are points which it must be left to the physiology of the future fully to determine.

The activities of which we have spoken involve sense-organs, nerves, nerve-centres, and muscles. It is the function of the sense-organs to be the recipients of stimuli. It is the function of afferent nerves to transmit the waves of stimulation from the sense-organs to the nerve-centres. It is the function of the nerve-centres to receive the waves of stimulation brought by the afferent nerves, to organise them, and to transmute them into outgoing waves of stimulation. It is the function of the efferent nerves to convey these outgoing waves to the muscles which are thus stimulated to orderly contraction. The common property of nerves and muscles is *irritability*. But the irritability is differently manifested in these two different tissues.

It is not necessary here to dwell longer on this aspect of general physiology. It will be sufficient to note that the whole surface of the body is, so to speak, in communication with the surrounding world through the sense of *touch* ; while the organism is to some extent made acquainted with the state of its own organs by means of stimuli arising within the body (*e.g.* hunger). To these are added special senses which might perhaps be termed *organs of prediction*. *Taste* predicts what materials taken into the mouth are suitable for allaying hunger. *Smell* still further predicts, even before they are taken into the mouth, what materials are fitted for this purpose. *Hearing* predicts or recognises from afar pleasurable or painful sensations, and thus enables the organism to seek the one or to avoid the other. Still more does *Sight* predict the pleasurable and the painful, and enable the organism to act betimes. It must be noticed that, in speaking of the pleasurable and the painful, we have drifted into the region of feeling. It is almost impossible, in speaking of the physiology of the sense-organs and nervous system, not to do so. And although we know nothing, except by inference, of the feelings of animals, we shall probably not be far wrong in assuming that, either as the result of evolution, or in some other manner at present unknown, pleasure is associated with such activities as are for the good of the organism and of the race to which it belongs, and that pain is associated with such activities as are harmful to the organism or the race.

Leaving this aspect of general physiology, let us now note that to enable sense organ, nervous arc, and muscle (which, to avoid repetition, we may call the *senso-motor arc*) to fulfil their physiological function many other processes must go on within the body. For they are living structures, and their life must be maintained. The body is so often likened to a cunningly-wrought piece of mechanism that there is some danger of losing sight of this fundamental distinction between organism and mechanism. In a machine, such as the steam-engine, fuel has constantly to be supplied, and it is by the combustion of the material thus introduced that work can be done by the mechanism. So too in an organism food has constantly to be taken into the body, and it is by the slow combustion of the material thus introduced that work can be done by the animal. So far there is analogy. But in the case of the engine, the combustion of the fuel is restricted to a certain part of the machine, and the other parts undergo no more waste of their substance than is caused by friction. In the case of the organism, on the other hand, combustion is not restricted to any spot, but goes on within the working parts, which parts, in and through their very activity, are constantly wasting away, and therefore need constant repair. So that if we liken a nerve to such a mechanical contrivance as a telegraph wire, we must not forget that in transmitting a wave, the nerve is partially destroyed, and therefore needs constant repair. In this it differs entirely from the telegraph wire; and in this difference we have a fundamental distinction between organism and mechanism.

The parts of the senso-motor arc are, therefore, constantly undergoing waste due to oxidation. Hence the necessity of a *blood circulation* (1) to renew the wasted substance, (2) to carry off the waste products, and (3) to supply the oxygen required for further activity through oxidation.

The circulating medium is the blood. In the living state it is warm and fluid. It consists of an almost colourless liquid, the *plasma*, in which float minute solid bodies, the *corpuscles*. These are of two kinds: (1) *red corpuscles*, minute, flattened, round, or oval discs (see p. 64); (2) colourless, white, or pale corpuscles, which change their form when living, and resemble a minute organism called the amœba (see Ch. xx.). Blood undergoes rapid change when dying, causing clotting or coagulation. This is due to the formation of fine filaments of *fibrin*. If the blood be allowed to stand, the clot contracts, and a thin fluid, termed *serum*, is expressed. This serum is not the same as plasma; nor is the fibrin during life merely held in solution in the plasma to separate out on death. The fibrin is generated from the plasma and the white corpuscles. These contain a substance, fibrinogen, which, under the influence of a second substance, probably also derived from the corpuscles, gives rise to fibrin.

The blood may be *venous* or *arterial*. Our blood, for example, contains about 60 per cent of gaseous constituents by volume, chiefly oxygen, nitrogen, and carbonic acid gas. The carbonic acid gas is almost entirely held, in a state of very loose combination, by the plasma. The oxygen is bound up in the *hæmoglobin* of the red corpuscles. The proportion of oxygen and carbonic acid gas varies according to the venous or arterial condition of the blood. In us venous blood contains 12 per cent. oxygen and 46 per cent. carbonic acid gas; arterial blood, 20 per cent. oxygen and 39 per cent. carbonic acid gas. Thus the carbonic acid gas in the blood is always in excess of the oxygen. The operations of giving up the carbonic acid gas and absorbing fresh oxygen seem to be quite independent. Let us note then that the red corpuscles are the carriers of oxygen throughout the system; while the plasma distributes the nutritive material and carries off the waste products.

The organ for propelling the blood is the heart, which thus provides for the continuity of the circulation. In the pigeon and the rabbit it has two sides completely separated off from each other without possibility (in the adult) of direct communication. Each side has two parts—(1) a receiver, the auricle;

and (2) a pump, the ventricle. Between the two in each case
are valves, which permit the blood to flow from the receiver into
the pump, but prevent its flowing from the pump into the
receiver.

In the course of the circulation, beside the senso-motor arc, are
(1) lungs, (2) kidneys, (3) the alimentary canal. In the lungs
the red corpuscles are freighted with oxygen, which they convey
to various parts of the body, returning to the lungs empty. The
blood which contains the laden corpuscles, and which is bright-
red in colour, is called *arterial*; that which contains the
exhausted corpuscles, and which is dull-red in colour, is called
venous blood. In the lungs also the plasma delivers up one
main product of tissue waste, the carbonic acid gas. So that the
lungs have a double function. In their relation to the blood
corpuscles they have the function of arterialisation · in their relation to the blood plasma they have an excretory function. In
the kidneys a further and special product of animal metabolism
(see p. x), the urea, is eliminated from the plasma. In the alimentary canal the plasma is enriched with nutritive material.

In each of these organs the blood-vessels branch and subdivide until they form a close meshwork of minute vessels
termed a *capillary plexus*. In each of the multitude of capillaries
which make up such a plexus the walls are so thin that tissue
nutriment and tissue waste is readily given up or received by
the blood ; and from the extreme fineness and the great number
of the tubes, the amount of surface thus bathed by the blood is
relatively enormous. The sub-division of arteries into capillaries,
and the union of capillaries into veins, may readily be observed
under the microscope in the web of a frog's foot. The frog must
be lightly tied to a piece of thin board or thick card-board, in one
edge of which a V-shaped notch has been cut. Over the notch
the web is placed, threads being tied to two of the toes to keep
them sufficiently apart. The frog should be kept moist by
placing a piece of wet blotting-paper over his body. In every
organ of the body the blood-vessels form such capillary plexuses
as are thus seen spread out like a network in the web of a frog's
foot.

We have seen that, in the pigeon or the rabbit, the heart is like a double pump, or a pump with two completely separate chambers placed side by side. One of these, the pulmonary pump, forces blood into the lungs; the other, the systemic

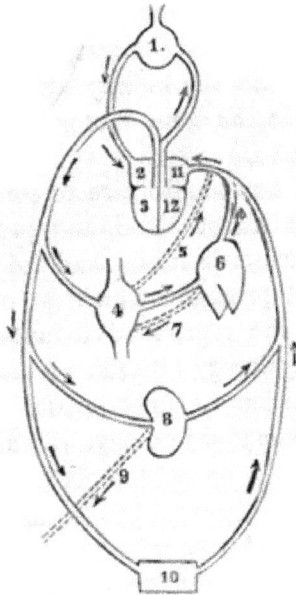

FIG. 23.—PHYSIOLOGICAL DIAGRAM.

1. Lung receiving blood from the right ventricle, 12. Here oxygen is absorbed into the system, and carbonic acid gas and water are got rid of by the system. 2. Left auricle receiving the arterialised blood from the lungs. Thence it passes to 3, the left ventricle, to be distributed by means of the arteries to the left of the figure. 4. Alimentary canal. Here tissue nutriment is absorbed into the system, and is carried (1) by the lacteal, 5, into the blood-stream returning to the heart; and (2) by the portal vein to the liver, 6, where bile is elaborated and poured into the alimentary canal by the bile-duct, 7. The kidney, 8, receives blood by a separate branch-artery. Here water and nitrogenous waste are eliminated and pass out of the system by the ureter, 9. Another branch artery carries blood to the senso-motor arc, 10. The blood returns by veins (to the right of the figure) to the right auricle, 11, whence it passes to the right ventricle, and so to the lungs, 1.

pump, forces to all parts of the body the blood delivered into its receiver from the lungs. From all parts of the body the blood collects in the receiver of the pulmonary pump. Thus all the blood has to pass through the lungs. But not all has to pass through either the alimentary canal to be enriched with tissue nutriment or through the kidneys to be deprived of nitrogenous tissue waste. This will best be seen by the diagram (Fig. 23). the blood which leaves the left ventricle may either pass to the alimentary canal, thence to the liver, where it undergoes complex and highly important changes, and thence to the right auricle; or it may go to the kidney, and thence to the right auricle; or it may go to the senso-motor arc, and thence to the right auricle.

Thus to enable the senso-motor arc to continue its functions, there must be a circulation of blood, and processes of digestion and absorption, of arterialisation, and of excretion, to say nothing of those vital processes which go on within the liver. These, together with the reproductive function, are the essentials of the general physiology of such an animal as a rabbit or a pigeon. But in addition to this essential physiology there is a vast amount of accessory physiology, of which more hereafter.

CHAPTER V.

GENERAL HISTOLOGY.

STUDIED under the microscope, the organs are found to be composed of *tissues* (muscular tissue, nervous tissue, glandular tissue, etc.), and the tissues to be built up of *cells*, together with certain cell products.

To make out the form and structure of the cells, a small piece of the tissue may be *teased* to pieces in a dilute solution (·75 per cent.) of common salt (normal saline) with two needles on a glass slide. The teased tissue must then be covered with a cover-glass and examined. But to make out the relations to each other of the cells in a tissue, thin *sections* of the tissue must be cut. To bring out clearly the structure of a cell **or** of a tissue, advantage is taken of the fact that different parts of **a cell,** and different cells in a tissue, are differently affected by certain *re-agents,* such as dilute acid or staining fluids. The cells or tissues so prepared must **be** *mounted* either **temporarily** for immediate study in normal saline solution or glycerine, or permanently in glycerine or Canada balsam.[1]

---

[1] For the guidance **of the student the** following hints on section-cutting **may** be useful :—

1. Cut a *small* fragment fresh from the **organ to** be examined.

2. Place in (*a*) osmic acid, 1 per cent., **for two** or three minutes ; or (*b*) in chromic acid, 5 per cent., for twenty-four hours ; or (*c*) in Kleinemberg's picric acid for eighteen hours. The object will thus be hardened.

3. Place in 50 per cent. alcohol for twenty-four hours, then 70 per cent. alcohol **for** twelve hours, 80 per **cent.** alcohol for six hours, and transfer to 90 per cent. alcohol. This removes **the water.**

4. Place in staining fluid **until** sufficiently deeply stained. (1) Hæmatoxalin, (2) picro-carmine, (3) eosin, and (4) borax-carmine, are the most useful re-agents.

1. Blood and Lymph.—Examined under the microscope (high power), the blood is seen to consist of a colourless fluid, the *plasma*, in which float a great number of yellowish *red corpuscles* and a few *white corpuscles*. The red corpuscles of the frog (Fig. 24, i. *b*) and the pigeon are flattened slightly bi-convex oval discs, and contain a *nucleus*. Those of the rabbit are circular bi-concave discs (Fig. 24, i. *c*) and contain no nucleus. The size of the corpuscles is, in the frog, $\frac{1}{900}$ inch; in the pigeon, $\frac{1}{2300}$; in the rabbit, $\frac{1}{3800}$. Treated with dilute acetic acid, the nucleus (when present) becomes more obvious, the colour of the surrounding parts of the corpuscle is gradually discharged, and eventually that of the nucleus also. If a needle be rapidly drawn across a drop of blood, some of the cells will be cut in two. There is no escape of the contents from the cut halves, showing that they are composed throughout of a viscid material, and not of a fluid contained within a membranous wall. The manner in which the corpuscles recover their shape after distortion by pressure may be seen in the capillaries of the living frog (see p. 60).

The white corpuscles (*leucocytes*) are of much the same size ($\frac{1}{2500}$ inch) in frog, pigeon, and rabbit, and are in all cases nucleated. On a cold slide they will probably appear round; but if the slide be warmed they will show *amœboid* movements, the proto-plasm flowing out into irregular processes and slowly creeping over the slide (Fig. 24, i. *a*). Sometimes they divide by *fission*. The nucleus first splits into two, and then the two halves of the

If borax-carmine be used, transfer it to 70 per cent. alcohol, to which a few drops of nitric or hydrochloric acid have been added.

5. After staining, place for an hour in 80 per cent. alcohol, and then in absolute alcohol for another hour.

6. Place in spirits of turpentine for an hour, and then transfer to paraffin (sold of proper hardness for this purpose of imbedding), which should be kept just melted. Leave for six or eight hours, and then cast in a mould. The small oblong vessels in which moist water-colours are sold answer well. Arrange the specimen with warm needles.

8. Pare away the paraffin so as to bring the object into view. Cut the thinnest sections you can get with a sharp razor.

9. Dissolve the paraffin in turpentine.

10. The section may now be examined or mounted permanently in Canada balsam.

corpuscle, each containing a nucleus, seem to creep off in opposite directions, the protoplasm between them getting more and more constricted, till its continuity at last is broken. In the capillaries of the frog's foot the white corpuscles will be seen to creep along the capillary walls.

The lymph is a colourless fluid containing, in the larger trunks, leucocytes quite similar to the white corpuscles of the blood In the *lacteals, e.g.* of the rabbit killed shortly after a full meal, the lymph is milky in appearance, and is called *chyle.* The milkiness is seen under the microscope to be due to multitudes of very minute fatty particles, which dissolve in ether, and are deposited as oil-drops on the evaporation of the solvent.

In the mesentery of the rabbit a number of *lymphatic glands* will be seen. A section of such a gland will show an outer portion or *cortex*, divided into compartments (*alveoli*), and an inner portion or *medulla*, formed of interlacing strands. In the alveoli of the cortex there are roundish masses or *follicles* of *adenoid* tissue formed of a network of connective fibres, the interspaces of which are crowded with leucocytes similar to the white corpuscles of the blood. The strands of the medulla are formed of similar tissue, also crowded with leucocytes. The Peyer's patches are composed of lymphatic follicles aggregated together and full of leucocytes. From such glands, and there are many such, leucocytes pass into the lymph.

2. Epithelium and Endothelium.—The *epithelium* is the lining membrane of such tubes or cavities as the mouth, stomach, intestine; the ducts and cavities of the salivary glands, testis, kidney, etc. They are for the most part in communication with the exterior. At the surface of the body the epithelium passes into the *epidermis.* The *endothelium* is a special variety of epithelium, and forms the lining membrane of such cavities or tubes as the peritoneal cavity, lymph sinuses, heart cavities, and blood-vessels. They are for the most part not in communication with the exterior.

The epithelium cells vary in form and arrangement in different parts of the body. Sometimes they form a single layer (*simple*

5

epithelium) as in the stomach and intestines, and in the tubules and finer ducts of the kidneys, salivary glands, etc. ; sometimes they form a tissue several layers deep (*stratified* epithelium) as in the mouth, bladder, and epidermis. Sometimes the individual cells are more or less *spherical*, as in the deeper layers of the epithelium of the mouth, and in the peptic glands of the stomach; sometimes they are flattened and *squamous*, as in the superficial layer in the human mouth; sometimes *polyhedral*, as in the external layer of the conjunctiva covering the eye; sometimes *ciliated*, as in the lining cells of the frog's mouth, and in the trachea of mammalia; sometimes *columnar*, as in the simple epithelium of the intestine. Sometimes the cells become *secreting* cells, as in the salivary glands or the pancreas; sometimes *horny* cells, as in the epidermis, in hair, and in nails; while in the tubules of the testis they become *seminal* cells, giving rise to spermatozoa.

Some of these varieties we shall become further acquainted

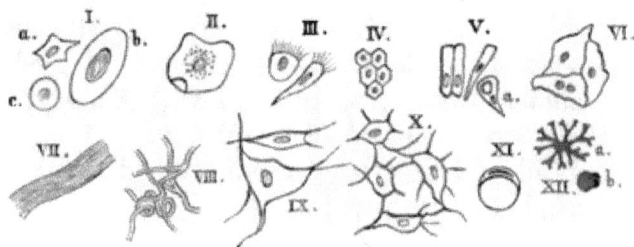

FIG. 24.—BLOOD, EPITHELIUM, AND CONNECTIVE TISSUE.

i. *a*. White blood-corpuscle. i. *b*. **Red corpuscle** (frog). i. *c*. Red corpuscle (rabbit). ii. Squamous epithelium, human mouth. iii. Ciliated epithelium, mouth of frog. iv. Corneal epithelium, rabbit. v. Columnar epithelium, frog's intestine. vi. Epithelioid cells of peritoneum of frog. vii. White fibrous **tissue**. viii. Yellow elastic tissue. ix. Branched connective tissue corpuscles. x. Corneal corpuscles. xi. Fat cell. xii. Pigment cells : *a*. passive; *b*. active.

with in this chapter. At present we may content ourselves with the following :—

(*a*.) Scrape gently the inner surface of your lip, and mount in saliva. **Flat nucleated** squamous cells will be seen (Fig. 24, ii.). **They form** the superficial layer of the stratified epithelium of the mouth. **The deeper cells are** rounder. These latter constantly multiply by division, and thus force **fresh layers of the squamous** cells to the surface, where they are **shed. In the skin a similar process goes on; but the external cells are more horny, and have lost their nucleus.**

(*b.*) Scrape gently the surface of the eyeball in a dead rabbit, and mount in normal salt solution. The cells are polyhedral and nucleated (24, iv.).

(*c.*) Scrape gently the inner surface of the roof of the frog's mouth, and mount in normal salt solution. The cells are ciliated along their free edges. The nuclei are readily seen (24, iii.).

(*d.*) Scrape gently the inner surface of the intestine of the frog, and mount in normal salt solution. The cells (24, v.) are columnar, and some of them pointed at one end. In some (mucous cells) a globule of mucus may be seen at the broader end. The nuclei are nearer the deep end.

Endothelial or epithelioid cells are flattened squamous cells of various shapes, polygonal, elongated, or irregular. They form a simple membrane one layer thick.

(*e.*) Remove carefully the transparent membrane that separates the pleuro-peritoneal cavity of the frog from the subvertebral lymph space. Treat it for two or three minutes with a dilute solution (·5 per cent.) of nitrate of silver, wash with distilled water, and expose to bright light until it is stained brown. The irregular endothelial cells of the peritoneum will be marked out with black lines (24, vi.), for the *cement substance* which lies between the cells is stained black by the silver. It may be that smaller cells surrounding minute orifices (stomata) will be seen. They are called *germinating endothelial cells,* and actively multiply by division. The stomata communicate with the coelom.

All of these cells (*a.-e.*) should be measured and drawn from Nature.

3. Connective Tissue.—This tissue is to be found in all parts of the body. It forms the framework or scaffolding, within and around which the various organs are built. Its fibres support the specialised cells of these organs.

A piece of *tendon*, by which a muscle is attached to a bone, and a piece of *inter-muscular tissue*, the filmy glistening material seen when two muscular bundles are separated, should be taken from a frog's hind leg, teased apart, and mounted (*a*) in normal saline solution, (*b*) in acetic acid. They may also with advantage be stained with magenta. In those mounted in normal saline, and especially in the tendon, a number of bundles of *white fibrous connective tissue* (24, vii.) will be seen. They are marked with longitudinal striations, and tend to break up into *fibrillæ*. In those mounted in acetic acid the fibrous tissue has been rendered almost or quite invisible. Great numbers (especially in the inter-muscular tissue) of fine wavy branching fibres of *yellow elastic tissue* will be seen (24, viii.) ; and here and there, with the nucleus and cell-substance more or less stained, will be *connective tissue corpuscles*, nucleated cells, elongated in the tendon, and branched in the inter-muscular tissue (24, ix.). These are the living formative cells. The white fibrous and yellow elastic tissue is to be regarded as *inter-cellular material*, analogous to the cement substance in endothelium and epithelium.

The *pigment cells* (24, xii. ; *a.* passive, *b.* active) of the frog's skin, which, as we have already learnt (p. 7), change their form in accordance with the intensity and quality of the light which falls on the frog's eye, are modified connective tissue cells, as are also the *corneal corpuscles* seen (24, x.) in a section of the cornea treated with gold chloride. *Fat-cells* (which may readily be obtained from the *corpus adiposum* of the frog) are also strangely modified connective tissue cells (24, xi.). In them the flattened nucleus and a small remnant of protoplasm are thrust to one side of the rounded cell, which contains a relatively huge globule of oil. Treat the cells with ether ; the oil will dissolve and the cells collapse.

4. Cartilage.—If the anterior or posterior cartilaginous expansion of the sternum be removed from a recently killed frog, and mounted in normal saline, rounded nucleated cells will be seen occupying cavities in a transparent matrix. The cells which occupy these cavities (*lacunæ*) will be seen in various stages of division.   In some cases the nucleus has divided so that there are two nuclei to one cell; in others the cell has also divided, so that two cells occupy one lacuna; in others a thin layer of matrix divides two cells which have but recently separated from one another (Fig. 25, i.).   The matrix is clearly an intercellular substance, answering to the cement substance in endothelium.

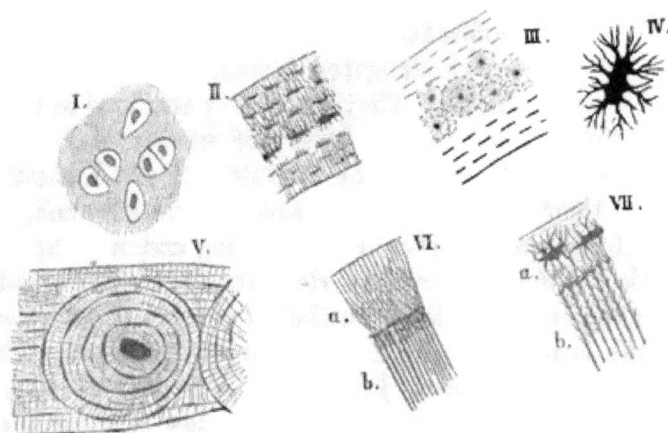

FIG. 25.—CARTILAGE, BONE, AND TOOTH.

i. Cartilage. ii. Dry femur of frog in transverse section. iii. Metatarsal of rabbit (transverse). iv. Lacuna of same. v. Haversian system (rabbit). vi. Transverse section of part of canine tooth of cat, showing, a. enamel, b. dentine. vii. Part of fang of a similar tooth, showing, a. cement, b. dentine.

The free surface of cartilage is invested by a membrane of fibrous tissue, the *perichondrium*.   Under favourable circumstances fine canals may be seen passing between the lacunæ, and from them to the perichondrium.   It may be that by them the cartilage is irrigated with lymph.

Elastic and *white fibro-cartilage* are varieties in which the matrix tends to pass into elastic or fibrous tissue.

5. **Bone.**—Bone sections may be prepared for microscopic examination in two ways; either (1) by grinding down (with file, pumice-stone, and hone) a piece of the solid and dry bone until it is sufficiently thin, and then mounting it dry; or (2) by

decalcifying the bone by dilute acid (chromic or hydrochloric), and then cutting sections in the usual manner.

We may begin by studying a section of a frog's femur. Externally the living bone is invested by a layer of fibrous connective tissue, the *periosteum*, on the inner surface of which is a layer of special bone-forming cells or *osteoblasts*. Within the living bone is a very vascular fatty substance—the *marrow*—also containing osteoblasts. A section of the dry bone (25, ii.) shows a number of dark elongated *lacunæ*, from which radiate a number of fine *canaliculi*. They are clearly arranged in two sets, an inner and an outer. The inner are smaller, and their fine canaliculi pass down towards the marrow cavity. The outer are larger, and their canaliculi passing towards the circumference of the bone, form branching tufts intersecting and uniting to form a network. The lacunæ run in rings round the bone, and thus mark out the *lamellæ*. Between the inner and the outer lamellæ is a *middle lamella*, in which there are no canaliculi. In the young growing bone each lacuna has a *bone cell* sending out processes into the canaliculi. There are a few Haversian canals near the nutritive foramina for the entrance of the blood-vessels·

A section of a metatarsal bone of the rabbit (25, iii.) shows us an inner and an outer set of *circumferential lamellæ* with small lacunæ. But between these two regions there is a wider space, in which there are a number of groups of lamellæ surrounding a small cavity. Between the lamellæ are lacunæ. The central cavity (from which fine radiating lines proceed outwards) is the cross-section of a *Haversian canal*, containing blood-vessels, the lamellæ round which are called *concentric lamellæ*. The Haversian canals are in connection either with the medulla or with the periosteum. One of the lacunæ is figured under a high power in 25, iv.

A section of the femur of a rabbit shows us in the main an aggregate of Haversian systems, one of which is shown in Fig. 25, v. There may be some *circumferential lamellæ* externally, and between the Haversian systems some *interstitial lamellæ.*

Close to the marrow cavity, especially towards the ends of the growing bone, the structure is much more open than that of the compact bone above described.   This is called spongy-bone. Many of the Haversian canals have been hollowed out into large *Haversian spaces*, which may show a tendency to run one into another.

It is clear that the ossified matter of the bone-lamellæ is an intercellular substance—a substance produced by the bone-cells— answering to that which is transparent and hyaline in cartilage, and fibrillated in fibrous connective tissue.   In a bone softened with hydrochloric acid some of the lamellæ may be torn off in shreds.   The matrix thus obtained shows crossing fibres; and from the lamella processes are given off which project at right angles to the lamella and pass through holes in adjacent lamellæ. The holes for the admission of such processes will also be seen.

The bone of the pigeon is essentially similar to that of the rabbit.   But a cross-section of the bone of a cod-fish shows a structure very different from that seen either in the frog or in the rabbit.   It is more like ossified or calcified cartilage, there being none of the lacunæ or lamellæ so characteristic of true bone.

6. Teeth.—Tooth sections may be prepared either from the dry tooth by grinding, or from the decalcified tooth by the ordinary process of section-cutting.

In the interior of the tooth, passing up from the bottom of the fang, is the *pulp-cavity*, empty in the dry tooth, filled with vascular and nervous pulp in the living state.   Around this, forming the mass of the tooth, is the *dentine*, through which there run a number of slightly waving or spiral *tubuli*, sometimes giving off fine branches.   In the fresh state each contains a fibre connected with the cells (*odontoblasts*) which line the pulp- cavity.   At their outer ends the tubules pass into a network of intercommunicating spaces.   Thus the dentine is an inter- cellular substance formed between the fibres of, and due to the vital activity of, the odontoblasts.

The material encrusting the dentine differs according as our section passes through the crown or the fang.   In the crown it consists of *enamel* (Fig. 25, **vi.**), composed of *enamel prisms*, set

more or less at right angles to the surface, or in gently curved
bundles, the curves at slightly different levels intersecting each
other. The individual prisms may be isolated by means of
hydrochloric acid. In the fang the encrusting material is called
cement, which resembles bone in structure. In it are large
lacunæ, the canaliculi of which communicate with the ultimate
endings of the tubules of the dentine (Fig. 25, vii.).

The tooth of the cod-fish differs markedly in structure from
the mammalian tooth above described. An inner zone has
large canals; a thinner outer zone fine-branched tubules.

7. Muscle.—Muscle is divided into (*a*) striped or voluntary
muscle, and (*b*) unstriped or involuntary.

(*a*.) *Striped.*—Of this kind are the most important muscles
of the trunk and limbs. The fibres are elongated, and are
surrounded and bound together by connective tissue (*endomysium*), in which run the capillaries of the blood-vessels. The
bundles into which the fibres are thus bound together are in
turn invested by stronger bands of connective tissue (*peri-
mysium*).

Fresh striped muscle of the frog (cp. also that of an invertebrate such as the water-beetle, hydrophilus, or the cockroach),
mounted in normal saline, is seen to be composed of glassy or
slightly opalescent fibres surrounded by a delicate sheath, the
sarcolemma. Each is transversely striated, and contains fusiform muscle corpuscles. Treatment with acetic acid renders the
striation indistinct, and the corpuscles more conspicuous. They
are embedded in the midst of the fibre in frog's muscle, but lie
just beneath the sarcolemma in that of the rabbit. Dead fibres
tend to split up into a number of separate *fibrillæ*, and a muscle
that is hardened in strong spirit very readily splits up into such
elements. Muscle fibres hardened in picric acid tend to break
up into discs.

In a fresh muscle fibre teased in normal saline and examined
under the highest powers of the microscope, the following structure may, under favourable conditions, and by careful focussing,
be observed :—The fibre shows alternate bands of lighter and

darker substance. The lighter bands are the *septal zones* (Fig. 26, iii.), the darker the *interseptal zones.* Within the latter very delicate longitudinal striations may be made out. In the midst of the septal zone is a delicate dotted line, the *septal line.* The space between the septal lines is termed a *muscle segment.*

In growing muscle spindle-shaped cells are active in the formation of muscular fibres, just as spindle-shaped or branched cells are active in the formation of fibrous connective tissue.

Fig. 26, i., shows a muscular fibre of the frog, the elongated black spots being the muscle corpuscles. Where the fibre is twisted the sarcolemma becomes obvious. 26, iii., is a

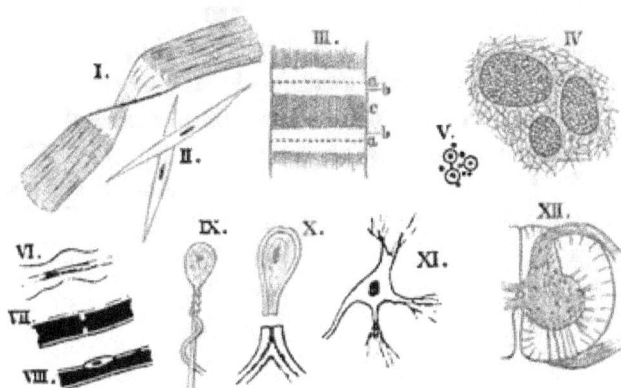

FIG. 26. — MUSCLE AND NERVE.

i. Striped muscle. ii. Unstriped muscle cells. iii. Striped muscle under highest power. *a.* Septal line. *b.* Septal zone. *c.* Interseptal zone. iv. Transverse section of sciatic nerve of rabbit. v. A portion more highly magnified. vi. Nerve fibre, after chloroform. vii. Nerve fibre with primary node, after osmic acid. viii. Nerve fibre with nerve corpusle, after osmic. ix. Nerve cell from sympathetic ganglion of frog. x. Unipolar nerve cell. xi. Multipolar nerve cell. xii. Section of spinal cord from cervical region of kitten.

diagram to show : *a.* septal line, *b.* septal zone, *c.* interseptal zone.

(*b.*) *Unstriped.*—Involuntary unstriped muscle occurs in the walls of the alimentary canal, of the bladder, and of the blood-vessels. The fibres are moderately elongated, flattened, and tapering at the ends (fusiform). Each has a readily-stained nucleus (26, ii.), which may also be made obvious by treatment with dilute acetic acid. They are embedded in a cementing substance.

8. Nerve.—The nervous elements are divided into (*a*) nerve fibres, and (*b*) nerve cells.

(*a.*) *Nerve fibres.*—Such a nerve as the sciatic consists of a number of nerve bundles bound together. Each nerve bundle (of which three are shown in 26, iv.) is surrounded by a fibrous sheath of connective tissue (*perineurium*), and is bound to other bundles by bands of similar tissue (*epineurium*), while the fibres within the bundle are bound together by more delicate *endo-neurium*. Within the bundle are a number of dots, many or most of them surrounded by a transparent ring (26, v.). Nerve fibres isolated at once and examined in normal saline appear as delicate glassy rods, without apparent external membrane, but with faint indications of a central thread. Under the influence of re-agents, however, we see the nerve fibre com-posed of three distinct parts—a central thread, the *axis fibre*, around this a fatty *medullary sheath*, and external to this again a *primitive sheath*. Chloroform dissolves the medullary sheath, and makes clearly visible the axis fibre, as in 26, vi. Osmic acid stains the medullary sheath black. This re-agent makes obvious certain breaks in the continuity of the medulla, the *primary nodes* (nodes of Ranvier) of which one is shown in 26, vii. By the subsequent addition of a staining fluid the *nerve corpuscles* (26, viii.) in or attached to the primitive sheath are rendered clearly visible. There is generally one (sometimes more) such nerve corpuscles between two primary nodes. Besides the primary nodes there are also much more numerous and less pronounced slanting breaks in the continuity of the medulla, constituting *secondary nodes*, and giving to this sheath an im-bricated appearance.

Near the peripheral termination of a nerve fibre the axis fibre loses its medullary sheath, and becomes a *non-medullated nerve-fibre* (Remak's fibre). The dots (in 26, v.) not surrounded by a clear ring are such fibres, the others being medullated. The olfactory nerve-fibres are of this non-medullated type, as are also most of the fibres of the sympathetic system. Such fibres still retain the primitive sheath. But eventually even this may be lost, and the axis fibre may split into fine fibrillæ which may terminate in specially-modified cells. In the optic and auditory nerves there is a medullary sheath, but no primitive sheath, and no primary nodes.

(*b.*) *Nerve cells.*—Traced inwards to the spinal cord a nerve is found to result from the union of two *roots*, a ventral or motor root, and a dorsal or sensory root. On the latter is a swelling or ganglion. Here there are a number of *unipolar nerve cells* (26, x.) Each is invested with an investing sheath continuous with the primitive sheath. An axis fibre may, in some cases, be seen to be in connection with the cell, and to bifurcate at some little distance from it. In the ganglia of the sympathetic system of the frog, some bipolar cells of the type figured in 26, ix. may be found. The straight axis cylinder process passes into a non-medullated fibre : the spiral fibre, after leaving the axis cylinder, acquires a medullary sheath and becomes a medullated fibre.

On the ventral root of the spinal nerve there is no ganglionic enlargement. The fibres pass inwards into the spinal cord, and there come into relation with multipolar ganglion cells such as that figured in 26, xi.

9. The Spinal Cord and Brain.—In Fig. 26, xii. is figured a section of the spinal cord (from the cervical region of a kitten), showing the origin of the dorsal and ventral nerve roots. In the middle of the cord is a canal (*canalis centralis*) lined with ciliated epithelium. Above and below this the cord is nearly cleft in twain by a *dorsal* and a *ventral fissure*, into the latter of which pass bands of connective tissue derived from the investing membrane or *pia mater*. The substance of the cord is composed of *white matter* externally, and *grey matter* internally. The white matter consists of medullated nerve fibres without any primitive sheath and interrupted by no nodes. They run longitudinally, but form also *commissures*, passing from one half of the cord to the other. The grey matter, besides having somewhat fewer medullated fibres and more blood-vessels, has also, as a characteristic, great numbers of nerve cells of various shapes and sizes, but generally multipolar and stellate, and larger in the ventral of the two divisions or *cornua* into which the grey matter of each side of the cord is divided. In both white matter and grey matter the fibres and cells are imbedded in a supporting connective tissue (*neuroglia*) which forms a very large proportion of the grey matter. This consists of (1) homogeneous semi-fluid matrix, (2) of a network of delicate fibrils, and (3) of branched nucleated cells.

The brain is composed of white matter and grey matter; but in the cerebral hemispheres and cerebellum the grey matter is external to the white.

10. Nerve Endings.—Nerve fibres terminate peripherally in various ways. In the epithelium of the skin the axis fibre breaks up into a network of fibrillæ lying in the interstitial substance between the cells. In striped muscle the network into which the axis fibre breaks up encloses nuclei of various shapes, and is known as an *end-plate*. And in connection with the special sensations there are specialised terminations of the sensory nerves.

(*a.*) *Touch.*—In sections of a rabbit's or a cat's lip, the deeply implanted hairs (mystaces) will be seen to terminate in enormously enlarged bulbs, with which nerve fibrillæ enter into close connection. Such hairs may be regarded as special sense organs. In the skin of the human finger the sensory nerves are in connection with *pacinian corpuscles*, large oval bodies, in which the central nerve thread is surrounded with a number of coats, onion-fashion (Fig. 30, iii.). Somewhat similar corpuscles are found in the tongue, mesentery, and elsewhere. They may be tactile, but their function is not certainly known. Much smaller corpuscles in the papillæ beneath the epidermis of the human finger are undoubtedly tactile.

(*b.*) *Taste.*—On each side of the tongue of a rabbit is a plaited patch, the so-called *papilla foliata*. Fig. 27, i. shows a portion of a section across the folds of this patch. On the adjacent sides of two folds are seen four *taste-buds* imbedded in the epithelium and connected with nerve fibres at their inner ends. At ii. one of these taste-buds is shown on a larger scale.

(c.) *Smell.*—In the olfactory membrane of the nose certain elongated spindle-shaped cells are wedged in between the more

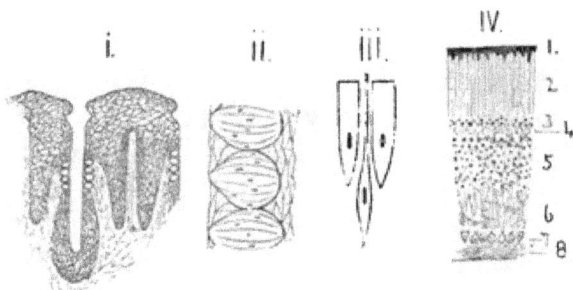

FIG. 27.—NERVE-ENDINGS.

i. Taste-buds of papilla foliata of rabbit. ii. A bud enlarged. iii. Olfactory cells (dog). iv. Retina of frog. 1. Pigmented epithelium. 2. Rods and cones; between this and 3 is a structureless limiting membrane. 3. Outer nuclear layer. 4. Outer molecular layer. 5. Inner nuclear layer. 6. Inner molecular layer. 7. Layer of nerve cells. 8. Layer of nerve fibres. The layer marked 8 is turned towards the source of light, and is bounded by a structureless limiting membrane.

ordinary epithelium cells as indicated in 27, iii.   These are in connection with the ultimate endings of the olfactory nerve.

(d.) *Hearing.*—Somewhat similar spindle-shaped cells produced into long *auditory hairs* are found in special regions of the organ of hearing, and are in connection with the auditory nerve.

(e.) *Sight.*—The most complex of nerve-endings are those of the optic nerve in the retina of the eye.   Fig. 27, iv. gives a section of the frog's retina under a high power ($\frac{1}{8}$).   There are eight distinct layers: 1. pigmented epithelium : 2. the layer of rods and cones ; 3. the outer nuclear layer ; 4. the outer molecular layer ; 5. the inner nuclear layer ; 6. the inner molecular layer ; 7. the layer of nerve cells ; 8. the layer of nerve fibres.   The last layer, that of nerve fibres, is the most internal, nearest the centre of the eyeball, and presented towards the rays of light. But the light vibrations are received by the ends of the rods and cones in the pigmented epithelium, and then transmitted to the

layer of nerve fibres, through the intermediate layers, and thence
by the optic nerve to the brain. Thus the light passes outwards
through the thickness of the retina mechanically as through a
transparent medium, and then passes backwards through the
retina organically as a wave of molecular change through the
nerve endings. The granular layers owe their dotted appearance
to the (stained) nuclei in the specially modified cells.

11. The Stomach and Intestine.—A cross-section of a frog's
stomach examined under a low power (Fig. 28, i.) shows a number of inwardly projecting folds, and if it be stained there

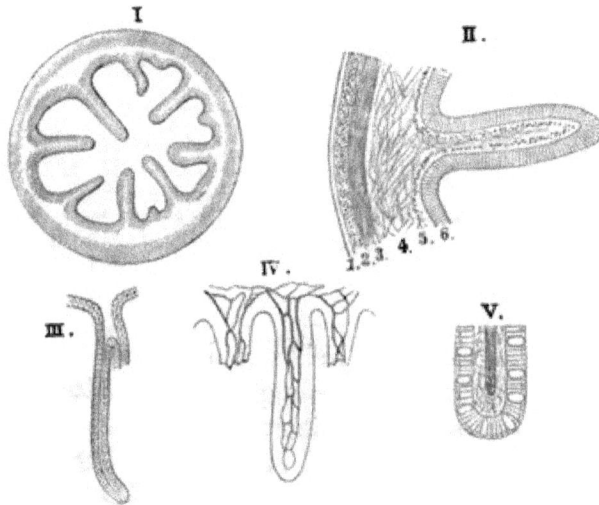

Fig. 28.—Stomach and Intestine.

i. Stomach of frog. ii. A portion of the same under higher power. 1.
Peritoneum. 2. Longitudinal muscular layer. 3. Concentric muscular layer.
4. Submucosa. 5. Muscularis mucosæ. 6. Mucosa containing peptic glands.
iii. Peptic gland (dog) under still higher power. iv. Villus (dog) injected to
show blood-vessels. v. Villus (rabbit) stained to show cells. *From sections.*

will be an internal lighter and an external darker zone. Under
a higher power (28, ii.) six tolerably distinct layers are seen :—
(1.) Externally the peritoneum, often lost from the section ;
(2.) external muscular layer with fibres running longitudinally ;
(3.) internal muscular layer with fibres running concentrically ;

(4.) submucosa of connective tissue ; (5.) muscular layer of the mucosa ; (6.) mucosa.　In the mucosa of the mammalian stomach great numbers of glandular tubes (the *peptic glands*) are seen. In each gland (28, iii.) there is a single layer of secreting cells continuous with the epithelial lining of the stomach.

In the intestine of the mammal or the bird there are a vast number of minute processes (*villi*) projecting inwards, and giving a velvety appearance.　Fig. 28, iv. shows one of these villi, in which the blood-vessels are injected, while 28, v. shows the end of a villus.　It is covered with columnar epithelium cells, many of which are converted into *goblet or mucous cells*, larger and clearer. Beneath this the tissue is lymphatic in structure, and is apt to be crowded with leucocytes.　The central blind tube in Fig. 28, v. is the chyle-vessel or terminal rootlet of the lacteal system of lymphatics.　The layers of the intestine are essentially similar to those of the stomach.

12. **The larger Alimentary Glands.**—In favourable sections of a salivary gland the *alveoli* (29 i. from the sublingual gland

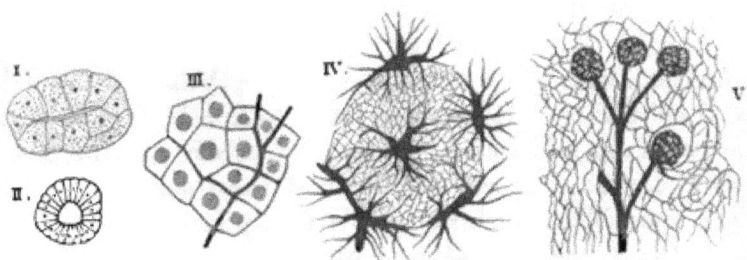

Fig. 29.—Salivary Gland, Liver, and Kidney.

Alveolus of sublingual gland of rabbit. ii. Duct of same. iii. Liver of rabbit stained, with bile canaliculi injected. iv. Liver (cat) with blood-vessels injected. v. Kidney of rabbit injected to show glomeruli. *All from sections.*

of a rabbit) or true secreting terminations of the ducts may be made out.　Their cavities are nearly or quite obliterated.　In the ducts, on the other hand, the cavity or lumen is generally obvious, and the cells more columnar (29, ii.).

A section of the pancreas shows a somewhat similar structure.

In the case both of the salivary glands and of the pancreas the alveoli are surrounded by delicate bands of connective tissue, and are grouped into lobules, surrounded by stronger bands, while the lobules are further grouped into lobes separated from each other by yet stronger connective tissue continuous with that of the capsule which encloses the whole gland.

The structure of the *liver* is very different from that of the salivary glands or pancreas. The substance of the organ is more or less distinctly divided by connective tissue partitions into a a vast number of lobules about $\frac{1}{20}$ inch in diameter. In a section of a liver in which the blood-vessels have been injected (29, iv.), there are seen round the lobule several *inter-lobular venules*, ultimate branches of the portal system. They send a fine network of capillaries inwards towards the centre of the lobule, where the blood is collected by an *intra-lobular venule*, one of the ultimate factors of the hepatic vein. In 29, iii. we have a section in which not the blood-vessels but the bile canaliculi have been injected. They form a network between the liver cells, which are thus clearly marked out, and seen to be polygonal and nucleated.

13. The Lung.—The trachea of the rabbit is lined with ciliated epithelium. It branches into two *bronchi*, each of which again branches, rebranches, divides, and subdivides into a vast number of ciliated tubes, the finer and more delicate of which are the *bronchioles*. Each bronchiole finally branches into wider non-ciliated *infundibula*, which, with their small branches, form the ultimate termination of this extensive system of tubes. Finally, the walls of each infundibulum are closely beset with minute bags, the *air-cells* or *alveoli*, which open into the infundibulum. The air-cells are lined with flattened non-ciliated epithelial scales, beneath which lies a dense network of pulmonary capillaries. Here it is that the plasma of the blood gives up its carbonic acid gas, and the exhausted red corpuscles receive a fresh supply of oxygen. Each lobule is separated from its neighbours by delicate bands of fibrous connective tissue, by the action of which the infundibula and air-cells have a tendency to resist inflation, and to expel the contained air during expiration.

In birds the minute lung structure is different. There are no infundibula and no air-cells, though the finest tubuli are minutely sacculated. These minute tubuli—which form a close plexus in the lung—are placed in communication with each other by minute perforations in their walls. It will be remembered (p. 49) that the lungs are in connection with a system of large air-sacs.

In the frog the lung is a hollow structure, pitted on its inner surface with sacculations, formed by the infolding of the walls. These sacculations are more numerous in the anterior part of the lung. In their walls is the capillary plexus. Their edges are often covered with ciliated cells, but the cavities of the sacculations are lined with flattened epithelium.

14. The Kidney.—The kidney of the rabbit is hollowed out on one side, its inner edge forming a depression or *hilus*. Here the ureter enters and enlarges into a funnel-like dilatation, the *pelvis*, which ends in a number of smaller dilatations, the *calyces*. Into the pelvis there projects a conical mass of kidney substance, the *urinary pyramid*. The substance of the kidney is divisible into an outer portion, the *cortex*, which has a dotted appearance, and an inner portion, the *medulla*, which is radially striated.

In a section of a kidney in which the blood-vessels have been injected, arterioles are seen running horizontally in the intermediate region between cortex and medulla. They send up vertical branches into the cortical layer, which in turn branch, each branch ending in a convoluted tuft of blood-vessels called a *glomerulus* (Fig. 29, v.) Each glomerulus is (as shown diagrammatically to the right of the figure) enclosed within a *capsule* (lined with flattened epithelial plates), which forms the dilated end of a *urinary tubule*. The blood passes into the glomerulus by an afferent vessel, and out by an efferent vessel. It then breaks up into a delicate capillary plexus within the cortex

6

around the urinary tubules, which are lined with columnar or polyhedral secreting cells. The course of the tubules is peculiar. After winding about in the cortex (so that in a section the cut ends of numerous coils will be seen), each tubule passes down into the medullary region ; but instead of passing at once to the pelvis it turns back, making a sharp loop (*Henle's loop*), re-enters the cortex, and there again becomes convoluted, after which it again dives down into the medulla, and so reaches the pelvis. The size of the tubule varies, being larger in the convoluted parts of its course. This course is special to the mammalia. In a stained section, therefore, the cortex is characterised by the coiled tubules and glomeruli, while the medullary portion shows the straighter parts of the tubules, and vascular bundles.

It seems probable that from the glomeruli there passes into the capsule little but water containing a certain amount of salts in solution, the amount of fluid thus filtered from the blood depending largely on the blood pressure. The urea is probably secreted in the convoluted portions of the tubule surrounded by the capillary plexus.

In the frog there are glomeruli, capsules, and convoluted tubules, but the latter have not the remarkable mammalian course, and there is no distinction into cortical and medullary regions. There is, it will be remembered (p. 28), a double blood supply. That from the renal artery passes to the glomeruli, that from the renal portal vein passes to the convoluted tubules. And it has been experimentally proved that here, and not in the glomeruli, is urea (injected into the blood) excreted.

15. The Skin.—A transverse section of the skin of the frog is shown in Fig. 30, i. The external layer of the *epidermis* is composed of flattened scales. A middle layer shows rounder nucleated cells. A deeper layer shows larger columnar cells, with their long axes vertical. Beneath this is the *dermis*, composed largely of fibrous bands supporting nerves and blood-vessels. In the uppermost part of the dermis are large pigment cells.

Imbedded in the dermis, with their narrow funnels passing

through the epidermis (out of focus in 30, i.) are large flask-shaped *cutaneous glands* lined with oblong nucleated cells, which secrete a fluid by which the skin is kept moist, and which (especially in the toad) is extremely bitter.

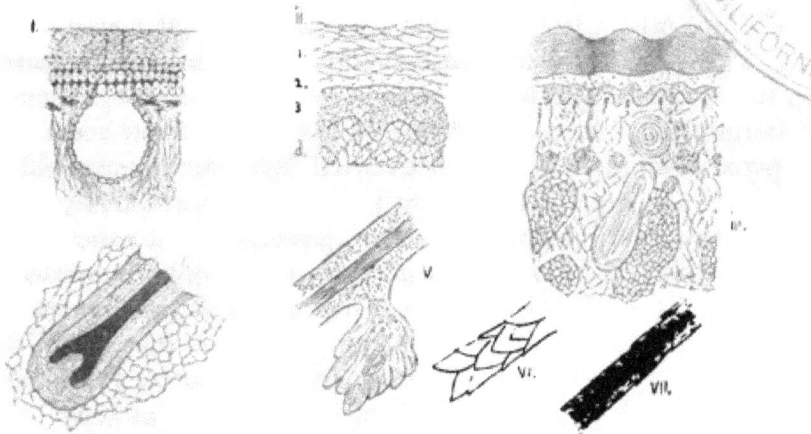

FIG. 30.—THE SKIN: HAIRS.

i. Skin of frog. ii. Skin of human arm. 1, 2, 3. Layers of epidermis. iii. Skin of human thumb. iv. Hair follicle (negro). v. Sebaceous gland (human). vi. External view, white hair of rabbit. vii. Darker hair of rabbit in optical section.

Fig. 30, ii. and iii. show sections of the human skin of the arm (ii.) and of the thumb (iii.). They show the three layers of the epidermis : (1) An outer layer of non-nucleated horny scales ; (2) a more transparent middle layer, in which nuclei can be seen ; (3) an inner darker layer of distinct and nucleated cells. It is by the constant multiplication of these lower cells that the overlying cells become thrust upwards towards the surface, becoming first flattened and then eventually mere horny scales, which are continually being shed.

Beneath the epidermis is the dermis. This is raised into *papillæ*, to which the epidermis is moulded, and in the papillæ of the thumb the *tactile corpuscles* may be readily seen under a high power. Deeper in the dermis lie the large oval *pacinian corpuscles* with their concentric coats. The tissue to the right of the oval corpuscle in iii. is fatty tissue, above which is the cut end of a blood-vessel.

Passing down through the epidermis into the dermis are spiral tubes, each ending in a coiled glandular knot. These are the *sweat glands.* One of these is shown in iii., while portions of three other glandular knots are cut in different planes.

Imbedded in the mammalian skin are the hairs characteristic of the order. Each hair is moulded on a little papilla of the dermis (iv.) Around the root of the hair is a *root-sheath* composed of an internal and an external layer continuous with the deeper and superficial layers of the epidermis respectively. Outside this is a fibrous dermic layer, passing up into the papilla. The whole is in iv. (from the scalp of a negro) imbedded in fatty tissue. The whole tubular depression in the skin containing the hair is called the *hair follicle.* Into it there open, near its external end, *sebaceous glands*, secreting an oily substance. The position of one of these is roughly indicated in 30, v.

Fig. 30, vi. and vii. show rabbit's hairs under a fairly high power. Fig. 30, vi. is from near the point of a white hair, and shows the imbricating scales of the *cuticle.* Fig. 30, vii. shows, besides these scales, the fibrous *cortex* and the internal irregular pigmented or air-containing *medulla.* In each case the proximal or root end of the hair is uppermost.

In the nails or claws the external epidermic scales become specially soldered together, and horny or fibrous in consistency. They are moulded in the rabbit on large papillæ, their roots being also covered over with a fold of skin so as to lie in a groove.

Attention may here be drawn to the different kinds of glands with which we have become acquainted in the skin and elsewhere.

1. Simple tube-like depressions, *e.g.* peptic glands.
2. Simple flask-shaped depressions, *e.g.* cutaneous glands of frog.
3. Tubes coiled at the end. Sweat glands.
4. Tubes with simple branches. Sebaceous glands.

5. Long tubes (ducts) with a vast number of branches ending in slight dilatations (racemose glands), *e.g.* salivary glands, pancreas.

6. Long tubes (ducts) ending in a network of inter-cellular canaliculi, which, in the rabbit, give off minute twigs that pass into the substance of the liver cells, and there terminate in vesicular enlargements.   Liver.

CHAPTER VI.

GENERAL EMBRYOLOGY.

THE vertebrate animal is developed from a fertilised ovum. The multitude of cells of which—together with cell products—the tissues of the adult organism are composed, is produced by the repeated division and sub-division of the original oosperm. Thus we may trace back the organism to the ovum with which a spermatozoon has entered into fertile union ; and the development of the ovum and spermatozoon in the parents we may trace back to the fertilised ovum from which they also were respectively developed. *Omnis cellula a cellulâ* is a law that, so far as we know, holds good for all animal life. It may be that in past ages, it may be that even now in the depths of the ocean, or in stagnant pools, the origin of a living cell from not-living matter has occurred or is taking place. No one has a right to assert that it has not and cannot. But as a matter of science such *abiogenesis* (or origin of living from not-living matter) has never yet been proved. And therefore, as students of science, we shall do well to accept the law of *biogenesis—omnis cellula a cellulâ*— as our working hypothesis.

The ovary and testis which in the parents produce the ovum and spermatozoon arise in the first instance, when the parent is still an embryo, as thickenings of the sheet of cells that lines the body-cavity. The thickened part of the sheet is called the *germinal epithelium* ; and for a while it is quite similar in male and female. By mutual ingrowth of cells, (1) from the germinal epithelium into the substance of the underlying body-walls, and (2) from the body-walls into the germinal epithelium, a thickened ridge, the *genital ridge*, is produced. But still there

86

is no distinction between ovary and **testis.** Ere long, however, differentiation sets in, **and** different lines of development are followed in the two sexes.

In the male the testis becomes a mass of minute tubes, the *seminal tubules,* twisted and convoluted in all directions, and lined with large cells, probably derived from the original germinal cells, which are destined to give rise to spermatozoa. They are termed *sperm cells.* From these a number of smaller cells arise by a process like budding. It is these smaller cells (*spermatoblasts*) which are converted into spermatozoa. The tubules in which they are developed open into the vas deferens, by which the ripe spermatozoa are carried out of the body.

In the ovary of the female there is no development **of tubules.** Certain large germinal cells are developed at the **expense of** other cells. These are the **ova.** The cells **adjoining each ovum** arrange themselves round it so as to form a special layer which forms the lining of a sort of bag, the *ovisac,* in which the ovum lies. When the ovum is ripe the **ovisac bursts,** and the egg is set free into the peritoneal cavity, or passes at once into the oviduct, and so out of the body.

The very young ovum, before it is mature, consists of a minute spherical mass of granular protoplasm. Within it **is a** specialised portion differing from the rest, and called **the** *germinal vesicle,* and within this again may be seen one or several *germinal spots.* The germinal vesicle is invested by a delicate membrane, and stretching from this membrane to the germinal spot or spots is a network of delicate fibres called the *reticulum.* The ovum is a *simple cell.* The germinal vesicle may represent the *nucleus* of the cell, and each germinal spot a *nucleolus.*

Such is the immature **ovum. From it** the **mature** ovum differs (1) in the possession of **an external membrane**; and (2) in the development within it **of a** greater **or less** quantity of *food-yolk* in the form of minute masses known as *yolk-spheres.* Both the amount and the mode of distribution of the food-yolk are matters of considerable importance. The amount differs markedly in our three types. The minute ovum of the rabbit

($\frac{1}{150}$th in.) has but little; the larger ovum of the frog contains a moderate amount; in the much larger fowl's egg (for we will here take the fowl as our typical bird) there is a relatively enormous quantity. The distribution of the yolk-spheres also differs in the two types in which they are freely developed. In the frog they are larger and more closely aggregated at one pole, than they are at the opposite pole, in which the yolk-spheres are smaller. This protoplasmic pole, with the smaller yolk-spheres, is termed the *animal pole.* In the fowl the yolk-spheres are of two kinds, smaller and earlier developed white yolk-spheres, larger and later developed yellow yolk-spheres. These are arranged in somewhat concentric layers around a central flask-shaped mass of the white yolk.

The germinal vesicle, which is relatively large in the immature ovum, occupies less relative space in the mature egg. In the rabbit's ovum it is near, but not at the centre. With the development of food-yolk it becomes thrust over towards the animal pole, near which it lies in the frog's ovum; while in the fowl's egg it travels to the periphery, where, together with the protoplasm round it, which remains relatively free from yolk-spheres, it constitutes the *germinal disc.*

The phenomenon now to be described, termed the *extrusion of the polar cells,* probably occurs in all three types. It consists in the division of the germinal vesicle of the ovum (or the nucleus of the cell) into two parts, of which one remains within the ovum, while the other, carrying with it a portion of the protoplasm, gives rise to a little protuberance, budded off from the ovum, and known as the *polar cell.* The polar cell is then constricted off, becomes separate from the ovum, and dies. A second polar cell is then budded off in a similar way. Thus two polar globules are produced by a process analogous to cell-division. It is supposed that they remove from the ovum matter detrimental to the further development of the egg. Each contains a share of the nucleus of the ovum. The share that remains to the ovum after the extrusion of the polar cells, containing perhaps only one quarter of the original nuclear matter, is called the *female pronucleus.*

After this has taken place the ovum is ready for impregnation. Here again we are forced to infer, from the results of investigations on other forms of life, the nature of the process of fertilisation. It would seem, however, that only one spermatozoon enters the ovum. This spermatozoon is essentially a cell, and its head contains the sperm-nucleus. When it enters the ovum the sperm-nucleus forms the structure known as the *male pronucleus.* The male pronucleus and female pronucleus now approach each other, and finally fuse, the fused product being known as the *first segmentation nucleus.*

Upon this the process of *segmentation* sets in, and does not cease until, by repeated cell-division, the fertilised ovum gives rise to the adult organism. It is obvious that during this process the cells must be nourished; (1) either by gradual absorption of food-yolk; or (2) by the individual efforts of the developing embryo; or (3) by absorption of nutriment elaborated by the mother. The frog, after the somewhat scanty stock of food-yolk is absorbed, is hatched, and forthwith has to obtain food for itself. The fowl undergoes a far larger proportion of its development within the egg at the expense of the abundant supply of food-yolk; then it fends for itself, with some maternal aid and advice. The rabbit is almost from the first supplied with nutritive material by the mother, within whose body embryonic development takes place; and even after birth feeds for some time on the mother's milk.

The process of segmentation is essentially the division of the primitive cell into two cells, each of these again into two, and so on until the one has become a multitude. In each case of cell-division there is also nucleus-division. And it has been found that the reticulum of the nucleus undergoes a remarkable series of changes constituting *karyokinesis.*

Fig. 31, i. shows diagrammatically a cell in which the nucleus is in the ordinary reticulum condition. In certain cases the following changes have been shown to occur. When karyokinesis sets in the capsule or limiting membrane of the nucleus tends to disappear, and the protoplasm of the nucleus takes on a striated appearance giving rise to a *nuclear spindle,* from the ends of which star-like striations extend into the protoplasm of the cell (Fig. 31, ii.). In the middle of the spindle the fibres of the

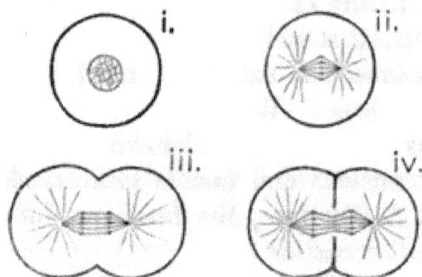

FIG. 31.—KARYOKINESIS.

reticulum aggregate so as to form the *equatorial plate.* This splits in the equatorial line into two plates which travel away from each other towards the poles, the protoplasm between them having a striated appearance (Fig. 31, iii.). Meanwhile the protoplasm of the general substance of the cell becomes constricted, in Fig. 31, iv., *e.g.* so far as nearly to divide the cell into two, and eventually to such an extent as to cause its complete division into two cells, each of which now contains a nuclear spindle, which may either relapse into a reticulum or straightway redivide. Watched under the highest power of the microscope the fibres of the equatorial plate are seen to group themselves into the form of a rosette, which then, with sundry other changes of form, separates into two rosettes, each of which forms a new equatorial plate. Such are the internal changes in the segmenting cells. We must now turn to the consideration of the segmentation of the ovum as a whole, and the formation of the so-called *germinal layers* to which this process gives rise, remembering that each case of cell-division is accompanied by karyokinesis. But before describing what takes place in frog, fowl, or rabbit it will be well to pave the way by taking an ideal case.

We will suppose that the ovum contains no food-yolk. It first divides into two equal halves, and then each of these again into two. We may call the points of intersection of these two planes the poles, and the planes vertical planes. We thus have four cells produced by two vertical planes, the segmentation being so far similar to that shown in Fig. 33. The next plane is equatorial midway between the poles. By this plane the four cells are subdivided into eight. Then follow two more vertical planes intermediate between the first two. By them the eight cells are divided into sixteen. Then follow two more horizontal planes midway between the equator and the poles. Thus we get thirty-two cells. So the process continues, until by fresh vertical and horizontal planes of division the ovum is divided into a great number of cells.

But meanwhile a cavity has formed in the midst of the ovum. This makes its appearance at about the eight cell stage, the eight cells not quite meeting in the centre of the ovum. The central cavity so formed is thus surrounded by a single layer of cells; and it remains as a single layer throughout the process of segmentation. So that the result of segmentation is the production of a hollow vesicle (Fig. 32, i.) composed of a membrane with a single layer of cells. Such a vesicle is called a *blastosphere*; the period during which this state of things continues is called the *blastula stage*; and the cavity of the blastosphere is called the *segmentation cavity* or *blastocœle*.

Now suppose [1] that one side of the blastosphere, which is represented in Fig. 32, ii., being composed of somewhat larger cells, is pushed in; just as one might take a hollow squash india-rubber ball, and push in one side so as to form a hollow two-layered cup. Suppose, moreover, that the mouth of the cup closes in so as to become much smaller. Then we have the state of things shown in Fig. 32, iii. This is the *gastrula* stage. Thus we have constituted, by a process termed *invagination*, an embryo with two layers of cells, an outer layer or *epiblast* (*ep.*), and an inner layer or *hypoblast* (*hy.*). They are the two primitive *germinal layers*. The cavity within the embryo is the primitive digestive body-cavity or *archenteron* (*a. e.*). The opening to the exterior is the *blastopore* (*bl. p.*).

A transverse section through the middle of the embryo at the gastrula stage shows a central cavity surrounded by a two-layered body-wall. This state of things does not long continue. Along a definite line on the surface of the embryo, marking off a region henceforth to be known as the dorsal, the epiblast, or

---

[1] The advanced student will ascertain for himself how far these suppositions are gratuitous.

outer layer becomes somewhat thickened, forming the *neural plate*. The edges of the thickened area rise up on either side so as to form a median groove, the *neural groove* (32, iv. *n. g.*); and then close over above the groove so as to convert it into a

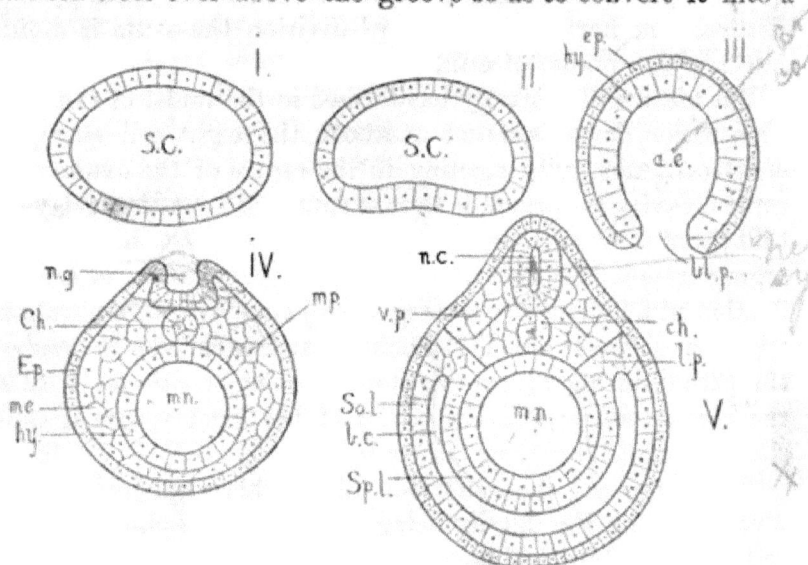

FIG. 32.—DIAGRAM OF EARLY STAGES OF DEVELOPMENT.

i. and ii. Blastula stage. iii. Gastrula stage. iv. Establishment of mesenteron, notochord and medullary groove. v. The closure of the medullary folds and cleavage of the mesoblast.

a. e. Archenteron. *b. c.* Body-cavity (cœlom). *bl. p.* Blastopore. *ch.* Notochord. *ep.* Epiblast. *hy.* Hypoblast. *l. p.* Lateral plate. *n. c.* Neural canal. *n. g.* Neural groove. *m. p.* Mesoblastic plate. *me.* Mesoblast. *mn.* Mesenteron. *s. c.* Segmentation cavity. *so. l.* Somatic layer (somatopleure). *sp. l.* Splanchnic layer (splanchnopleure). *v. p.* Vertebral plate.

neural canal (32, v. *n. c.*). **This** is the first indication of the central nervous system. It lies in the mid-line of the dorsal region, and is **formed from** epiblast. Immediately beneath it **there is** formed a linear rod of cells derived from the hypoblast of the mesenteron, and called the *notochord* (*ch.*).

Meanwhile a third germinal layer (Fig. 32, iv.) is being formed between the epiblast and the hypoblast. It is the *mesoblast*. The **cells** of which it **is** composed are probably derived from the hypoblast; but in any case it soon forms a separate layer, divided by the notochord into two separate sheets, the *mesoblastic plates* (*m. p.*) passing round and meeting below the digestive cavity (*mesenteron, mn.*).

We have yet to note some further changes in the mesoblastic plates. (1.) Each splits[1] into two layers, of which the inner layer adjoining the mesenteron is called the *visceral* or *splanchnic layer* (*sp. l.*), while the outer, that adjoining the epiblast, is the *parietal* or *somatic layer*. The space between them is the body-cavity (*b. c.*). (2) On each side of the neural canal and the notochord longitudinal masses of the mesoblast are separated off from the mesoblastic plates. These median dorsal parts of the mesoblastic plates are now known as *vertebral plates* (*v. p.*), while the remainder of the mesoblastic plates are now termed *lateral plates*. In the vertebral plates the cleavage of the mesoblast and resulting body-cavity are eventually obliterated. (3.) The vertebral plates become segmented by transverse constrictions into a series of more or less cubical bodies (for which in the chick see Fig. 37, *m. s.*), the *mesoblastic somites*. (4.) In the lateral plates the cleavage of the mesoblast is obliterated only at the dorsal end, near the vertebral plates. Below this the body-cavity is persistent. The splanchnic layer eventually forms the muscular coat of the alimentary canal, while the somatic layer gives rise to the endothelial lining of the body-walls.

The blastopore has probably by this time closed, so that the mesenteron has neither anterior nor posterior openings to the exterior, though posteriorly it may open into the neural canal, by a passage known as the *neurenteric canal* (*n. e. c.* in frog, Fig. 36, iii.). External openings are formed, often comparatively late in embryonic life, by the formation of external depressions which meet hollow outgrowths from the mesenteron. The anterior depression, that for the mouth, is called the *stomodæum*; the posterior depression, that for the vent, is called the *proctodæum*. The former is always in vertebrates a new invagination; but the latter may be in some types (perhaps in the

---

[1] The cleavage of the mesoblast begins *before* the mesoblastic plates meet below the mesenteron. The two processes are kept separate in the figures and description for the sake of clearness.

frog) the persistent blastopore. The lungs, and the glands of the alimentary canal arise as hollow outgrowths therefrom. The heart, and all the blood-vessels arise as tubes within the meso-blast, from which layer are also formed all the muscles, cartilages, and bones of the body. The hypoblast gives rise to the epithelial layer of the alimentary cavity and of the glands connected there-with. The epiblast forms the epidermic layer of the skin ; but from it are also formed the central nervous system and certain important structures connected with the organs of special sen-sation.

We have thus seen how the three germinal layers, the ali-mentary canal with its mouth and anus, the body-cavity, the central nervous system, and the notochord may be formed in the case of an ideal, but not wholly imaginary, animal. The student is clearly to understand that both Fig. 32 and this descrip-tion of the mode of formation of the leading structures in the body are, so to speak, diagrammatic, and introduced merely to aid him in comprehending how these structures are formed in the frog, the fowl, and the rabbit, to which we may now turn.

Segmentation of the Ovum.—(1.) In the frog the mode of seg-mentation does not differ very markedly from that described. The segmentation is complete (*holoblastic*), that is to say, the planes of division run right through the whole thickness of the ovum, all its substance being segmented or broken up into cells ; but owing to the inequality of the distribution of the food-yolk it is unequal or irregular. Where the protoplasm is most concen-trated, namely, at the animal pole, segmentation is more rapid. Thus the first two furrows form rapidly in the upper half of the ovum, but more slowly extend to the lower half. The first horizontal furrow is not equatorial, but near the animal pole (Fig. 33). And throughout segmentation cell-division goes on more rapidly in the animal hemisphere, so that the cells are smaller there than in the opposite hemisphere, It is usually found that wherever large quantities of food-yolk occur, segmen-tation is more or less unequal, the velocity of segmentation being directly proportional, and the size of the cells inversely proportional to the concentration of the protoplasm. At the close of the segmentation of the frog's egg (Fig. 34, i.) the ovum forms a blastosphere containing an excentric segmentation cavity, roofed over with several layers of small cells, and resting on a continuous mass of larger yolk-containing cells.

(2.) In the rabbit the segmentation is complete and nearly regular, but presents some peculiar features. Two processes seem to go on together. In addition to the formation of a blastula with a segmentation cavity, there goes on a slipping inwards of some of the cells to fill up this cavity. And since the latter process keeps pace with the former, the cavity is filled up as rapidly as it is produced. Thus at the close of segmentation there is an outer layer (*ou.*, Fig. 34, ii.) of clear cells surrounding, except at one spot (which we may term the *pseudo-blastopore*, *v. b.*), an inner mass of granular cells (*In.*).

FIG. 33.—CLEAVAGE IN FROG'S OVUM.

i.-iv. Stages with 2, 4, 8, and 16 blastomeres. v. A later stage when the smaller blastomeres cover half the ovum. vi. A still later stage when the smaller blastomeres have enveloped the whole ovum, except at one spot, the blastopore, *bl.*

(3.) In the fowl, owing to the enormous quantity of food yolk and the consequent concentration of the protoplasm in the germinal disc, the segmentation is not only very irregular, but also incomplete or partial (*meroblastic*). That is to say, the planes of division do not extend throughout the ovum, but are confined to that comparatively small area of its surface known as the germinal disc. The first two furrows are at right angles, but do not cross each other quite in the middle of their length. Four more radial furrows follow intermediate between these, thus dividing the germinal disc into eight somewhat irregular cells. The next furrow is irregularly circular (or polygonal), giving rise to eight smaller central and eight larger peripheral cells. Henceforth the central segments divide more rapidly

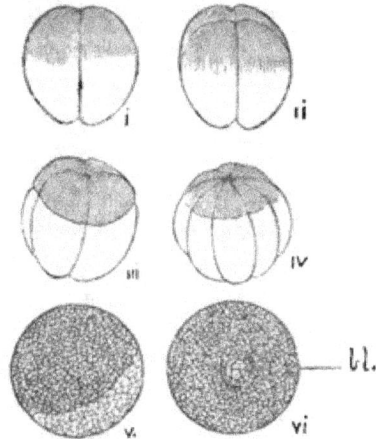

than the peripheral, while furrows beneath and parallel with the surface separate the superficial from the deeper cells. Thus at the close of segmentation (*i.e.* before incubation) we have a

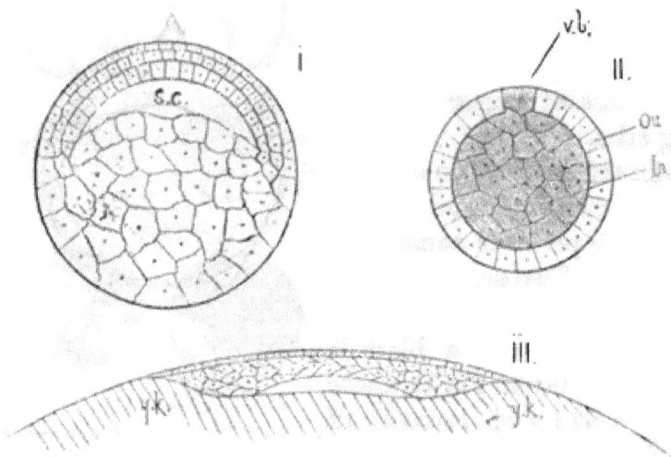

FIG. 34.— AT THE CLOSE OF SEGMENTATION.

i. In frog.   ii. In rabbit.   iii. In fowl.

In. Inner cells.   *Ou.* Outer cells.   *v. b.* Pseudo-blastopore.   *yk.* Yolk.

superficial layer of small columnar cells forming a distinct membrane, and a deep layer of larger cells irregularly disposed, but deeper around the edges of the disc than at its centre (Fig. 34, iii.).

The Egg Membranes.—The frog's egg is closely invested by a delicate *vitelline membrane.* External to this is the albuminous envelope secreted by the walls of the oviduct.

The fowl's egg is also closely invested by a vitelline membrane. In passing down the oviduct (after impregnation) it receives a thick albuminous coating of white-of-egg. At either end of the egg this assumes a denser condition, and forms a twisted knotted cord (*chalaza*). A two-layered shell-membrane, and finally the shell itself, are added in the lower parts of the oviduct.

The rabbit's egg is surrounded by a radially striated membrane (*zona radiata*), which probably does not answer to the vitelline membrane of the other types, though this membrane may exist in the rabbit's ovum previous to the rupture of the follicle. As it passes down the Fallopian tube the egg becomes invested with an albuminous envelope.

The Establishment of the Germinal Layers.—(1.) We have seen that in the frog there is at the close of segmentation a blastosphere containing a comparatively small segmentation cavity placed excentrically, roofed over with several layers of smaller cells, and resting on a considerable mass of yolk-containing cells (Fig. 34, i.). So far there is no differentiation into epiblast and hypoblast. The large amount of food-yolk and the mass of lower cells prevents invagination from taking place in so simple a manner as in our illustrative case. Still invagination does take place. The small roofing cells gradually creep over the rest of the sphere, by the conversion at the edges of the roof of large yolk-containing cells into small roofing cells. But having reached the equatorial line, they grow in over a small arc, forming a sort of lip. This marks the hinder end of the future embryo. The lip extends further and further inwards and upwards so as to run parallel, or rather concentric, to the dorsal surface. This extension of the lip is due partly to a true ingrowth of the external roofing cells, but perhaps partly also to a differentiation of the yolk-containing cells along its line. As it extends inwards and upwards a slit-like cavity opens beneath it (Fig. 35, i.). Meanwhile the smaller roofing cells creep further and further round the rest of the sphere, until only a small patch of the larger yolk-containing cells remains uncovered. As the slit-like mesenteron enlarges, the original segmentation cavity is gradually obliterated. Eventually the plug of yolk-containing cells beneath the inturned lip disappears, and the mesenteron communicates with the exterior by a small circular opening.

Here, then, we have the gradual formation of an external layer of cells which turns inwards so as to line the roof of a cavity which eventually opens to the exterior by a posterior orifice. The external layer is the epiblast; the internal layer continuous with it at the lip is the hypoblast; the slit-like cavity is the mesenteron; and the posterior orifice is the blastopore. The

7

whole thing is comparable with what we had in the illustrativ
case; but it is much modified by the presence of yolk-containin
cells. The hypoblast, for example, only forms the roof of th
mesenteron, and only gradually grows downwards to line th
sides.

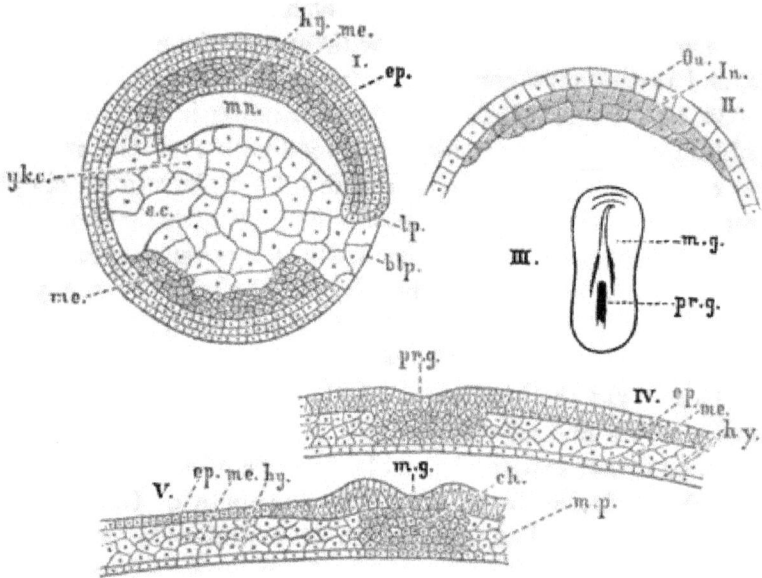

Fig. 35.—Formation of Germinal Layers.

i. Section of frog's ovum. ii. Section of upper part of blastodermic
vesicle of rabbit. iii. Supervicial view of blastoderm in fowl. iv. Section
through, and at right angles to, the primitive groove in rabbit v. Section
through, and at right angles to, the medullary groove.

blp. Blastopore. ch. Notochord. ep. Epiblast. hy. Hypoblast. In.
Inner cells. lp. Lip of blastopore. m. g. Medullary groove. m. p. Meso-
blastic plate. me. Mesoblast. mn. Mesenteron. Ou. Outer cells. pr. g.
Primitive groove. s. c. Segmentation cavity. yk. c. Yolk-containing cells.

So far we have only got epiblast and hypoblast. But th
invaginated layer which gives rise to hypoblast gives rise als
to mesoblast. For it is not composed of a single layer of cell
but of several layers, of which the lowest, that which lines th
roof of the mesenteron, is hypoblast, while the others give ris

to the mesoblast of this region. On the ventral side of the blastopore, mesoblast either grows inward from the epiblast, or arises by differentiation of the larger yolk-containing cells.

Eventually all the yolk-containing cells are converted into mesoblast, or hypoblast, except a certain number, which eventually become enclosed in the mesenteron, by the growing of the hypoblast round them above the mesoblast (*me.* in Fig. 36, iii.), so as to bring them inside the digestive cavity, where they form a small internal yolk-sac, the cells of which are gradually absorbed.

The formation of the notochord, and the differentiation of the mesoblastic plates into vertebral and lateral plates ; and the permanent cleavage of the mesoblast in these lateral plates, so as to form a splanchnic and a somatic layer, with a body-cavity between them ; all this takes place in a manner substantially similar to that described in the illustrative case.

(2.) We have seen that in the fowl's egg segmentation converts the germinal disc into a two-layered *blastoderm*, the cells of which are smaller in the superficial than in the deeper stratum. The central part of the blastoderm (*area pellucida*) rests on a clear fluid contained in a shallow space (Fig. 34, iii.) ; its edges (*area opaca*) rest on unsegmented white yolk (*yk.*), and appear as an opaquer rim.

The hypoblast is not, in the fowl, produced by invagination. The two layers of the original blastoderm are directly converted into epiblast and hypoblast. The former (epiblast) is constituted from the first (*i.e.* before incubation) of smaller cells arranged as a definite membrane. The lower layer is rapidly converted, under the influence of incubation, into a definite hypoblastic membrane of flattened cells. These form the roof of the shallow space in the area pellucida, which may therefore be regarded as the mesenteron. A segmentation cavity is only found at a very early stage, and is soon obliterated by the rapid growth of the lower layer cells. Between the hypoblast and epiblast there remain some undifferentiated cells, but they do not form a distinct intermediate layer.

In the fowl, then, we seem to see carried further a process

that was, so to speak, begun in the frog. In our illustrative case, we saw the hypoblast produced entirely by invagination. In the frog we saw it partly produced by invagination, partly by differentiation of cells which lie beneath the epiblast. Here we find the hypoblast entirely formed by such differentiation, apparently without any invagination.

At first the blastoderm covers only a small area. But it gradually spreads outwards in all directions, creeping over the yolk beneath the vitelline membrane, until at last (and this does not take place until comparatively late in development) it envelopes the whole mass of yolk.[1] This is similar to what we saw in the frog; but it is carried further. The investment of the yolk is complete. There is no circular patch of yolk cells left uncovered to mark the position of the blastopore. Thus the blastoderm creeps over the whole surface. But the embryo is entirely formed in the small area pellucida.

At the hinder end of this area a linear opacity makes its appearance. This is known as the *primitive streak.* After a while it becomes grooved (Fig. 35, iii., *pr. g.*). It is produced by a thickening of the blastoderm and the formation of rounded cells which grow downwards, and soon spread outwards in wing-like masses at the sides of the primitive groove. This is shown diagrammatically in Fig. 35, iv. These lateral masses are henceforth to be regarded as mesoblast. The process by which they are formed is not dissimilar to that by which the mesoblast is formed in the frog at the inturned lip. There we saw a hypoblast and a mesoblast layer thus formed. Here we find that only the mesoblast is thus produced. This takes place in the hinder third of the area pellucida.

In the anterior two-thirds the mesoblast is formed in a different way. Here (as seen in Fig. 35, iii. and v.) the medullary groove is beginning to appear. The hypoblast here breaks up into two layers—a flattened lower layer, which retains the character and name of hypoblast, and a more indefinite

---

[1] The hypoblast and mesoblast are said to be here formed mainly through the instrumentality of the nucleated protoplasm of the *germinal walls* at the edge of the area opaca.

upper layer of cells, which constitute the mesoblastic layer. In this region, the cells in the median line (which longer retain their primitive connection with the hypoblast, Fig. 35, v.) become concentrated into the notochord (*ch*), while the two mesoblastic plates, one on each side, become more and more completely differentiated. The upper parts of these plates give rise subsequently to a vertebral plate segmenting into mesoblastic somites, the lower parts, or lateral plates, being cleft into two layers, enclosing the body-cavity.[1]

(3.) We have seen that in the rabbit, at the close of segmentation, an outer layer of clear cells surrounds, except at one spot, the pseudo-blastopore, an inner mass of granular cells. It is possible, as above suggested, that the slipping inwards of the inner cells is analogous to the process of invagination. But if so, this process is a secondary one, and the pseudo-blastopore is probably not homologous with the blastopore.

The outer cells ere long grow over the pseudo-blastopore, and completely enclose the inner cells. A fissure then makes its appearance between the inner cells and the outer cells, on that side of the ovum which is furthest from the pseudo-blastopore. The cells separate more and more, until a relatively large cavity is formed. And this cavity increases more and more, until the inner cells form merely a lens-shaped mass attached to the inner side of the outer cells (Fig 35, ii.). Thus, by an internal separation of the cells, is formed a blastodermic vesicle The greater part of its walls consists of a single layer of flattened outer cells. It rapidly enlarges, the cavity being filled with an albuminous fluid ; and as it continues to enlarge, the lens-like mass of inner cells spreads out and becomes flattened, so that there is formed a central area where the cells are two rows thick, and a peripheral area where they are one row thick. In the central, or *embryonic area*, the lower layer of inner cells forms the flattened hypoblast, while the upper layer of inner cells fuses with the layer of outer cells to

---

[1] The cleavage of the mesoblast at first extends into the vertebral plates, but is here, for the most part, obliterated, though even here it may subsequently reappear.

form the epiblast. In the peripheral area around this the outer cells form flattened epiblast, and the inner cells form hypoblast. The lower part of the vesicle is constituted by a single layer of outer cells forming flattened epiblast. In the course of development the hypoblastic layer grows outwards, so as to line this part also. So that we have a state of things somewhat analogous to that which obtains in the fowl; but with this difference, that in the fowl the blastoderm grows round and encloses a mass of yolk, while in the rabbit it grows round and encloses a cavity containing an albuminous fluid. And just as in the fowl it is only the pellucida area which gives rise to the embryo, so in the rabbit it is only the embryonic area that gives rise to the developing organism.

In this area the primitive streak is formed in a manner very similar to that which obtains in the fowl; its surface becomes furrowed with a primitive groove, and from it the mesoblast of the posterior part of the embryonic area is produced (Fig. 35, iv.). In the anterior part of this area, where the medullary groove appears, the mesoblast is derived, as in the chick, from the hypoblastic layer (Fig. 35, v.). From the hypoblast of the axial line the notochord is developed; and on either side are formed the mesoblastic plates, separating into (1) a vertebral plate with its mesoblastic somites, in which the body-cavity, arising, as in other types, by cleavage of the mesoblast, is obliterated, and (2) lateral plates, in which the cleavage of the mesoblast and resulting body-cavity is persistent.

The Neural Tube.—In all three types the mode of formation of the neural tube differs in no essential from that described in our illustrative case. A longitudinal neural plate is formed in the mid-dorsal line. On either side neural folds (*laminæ dorsales*) grow up, and, eventually meeting above, enclose a canal. In the frog this takes place along the whole dorsal line (Fig. 36, i.), and the canal opens into the blastopore. In the fowl and rabbit it is formed in the anterior two-thirds of the embryonic area, and the two neural folds diverge posteriorly, so as to enclose the front end of the primitive streak

(Fig. 35, iii.). Thus the primitive streak occupies the same position as the blastopore of the frog, and is very probably homologous with that structure. On the dorsal side of the blastopore of the frog hypoblast and mesoblast come into close relation with the inflected epiblast. From the primitive

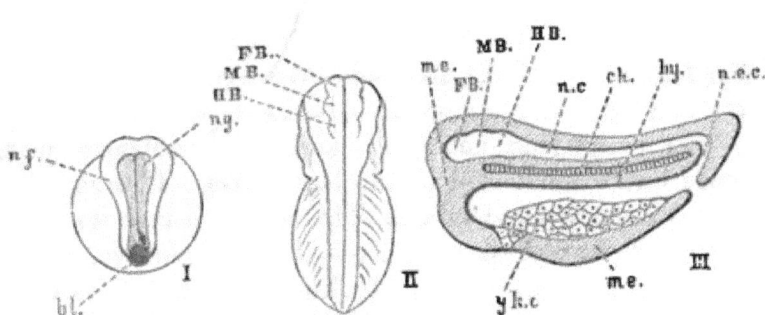

FIG. 36.—THE NEURAL TUBE.

i. and ii. Dorsal views of frog embryo. iii. Longitudinal section of embryo in about the same stage as ii.

bl. Blastopore. *ch.* Notochord. *FB.* Fore-brain. *HB.* Hind-brain. *hy.* Hypoblast. *MB.* Mid-brain. *nc.* Neural canal. *nf.* Neural fold. *me.* Mesoblast. *n. g.* Neural groove. *n. e. c.* Neurenteric canal. *yk. c.* Yolk-containing cells.

streak of the fowl and rabbit the mesoblast grows inwards (Fig. 35, iv.), and at the anterior end of the primitive streak this mesoblast meets that produced in the neural region from the hypoblast (Fig. 35, v.), thus effecting at this point an indirect junction of epiblast and hypoblast. In some of the lower vertebrate types, moreover, among those which possess a primitive streak, a passage here perforates the blastoderm, and leads into the primitive digestive cavity. So that it is probable that the primitive streak is homologous with at least the dorsal part of the blastopore. In the fowl, the point where the spreading blastoderm finally coalesces, so as to enclose the yolk (p. 100), possibly marks the ventral portion of the blasto-pore. What relation, if any, the pseudo-blastopore of the rabbit's ovum bears to the true blastopore is not known. In any case, the blastopore of the frog, and the anterior end of the primitive streak of the fowl and the rabbit, mark the posterior end of the young embryo, its anterior end being indicated by the anterior end of the neural canal.

At this anterior end the neural canal enlarges into three so-called cerebral vesicles, the precursors of the fore-brain, mid-brain, and hind-brain. They are seen in Figs. 36 and 37, *FB.*, *MB.*, *HB.*

The folding off of the Embryo.—We have seen that in the rabbit and the fowl the embryo, marked out by the neural tube, the underlying notochord, and the mesoblastic plates, occupies but a small area of the blastoderm. In front of the neural tube a crescentic groove (Fig. 35, iii.) makes its appearance, bending round the anterior end of the embryo, with the horns of the crescent pointing backwards. This marks off the *head-fold.* Behind the neural tube a similar groove is subsequently formed, with the crescentic horns directed forward, giving rise to the *tail-fold*, and soon the horns of the crescent become connected at the sides by lateral folds. All these grooves gradually become constricted beneath the embryo.

Take your knife or pencil-case and lay it upon the table. Spread over it your handkerchief, and tuck the handkerchief under the ends. Pass a piece of string round the ends in the tucks, single knot it, and gradually draw it tight. The string will form curved grooves under the ends, and straight lateral grooves along the sides; and as you pull it tight the grooves at the ends become deeper and deeper, until the part of the handkerchief containing the knife is folded off from the rest, and only connected by a puckered stalk. This will illustrate in a clumsy but comprehensible fashion the beautiful process by which the embryo is folded off from the rest of the blastoderm, which in the case of the fowl envelopes the whole of the yolk, and is called the yolk-sac (Fig. 38, i. *y. s.*), and in the case of the rabbit contains an albuminous fluid, and is called the umbilical vesicle.

It must be remembered that in both these types the hypoblast covers a cavity—the primitive alimentary cavity. During the process of the folding off of the embryo this cavity is not obliterated, but the walls of the embryo folding in beneath it,

and at either end, and at the sides, convert it into a tube
(Fig. 38, iv.). So that we have a small tubular embryo connected by a hollow stalk with the large yolk-sac or the umbilical
vesicle.

In the meanwhile the cleavage of the mesoblast is in progress.
The ventral and lateral plates split into two layers, of which
one closely adheres to the hypoblast of the digestive tube. This
is the splanchnic layer (*sp. l.*), which, with the adjoining
hypoblast, forms the *splanchnopleure.* The other forms the inner
lining of the body walls. This is the somatic layer (*so. l.*),
which, with the adjoining epiblast, forms the *somatopleure.* The
space between them is that of the primitive body-cavity (*b. c.*).
These are seen in Fig. 38, iii., which is a quite diagrammatic
transverse section of the embryo.

In the frog there is no folding off of the embryo, and no
yolk-sac or umbilical vesicle, though the yolk-cells beneath the
mesenteron are homologous with these structures. The cleavage
of the mesoblast takes place as in the other types.

The Heart, Cerebral Vesicles, and Mesoblastic Somites.—It
will be well to describe here the appearance of embryo chicks of
about 24, 36, and 48 hours incubation. They may be obtained
either from the hen or by means of an incubator kept at a temperature of from 38°-40° C. To prepare the embryo for observation the
student should place the egg in a basin of normal saline warmed
to 38° C., and carefully break away the shell. On exposing the
yolk the embryo will be clearly visible on its upper surface. It
lies in the midst of the area pellucida, which is surrounded
by a mottled vascular area containing embryonic blood-vessels.
With sharp scissors an incision should be rapidly carried round
the vascular area. The blastoderm may then be floated off from
the yolk, and the vitelline membrane which covers it may be
removed. It should then be floated on to a glass slide and
covered with a cover-glass, taking care to have plenty of normal
saline.

In Fig. 37, i., the embryo (of about 24 hours) is seen, from
the ventral aspect, in the midst of the area pellucida. Anteriorly,

in the middle line, is seen the first cerebral vesicle or fore-brain (*FB.*), followed by the mid-brain (*MB.*), and the hind-brain (*HB.*), the latter passing into the myelon, still unclosed behind. The line marked *a* is the amnion fold (see p. 108). *Pr. g.* marks the remains of the primitive groove. The noto-chord (*ch.*) is seen through the neural tube. On either side

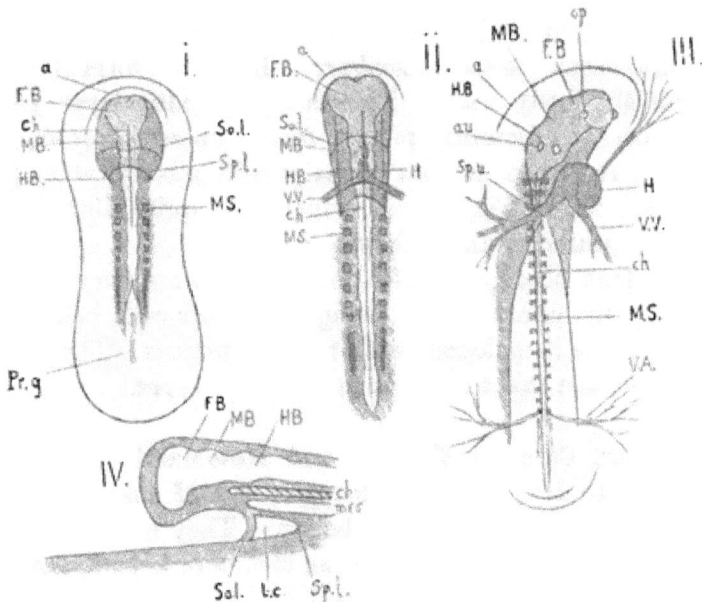

Fig. 37.—CHICK EMBRYOS.

i., ii., iii. Chicks of 24, 36, and 48 hours incubation. iv. Diagrammatic section through anterior end of i.

a. Amnion fold   *au* Auditory vesicle.   *b. c.* Body-cavity (cœlom). *Ch.* Notochord.   *FB.* Fore-brain.   *H.* Heart.   *HB.* Hind-brain.   *MB.* Mid-brain.   *MS.* Mesoblastic somites.   *mes.* Mesenteron.   *op.* Optic vesicle. *Pr. g.* Primitive groove.   *So. l.* Somatic layer.   *Sp. l.* Splanchnic layer. *Sp. u.* Union of splanchnic layers.   *V. A.* Vitelline artery.   *V. V.* Vitelline vein.

of the spinal cord (*myelon*), are seen the cubical masses (*meso-blastic somites*), into which the vertebral plate is segmenting. The lines *So. l.* and *Sp. l.* mark the limits of the somatic and splanchnic layers of the head-fold.

As this may present a difficulty to the student the diagram

(37, iv.) is introduced to aid him in overcoming it. It represents the head end of an embryo, the amnion-fold (see p. 108) being omitted to avoid confusion. *FB. MB. HB.* are the three cerebral vesicles ; *Ch.* the notochord which underlies the second and third. Beneath it lies the blind anterior end of the primitive digestive cavity (*mes.*). The shaded part in front of and beneath it is the mesoblast, which has cleft here, just as it cleaves at the sides of the embryo (Fig. 38, iii.), into a somatic layer (*So. l.*), and a splanchnic layer (*Sp. l.*), with the body-cavity (*b. c.*), between them. When it is remembered that the embryo is being folded off at the sides, as well as in the region of the head, the true meaning of the lines *So. l.* and *Sp. l.* in 37, i., will be seen.

In Fig. 37, ii., an embryo of about 36 hours, it will be seen that the number of mesoblastic somites has increased, and that the heart (*H.*), has now become obvious. It lies at the point of union of the two vitelline veins (*V. V.*). The development of this organ is illustrated in Fig. 38A, ii., and is described in Ch. vi.

In Fig. 37, iii., an embryo of about 48 hours, there is considerable change. The mesoblastic somites have increased in number. The heart is more definite, and appears as a coiled tube. Its rhythmic contraction will no doubt be seen in the embryo of this age, and of the age figured in ii. Besides the vitelline veins (*V. V.*), the vitelline arteries (*V. A.*), are clearly visible, but their connection with the heart is obscured by the mesoblastic somites. In the fore-brain two rounded optic vesicles (*op.*) are seen to have budded out ; and on either side, in the region of the hind-brain, are the auditory pits (*au.*), from which the auditory capsules will be formed. They are not, like the optic vesicles, parts of the brain, but involutions of the external epiblastic layer. *Sp. u.* marks the point where the splanchnic layers of the mesoblast of the two sides have united beneath the blind end of the primitive digestive cavity. Anterior to this the splanchnic investment of the mesenteron is complete. Posterior to this the mesenteron has not yet been converted into a tube, and the splanchnic layers diverge.

The Embryonic Appendages and Fœtal Membranes.—We have now to consider certain important structures which characterise the fowl and the rabbit, but which are absent in the frog. The first of these is the *amnion.*

Revert for a moment to our rough and ready illustration, with knife and handkerchief, of the manner in which the embryo is folded off. Suppose that all around our mimic embryo we fold the handkerchief in a ridge, and then, threading a piece of cotton along the summit of this ridge, draw the ends of the string together, and so pull the folds over the embryo until they meet above it. This again illustrates, in a rough but comprehensible fashion, the way in which, in the chick embryo, the amnion folds grow up all round, and eventually meet and coalesce above the embryo. This illustration, however, fails to mark an important distinction. It shows, indeed, that in the folding off of the embryo the folds are inwards, while in the formation of the amnion they are outwards. But it does not show that in the former process the folds affect the whole thickness of the blastoderm, while in the latter, the formation of the amnion, only the somatic layer is affected. This will, however, be readily seen in Fig. 38, ii. and iii., which are diagrammatic cross-sections of the embryo. In ii., in which the folding off of the embryo is omitted for the sake of clearness, the amnion fold (*am. f.*), is seen at the edge of the somatic layer. In iii., in which the embryo is represented as folded off, the amnion folds have nearly coalesced above the embryo. The folds are hollow, and enclose a space (*s. s. c.*) which is continuous with the body-cavity.(*b. c.*). It must be remembered that such folds extend all round the embryo, the head-fold being very well marked. The edge of its outer limb is shown in Fig. 37, i., ii., iii., the inner layer being so closely applied to the head as not to be distinguished. At last all these folds unite and coalesce above the embryo; the outer layers unite and become continuous; the inner layers do the same; and thus the space (*s. s. c.*, Fig. 38, iii. and iv.) between them becomes continuous. The inner layer is known as the true amnion, the outer layer as the false amnion. The false amnion lies beneath and in close contact with the investing membrane (zona radiata of the rabbit, and vitelline membrane in the fowl), and is known as the *serous membrane* (*s. m.*). The space beneath it may be called the *subserous space* (*s. s. c.*); it is

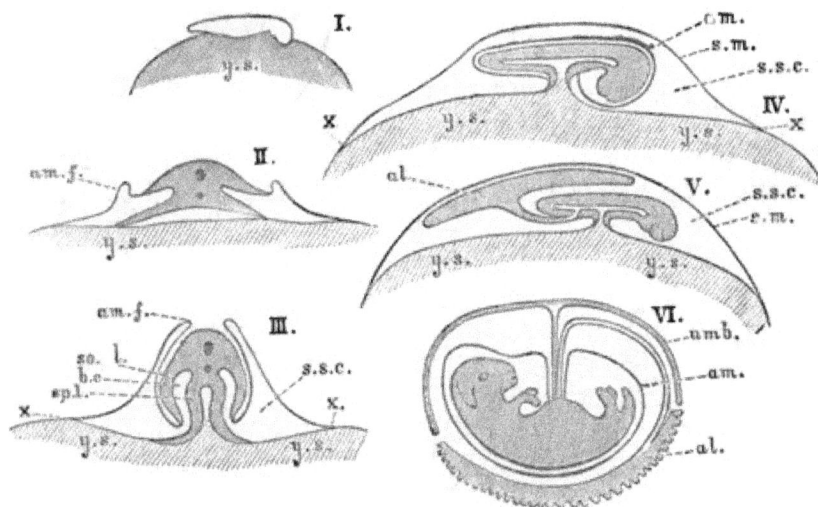

FIG. 38.—FŒTAL MEMBRANES.

i. Embryo folded off from yolk-sac. ii. Transverse section through embryo, showing cleavage of mesoblast and amnion fold. iii. Transverse section through older embryo. iv. Longitudinal section at a stage a little later than iii., the amnion folds having united. v. Longitudinal section to show formation of allantois. vi. Embryonic rabbit with fœtal membranes.

al. Allantois. *am.* Amnion. *am. f.* Amnion fold. *b. c.* Body-cavity (cœlom). *s. m.* Serous membrane. *s. s. c.* Subserous cavity. *so. l.* Somatopleure. *sp. l.* Splanchnopleure. *umb.* Umbilical vesicle. *x.* Point where serous membrane joins yolk-sac. *y. s.* Yolk-sac.

continuous with the body-cavity (38, iii.). Fig. 38, iv., is a diagrammatic longitudinal section (the body-cavity being omitted), in which the amnion folds have coalesced, the inner layer forming the true amnion (*am.*), the false amnion forming the serous membrane (*s. m.*). Within the inner layer or true amnion a fluid collects, in which the embryo lies as in a water-bed.

In the rabbit the head of the embryo is sunk into a pit or depression of the blastodermic vesicle, from the edge of which the amnion fold takes its origin. In this depression, termed the *proamnion*, the mesoblast is absent. It is shown in 38A, iii. The illustrations in this figure exhibit the relations of epiblast, hypoblast, and mesoblast in the fœtal membranes.

Meanwhile—beneath the embryo—the yolk-sac, surrounded by a thin blastodermic layer, is in contact with the investing membrane. But by the continuation of the cleavage into this region the blastoderm is split into two layers, of which one (continuous with the splanchnic layer) remains adhering to the yolk-sac, while the other (continuous with the somatic layer outside its amnion fold) adheres to the investing membrane, and forms the serous membrane of this region. This will, perhaps, be sufficiently clear by a study of iii. and iv., Fig. 38. Note

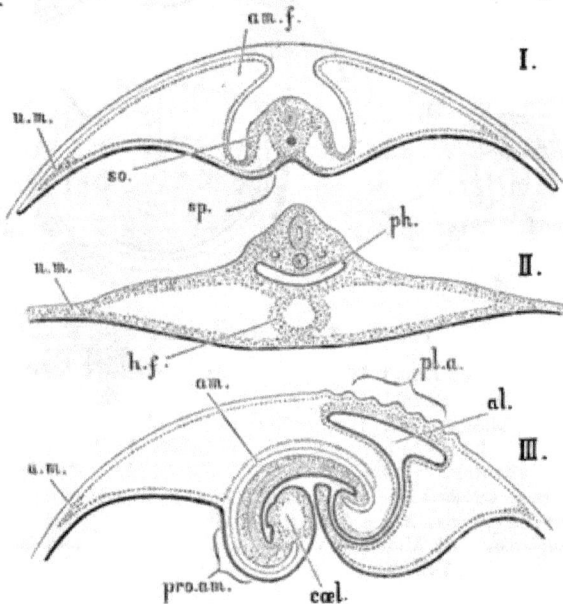

FIG. 38 A.—RELATIONS OF EPIBLAST, HYPOBLAST, AND MESOBLAST.
  I. Transverse section of Embryo before the Amnion folds have met.
 II. Transverse section of Embryo (Amnion omitted) through Pharynx and
       Heart-folds.
III. Longitudinal section of Rabbit Embryo to show Proamnion.

Thin line, epiblast; thick line, hypoblast; dotted, mesoblast.

al. Allantois. *am.* Amnion. *am. f.* Amnion fold. *cœl.* Cœlom. *h. f.* Heart fold. *ph.* Pharynx. *pl. a.* Placental area. *so.* Somatopleure. *sp.* Splanchnopleure. *u. m.* Uncleaved mesoblast.

first that in iii. the whole of the amnion fold down to *x* is continuous with the somatic layer (*so. l.*). Then note that in iv. the serous membrane, derived from the outer layer of the amnion fold, forms a distinct membrane as far as *x*. Now, picture the yolk-sac continually **shrinking as** its contents are absorbed into, and minister **to** the growth of, the embryo. But owing to the shrinkage **of the** yolk-sac (*y. s.*) the sub-serous space (*s. s. c.*) gradually extends round the yolk to the lower pole of the egg. And as it does so the serous membrane is split off from it, and forms a complete layer beneath the vitelline membrane (**which** is not represented in the figure, but closely adheres to **the** serous membrane).

Another important structure, the *allantois*, has still to be described. From the hinder-end of the digestive tube there grows out, even before the amnion folds have met, a hollow outgrowth. It is lined with hypoblast, but covered externally by a layer of splanchnic mesoblast. It spreads out (Fig. **38,** v.) **as a** flat sac lying over the dorsal and posterior part of the embryo between the true amnion (omitted in the fig.) and **the** serous membrane. In the fowl a rich network of blood-vessels is developed in it, and it performs the functions of embryonic respiration. **Furthermore, in the later** stages **of** embryonic growth it extends outwards and downwards until its lower folds **meet** below, overlap, and fuse together, thus enclosing the albumen in a space lying between the allantois and the yolk-sac, which has been termed the placental sac. Simple villi developed on the walls of this sac absorb the enclosed albumen partly into the blood-vessels of the yolk-sac, partly into those of the **allantois.** In the rabbit (Fig. 38, vi.) its mesoblastic layer unites and fuses with the serous membrane. The umbilical vesicle, which has become flattened out (*umb.*), also fuses with the serous membrane. These **two membranes,** which become practically continuous, are known as the *chorion*. That **in the** dorsal region, formed from the allantois, is **called** *true chorion*, that in the ventral region, formed out **of** the flattened umbilical vesicle, is called *false chorion.* From the outer surface of **the** true chorion finger-like processes (*villi*) grow out and fit into crypts **or** depressions formed in the wall of the uterus of the mother. Both walls become highly vascular, and nutritive fluid passes from the mother **to** the fœtus, each villus being provided with an artery, a vein, and a capillary plexus connecting the two. There is no transfusion of blood from the mother to **the** offspring, **but** a diffusion of nutritive material from the mother, and of waste products from the fœtus. These maternal and fœtal structures taken together constitute the organ known as the *placenta*.

While the placenta is being developed the folding off of the embryo becomes more complete, there being (as seen in Fig 38, vi.) only a narrow *vitelline duct* connecting the remnant of the umbilical vesicle with the embryo. Near the vitelline duct is seen the duct of the allantois, and around both is the somatic stalk of the amnion. The common stalk, consisting of vitelline duct, allantoic duct, and enveloping amnion, is now known as the umbilical cord. The mesoblast of the investing amnion stalk develops a gelatinous tissue which cements together the whole of the contents. The allantoic arteries in the cord in some mammals wind in a spiral manner round the allantoic vein.

At birth the fœtal membranes, together with the uterine mucous membrane with which they are so closely interlaced, are delivered as the *after-birth*, such a placenta being called *deciduate*, as opposed to non-deciduate where the fœtal membranes separate from the maternal. The mother gnaws through the umbilical cord close to the embryo, and probably eats the placenta.

The vitelline duct completely atrophies; but the allantoic stalk gives rise to two structures, (1) the *urinary bladder*, formed by a dilatation of its proximal extremity; (2) a cord known as the *urachus*, connecting the bladder with the wall of the body at the umbilicus.

We have thus traced the rabbit to its birth. Let us now return to the fowl. We have seen that the allantois grows out at the posterior end of the embryo, and, applying itself closely to the serous membrane, has a respiratory function. Thus does the growing chick obtain the necessary oxygen. The necessary food is obtained from the yolk-sac, the contents of which, however, do not pass directly into the alimentary canal, for the duct of the splanchnic stalk becomes early obliterated, but are absorbed indirectly through the blood which circulates freely in the vitelline arteries and veins. Owing to this absorption of the yolk the yolk-sac gradually shrinks, and eventually, while still of considerable size, is withdrawn into the body-cavity, the walls of which gradually grow together at the umbilicus.

At the broad end of the egg there is, between the two shell

membranes, a space which contains air. As development proceeds it increases in size, and on the fourteenth day of incubation the chick moves so as to lie lengthways in the egg, with its
beak touching the inner wall of the air-space. On the twentieth
day the beak pierces this wall, and the chick begins to breathe
the contained air. The blood, which has hitherto been oxygenated in the embryonic respiratory organ (the allantois), is
now arterialised in the adult respiratory organ, the lungs. The
functionless allantois dries up and shrivels away, as do also the
serous membrane and the amnion. The chick breaks the shell
with its beak and is hatched.

In the frog there is neither amnion nor allantois. The
urinary bladder, however, arises as an outgrowth of the
ventral wall of the posterior end of the cloaca, and has therefore been regarded as homologous with the similar structure in
the rabbit.

Summary.—The frog, the fowl, and the rabbit pass through
embryonic development under very different conditions. The
primitive embryo, such as that of *Amphioxus* (p. 114), is
free at the gastrula stage. It moves through the water
in which it is hatched by means of the cilia with which its
surface is freely covered. And from that time forward it has
to obtain its nourishment by its own individual exertions.
The frog embryo is provided with a supply of food-yolk sufficient to carry it over the early stages of development. It
is, however, hatched comparatively soon, long before it has
assumed anything like its adult form. There is a certain
amount of transformation within the egg membrane, but a large
proportion of the embryonic development takes place after the
organism is hatched, and constitutes metamorphosis. And
during metamorphosis the organism is absolutely dependent on
its own exertions for the necessary nutriment. The fowl
embryo is provided with an abundant supply of food-yolk,
sufficient to carry it over, not only the early, but also the later
stages of embryonic development. The mother, however, supplies under natural conditions the necessary warmth, without
which developmental changes cannot proceed. And when the
chick is hatched, although the hen affords it protection, and

aids it in its search for food, she does not prepare for it food-stuff by her own vital processes. The rabbit embryo is provided with but little food-yolk. It is dependent on the mother from the first, not only for warmth, but also for nutriment. It is developed within the maternal organism, with which it is con-nected in such a way that diffusion of nutrient fluid can take place from the one to the other. And when it is born it is for a while dependent on the mother, not only for protection and aid, but also for food-milk, which is elaborated by her own vital processes.

In all three types the essential nature of the process of fertilisation is the same. In all, development results from, and is dependent upon, the union of ovum and spermatozoon. And, in all, this union gives rise to a process of cell-division or segmen-tation. But the character of the segmentation differs. In the simple *alecithal* ovum, in which there is no food-yolk, or in which the little that there is is distributed uniformly, the segmentation is complete, or *holoblastic*, and *regular*. In the frog's ovum, in which the yolk is arranged in polar fashion (*telolecithal*), the segmentation, though still holoblastic, is *unequal*. In the fowl's ovum, where the yolk is also arranged in polar fashion (*teloleci-thal*), but is in enormous quantity, the segmentation is partial (*meroblastic*) and irregular. In the rabbit's ovum there is very little food-yolk, and that arranged uniformly (*alecithal*), and the segmentation is complete, or holoblastic, but slightly unequal.

At the outset we made use of a supposititious case of simple embryonic development for purposes of illustration. The sup-posed organism possessed an alecithal ovum, which underwent holoblastic segmentation in a quite regular fashion. A hollow blastosphere was formed which underwent simple invagination, by which the segmentation cavity was obliterated. Thus a two-layered gastrula was produced, the inner layer being hypoblast, the outer layer epiblast. The gastrula-cavity was the archen-teron; the opening of the gastrula-cup the blastopore. Meso-blast was formed between the two primitive layers; and the body-cavity, or cœlom, arose by cleavage of the mesoblast. Along the dorsal surface of the embryo the medullary axis was differentiated from the epiblast. We may now notice that, with certain modifications, the little Lancelet (*Amphioxus*) presents us with a mode of development similar to that which was there described.

Let us now summarise the more important events in the course of development :—

1. The immature ovum consists of naked protoplasm containing a germinal vesicle (nucleus) and germinal spots (nucleoli).

2. The mature ovum contains a variable amount of food-yolk, but little in *Amphioxus*, and that little uniformly distributed (alecithal) ; more or less in fowl and frog, concentrated at one pole (telolecithal) ; but little in the rabbit (alecithal).

3. The detached ovum is surrounded by primary and secondary membranes. The primary membranes (zona radiata, rabbit ; vitelline membrane, fowl and frog) are formed in the ovi-sac. The secondary membranes (albuminous layer, shell membranes) are added in the oviduct.

4. The nucleus, previous to fertilisation, divides twice, probably by karyokinesis ; one half is extruded in each case in a polar cell ; the portion that remains is the female pronucleus.

5. Fertilisation is probably effected by a single spermatozoon, the nucleus of which, within the ovum, forms the male pronucleus. Fertilisation is effected after the ova are laid in the frog, but within the body of the parent fowl and rabbit.

6. Female and male pronucleus fuse to form the first segmentation nucleus.

7. Segmentation is probably in all cases accompanied by karyokinesis.

> In *Amphioxus*, holoblastic and nearly regular.
> In frog, holoblastic and unequal.
> In fowl, meroblastic and irregular.
> In rabbit, holoblastic and slightly unequal.

8. Epiblast and hypoblast produced—

> In *Amphioxus* by simple invagination.
> In frog partly by invagination, partly by differentiation.
> In fowl and rabbit by differentiation.

9. Archenteron produced—

> In *Amphioxus* by invagination.
> In frog partly by invagination.
> In fowl and rabbit beneath the blastoderm of the embryonic area.

10. Mesoblast produced—

> In *Amphioxus* from paired outgrowths of the archenteron.
>
> In frog partly from hypoblast at point of invagination, partly by differentiation of yolk-containing cells.
>
> In fowl and rabbit partly by down-growth of cells derived from the epiblast at primitive streak, partly by differentiation from hypoblast.

11. Neural plate, groove, and tube formed in all four types from epiblast of mid-dorsal line, the process being somewhat modified in *Amphioxus.*

12. Notochord formed in all four types by special up-growth of cells of hypoblast in the mid-line beneath the medullary axis.

13. Mesoblastic plates in all three types separate into vertebral plates and lateral plates, of which the former by transverse segmentation give rise to mesoblastic somites, and the latter, by cleavage of the mesoblast, to the body-cavity.

14. The embryonic body is formed in the frog by simple elongation, there being no yolk-sac or umbilical vesicle. In the fowl and rabbit it is folded off from the yolk-sac or from the umbilical vesicle.

15. In the fowl and rabbit a protective sac, the amnion, is formed over the embryo. This is absent in the frog.

16. In the fowl and rabbit an embryonic respiratory organ, the allantois, is developed as an outgrowth from the hind-gut. In the rabbit this comes into close relation with the uterine tissues of the mother, and thus acquires also a nutritive function. · It is absent or rudimentary in the frog.

17. Tertiary investing membranes (serous membrane in fowl, serous membrane and chorion in rabbit) are produced in the higher types.

CHAPTER VII.

THE GENESIS OF TISSUES AND ORGANS.

IN the last chapter we considered the General Embryology of the frog, the fowl, and the rabbit. We saw how the so-called germinal layers were constituted, how the body-cavity was formed, how the vertebral plate was segmented into mesoblastic somites, how the notochord and neural axis were formed, and how the latter expands anteriorly into three cerebral vesicles foreshadowing the fore-, mid-, and hind-brain. This, however, carries us but a little way towards understanding how the tissues and organs of the body take their origin. Their genesis, so far as the space of one short chapter will permit, will now be considered.

The cells of which the early embryo is composed are minute nucleated masses of protoplasm. They are at first for the most part undifferentiated; but those of the epiblast and hypoblast very early show signs of differentiation in the direction of flattened or of columnar epithelium cells. The cells of the notochord differentiate in a direction of their own, becoming net-like and vacuolated; while those of the mesoblast become, in the course of development, differentiated into blood, blood-vessels, muscle, connective tissue, cartilage, and bone, or are instrumental in the production of these tissues. For it must be remembered that, as we have seen in the chapter on General Histology, there is in most tissues, besides the actual cells, a greater or less amount of intercellular or of cementing substance, which may, as in epithelioid tissue (endothelium), be comparatively insignificant, or may, as in cartilage and bone, be important.

117

Blood.—In the vascular area of the blastoderm of fowl and rabbit some of the nucleated mesoblast cells become united one with another by protoplasmic processes, so that an irregular network of nucleated cells is produced. The nuclei of the cells rapidly multiply by division. Some remain central, and the protoplasm which surrounds each of these acquires a yellowish-red colour. Vacuoles are formed in the protoplasmic network in which they lie, and the vacuoles running together, there results a system of branched canals enclosing a fluid in which the nucleated corpuscles float in irregular clusters. Eventually these clusters break up into a number of red corpuscles, each with an internal nucleus and surrounding red-tinted protoplasm. These primitive red corpuscles seem to be capable of amœboid movements like the white corpuscles, and of multiplication by division. Thus is formed a network of canals—the primitive blood-vessels of the blastoderm—containing red blood-corpuscles. It is probable that within the embryo itself primitive blood-corpuscles and blood-vessels are developed in a similar way. The protoplasmic walls of these primitive blood-vessels at first contain only a few nuclei irregularly imbedded in the protoplasm. But subsequently the protoplasm around these nuclei and their descendants becomes differentiated into the flattened cells which compose the walls of the blood-capillaries, and which form the lining membrane of the larger vessels in the adult. Around this epithelioid layer in the larger vessels, connective tissue and muscular coats are subsequently formed.

The primitive nucleated red blood-corpuscles of the mammal are subsequently converted into the non-nucleated blood-discs, losing their nucleus and shrinking in size. Thus the nucleated red corpuscles are, during embryonic life, gradually replaced by non-nucleated red discs. But there can be no doubt that red discs are also produced in other ways. They are formed, for example, by the separation of masses of protoplasm from the large cells of the red marrow of bones. It has also been shown that white corpuscles develop into nucleated red corpuscles, and these into red discs. The white corpuscles would seem to be formed during early embryonic life in the liver, but subsequently in the spleen, and lymphatic glands.

Connective Tissue.—The mesoblast cells between the developing blood-vessels become somewhat separated from one another, but at the same time united by processes passing from one to the other. In the interspaces there collects an albuminous fluid, which will be converted into the ground substance of the tissue. Whether the fibrillated bundles of white fibrous tissue are (1) separated off from the outer portions of the cells, the residue of the cell remaining as a connective tissue corpuscle ; or (2) developed in the intercellular ground substance, the cells remaining as the corpuscles, is still undetermined. It is probable that elastic fibres are formed in the intercellular substance between fusiform cells.

Nerve.—The mode of development of nerve-fibres and nerve-cells requires further investigation. It is now generally believed, however, that all the nerves arise as outgrowths of epiblast cells from the central nervous system. These cells elongate and arrange themselves end to end with the consequent production of axis fibres. If during life an axis fibre is cut, the part furthest from the central nervous system dies and decays, and is replaced by the outgrowth of the cut end in living connection with the central system.

The medullary sheath is said to be formed from mesoblast. The cells are said to develop around the axis fibres. Each internode is supposed to be formed from a single cell, the greater part of which becomes converted into fatty matter, the nucleus and residuary protoplasm remaining as the nerve-corpuscle (see p. 74) The primitive sheath is regarded as the cell membrane of the medullary cell.

Muscle.—Unstriped muscular fibres are developed from mesoblastic cells (for the most part from the splanchnic layer of the lateral plates), which become elongated, fusiform, and flattened.

Striped muscular fibres, notwithstanding their extreme length (up to 2 inches), would also seem to be derived from single cells. At first oval, the fibre becomes an elongated spindle in which the nucleus subdivides, the new nucleus shifting along the length of the fibre. **Thus is** formed an elongated fibre containing **several** nuclei. The protoplasm of the fibre, either alongside or around the nuclei, becomes differentiated into **the** characteristic muscular fibrils, only a small remnant of indifferent protoplasm remaining around the nucleus as the muscle corpuscle.

We have seen that the involuntary (unstriped) muscle of the alimentary system is derived from the splanchnic layer of the lateral plate. The main muscles of the trunk are derived from the mesoblastic somites of the vertebral plates.

Each vertebral plate originally contains a cavity, part of the primitive body-cavity, by which it is divided into an inner and an outer portion. That half of the inner portion which lies next to the notochord and neural canal is utilised in **the** formation of the permanent vertebræ. The **other** half of this portion early **develops into longitudinal muscles** running throughout the length **of each mesoblastic** somite ; and this, together with **the** whole **of the outer** portion (which is also converted into **muscles**), is known **as the** *muscle plate.*

Cartilage.—This tissue is developed **from** mesoblastic cells by **the** formation of a capsule round **each.** A second capsule is formed within the first, a third within that, and so on in orderly succession, until a concentric system of capsules surrounds each cell, the capsules first formed expanding as fresh internal capsules are formed. The capsules so developed fuse together and blend into the hyaline matrix of cartilage. The capsules are either (1) secreted by the cells, or (2) formed by the differentiation of the superficial layer of **the** protoplasm of the cell, or (3) developed in the intercellular substance around the cells. In the case **of elastic cartilage, elastic** fibres are subsequently developed.

Bone.—While many of the bones **take** their origin by ossification in cartilage, others arise by independent ossification in connective tissue. The **former** are termed *cartilage bones*, the latter *membrane bones.*

The *parietal* of the rabbit, one of the roofing bones of the skull, is a good example of a membrane bone. The first sign of ossification here is the formation of a network of minute bars or spicules. The bars are fibrous, and composed of *osteogenetic fibres*, which are at first soft and pliant, but become hardened by the deposition within them of calcareous salts. Between the fibrous bundles are large fusiform cells, the *corpuscles.* These cells are instrumental in forming the bone, whence they have been termed *osteoblasts.* As ossification proceeds some of these cells become completely surrounded by the osseous tissue, and thus give rise to the bone corpuscles, each lying in a hollow space or lacuna.

Meanwhile the bone has become invested by a definite connective tissue membrane, the *periosteum.* The inner layer of this (*osteogenetic layer*) is largely composed of osteoblasts, which deposit successive lamellæ of bone on the surface of the growing parietal. At the same time lamellæ are being deposited in the meshes of the original network of minute fibrous bars. Thus the interstices of the network are converted into narrow channels which contain blood-vessels. And thus the whole bone is converted into a more or less compact structure. Finally, as the bone increases in thickness, its middle layer becomes hollowed out by absorption of the bony matter, so as to become spongy in structure. This internal spongy bone is called the *diploë.*

We may take the *femur* of the rabbit as our example of a cartilage bone. In a still young rabbit this bone consists of a shaft (*diaphysis*) containing a central space in which is lodged the marrow. At each end, separated from the shaft by a cartilaginous interspace, is a terminal ossification (or *epiphysis*). In the adult rabbit the epiphyses coalesce with the shaft by ossification of the intermediate cartilage.

In the embryo the minute femur consists of a rod of fœtal cartilage invested by a sheath of vascular perichondrium. Within the substance of the cartilage the cells arrange themselves in columns near the growing ends, and in groups in the more central portions. In the groups, and at the deeper ends of the columns, the cell cavities are seen to be enlarged by the swelling of the cells, which then tend to break down and atrophy. Meanwhile the tissue becomes vascularized by the introduction of blood-vessels from the vascular perichondrium. The channels occupied by these vessels, uniting with the enlarged and coalescent cell cavities, convert the tissue into a vascular sponge-work, which becomes calcified, and thereby strengthened by the deposition of granules of calcareous matter in the remaining strands or trabeculæ of cartilage.

So far we have merely calcified cartilage, not true bone. But along the edges of the *primary marrow cavities*, as they are termed, of the sponge-work, a number of osteoblast cells may now be seen. Each secretes, and may be seen to be surrounded by, a droplet of osteogen, from which true bone is formed ; and as this true bone replaces the cartilage, the tissue is converted into " spongy endochondral bone."

Meanwhile, around the blood-vessels of and beneath the perichondrium, osteoblasts are developed, which here, too, form spongy bone. The perichondrium thus becomes periosteum ; and the spongy bone developed in this region is termed periosteal spongy bone.

If now we transfer our attention to the centre of the shaft, we shall find that the spongy endochondral bone is there being absorbed or eaten away through the instrumentality of specialized cells (osteoclasts), some of which are very large (giant cells). The spongy bone in the centre of the shaft is thus completely absorbed, and a marrow cavity produced.

Around this cavity the two processes—bone formation by osteoblasts, bone absorption by osteoclasts—may be seen going on side by side. Large spaces or marrow cavities are excavated, and are then gradually filled in with bone, or narrowed to the dimensions of a small canal for the passage of a blood-vessel. And thus are formed embryonic Haversian systems, which differ from those of the completed bone in their greater irregularity, and the much looser texture of the lamellæ.

Finally, we have to note that beneath the periosteum a delicate tube of bone forms round the middle of the shaft, and this tube gradually spreads towards either end of the bone until it becomes a delicate but compact osseous sheath. It is produced by an osteogenetic layer (with osteoblast cells), which is developed on the inner face of the membrane. This layer of compact bone, with its well-defined Haversian systems, gradually increases by exogenous growth beneath the periosteum. And as, with the general growth of the bone in thickness, the marrow cavity widens all the spongy bone is here absorbed, and in the

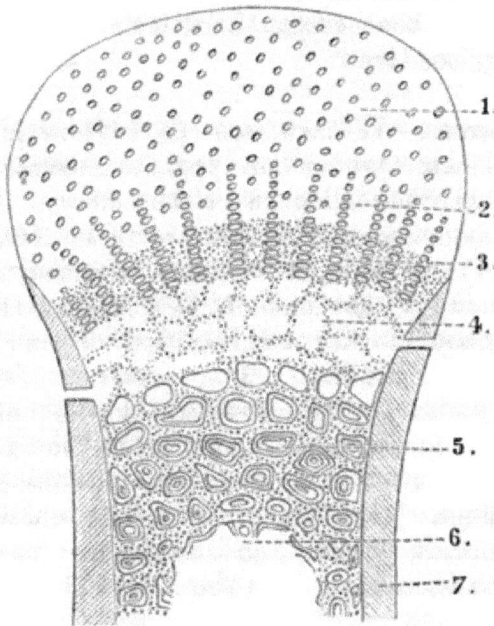

Fig. 39.—Diagram of Bone Formation.

Longitudinal section through head of bone.
1. Cartilage. 2. Columnar cartilage. 3. Vascularized cartilage. 4. Spongy calcified cartilage. 5. Region of deposition of bone lamellæ. 6. Reabsorption of bone to form medullary cavity. 7. Compact periosteal bone.

completed femur the central shaft is entirely composed of compact periosteal bone. At the ends of the growing bone, however, endochondral bone is still in process of formation, so that the growth of the bone in thickness is exogenous and periosteal, but in length is endogenous and endochondral.

Independent ossifications taking place at the growing ends of the bone give rise to the epiphyses, which remain separated from the shaft by a pad of cartilage. Growth then takes place mainly in this pad; its outer (epiphysial) half increasing by cartilaginous growth as fast as its inner (diaphysial) half is invaded by the growth of bone from the shaft. In the adult the whole pad becomes ossified and growth ceases. The whole femur has become bone, except the articular surfaces which still have a cartilaginous layer.

The Mesenteron.—We have seen (p. 105) how, in the chick, which we will here take as our type, the mesenteric tube is constituted during the folding off of the embryo. By the end of the third day of incubation there are three divisions of the mesenteron—(1) an anterior division, comprising the tubular portion produced by the union of the splanchnic layers of the cleaved mesoblast in the neighbourhood of, and somewhat posterior to, the head-fold; (2) a posterior division, comprising the tubular portion produced by the union of the splanchnic layers in, and slightly anterior to the tail-fold; and (3) a mid division, not yet converted into a tube, and resting upon the yolk-sac. The first will constitute the œsophagus, crop, proventriculus, gizzard, and duodenum; the second the large intestine and rectum; the third or mid division the small intestine.

In the anterior part of the first division four clefts on each side open outwards to the exterior. They are the visceral clefts, similar to those which are seen in the tadpole. Bounding them are five pairs of visceral arches, the anterior pair nearly meeting in the middle line, the succeeding pairs being successively shorter. The last three clefts soon completely close. But the inner ends of the first pair persist as the Eustachian tubes of the adult.

Immediately behind the region of the visceral clefts a ventral diverticulum is formed in the primitive digestive tube, from the hinder end of which[1] two diverticula grow out backwards into the thickened mesoblastic tissue. From each diverticulum numerous branches are given off, and these, in turn, branch and subdivide in the thickened mesoblast, which has now assumed the form of a separate lobe, and in which a rich network of capillaries make their appearance. Thus are formed the *lungs.* Finally, at the extremities of the diverticula, and some of their main branches, the characteristic *air-sacs* of the bird become differentiated.

In the region posterior to this the crop, proventriculus, and gizzard are differentiated.

Just behind the gizzard, in the short space between it and the point of union of the splanchnic layers, a pair of diverticula are formed, and grow out into the thickened mesoblast. From them solid cylindrical cords of hypoblast cells push their way in all directions through the mesoblastic tissue, and eventually fuse into a network, between the strands of which blood capillaries are formed in abundance. Thus are formed the right and left lobes of the *liver,* median outgrowths from which, meeting and fusing in the mid-line, form the wedge-like bridge between them. The bile canaliculi are produced either by the development of a lumen in the solid cords of hypoblast, or arise primitively as interspaces between the hypoblastic cells. The gall-bladder is a special pouch developed from the right primary diverticulum.

Slightly posterior to the hepatic diverticula a dorsal outgrowth gives rise to the posterior duct of the *pancreas.* The numerous branches formed in connection with it do not form a network as in the case of the liver, but give rise to the tubules of the pancreas. The other ducts of the pancreas are subsequently formed in a similar manner.

The mid-region of the mesenteron, that which at the end of the third day is not yet closed as a tube and is still widely open to the yolk-sac below, gives rise to the small intestine posterior to the duodenum. Lying immediately beneath the notochord

---

[1] The ventral diverticulum is partly formed by a pinching in of the primitive tube so as to cause it to be, so to speak, double-barrelled.

and mesoblastic somites, its dorsal attachment is at first very broad. As development proceeds the body-cavity extends further and further round its sides and dorsal aspect until it is only attached dorsally by a thin sheet of mesoblast. Meanwhile it has been, so to speak, pulled downwards away from the vertebral column, and has also increased in length to such an extent as to be thrown into coils. The thin suspending sheet of mesoblast has, during this process, been pulled downwards and folded and thus converted into the *mesentery*. At the same time, during the progress of these changes, the splanchnic layers gradually close in so as to convert this region also into a tube.

The posterior region of the mesenteron is at first in communication with a neurenteric canal similar to that of the frog (Fig. 36, *n. e. c.*). Ere long the communication of the neurenteric canal with the medullary canal is obliterated. Meanwhile the proctodæum is being formed by special invagination from the exterior to constitute the vent; and the portion of the mesenteron and neurenteric canal posterior to the vent is known as the *post-anal gut*. This part subsequently atrophies. The greater part of the remainder of the posterior division of the mesenteron forms the rectum. But its posterior portion—that which meets the proctodæal invagination—gives rise to that inner part of the cloaca which receives the generative ducts, and at this period of embryonic life, the allantois.

The Proctodæum.—The remainder of the cloaca, that which constitutes its outer moiety into which the Bursa Fabricii opens, is formed by the proctodæal invagination. This part is therefore lined with epiblast, while the inner division of the cloaca is lined with hypoblast.

The Stomodæum and Nares.—During the third day of incubation two small depressions are formed at the anterior end of the embryo. They are the *nasal pits*. During the next day the mouth begins to be marked out as a deep stomodæal depression, somewhat behind and ventral to these pits. Fig. 40, A., shows the changes which have taken place by about the sixth day. The mouth is bounded above by the *fronto-nasal process* (*f. n. p.*), at the outer edges of which are *nasal grooves* passing from the nasal pits (*n. p.*) to the mouth (*m.*). The lateral edges of the mouth lie between the somewhat divergent *superior* (*s. m. p.*) and *inferior maxillary processes* (*i. m. p.*), the latter processes forming also its lower boundary. As development proceeds,

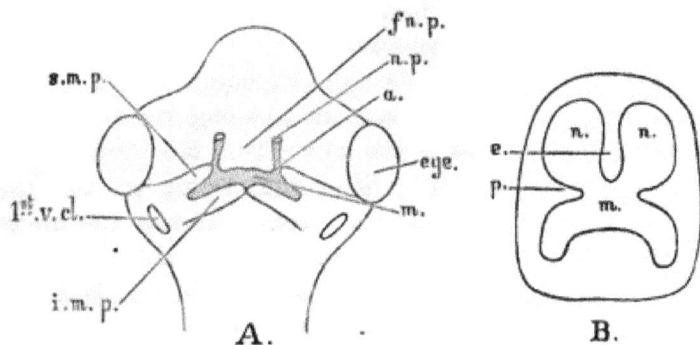

FIG. 40.—THE MOUTH.

A. Ventral view of head of chick embryo. B. Transverse section of mouth (after Gegenbaur).

a. Point where fronto-nasal process overarches the nasal groove. *e.* Internasal septum. *fn. p.* Fronto-nasal process. *i. m. p.* Inferior maxillary process. *m.* Mouth. *n.* Nasal chambers. *n. p.* Nasal pit. *p.* Incomplete palate. *s. m. p.* Superior maxillary process. *1st v. cl.* First visceral cleft.

the fronto-nasal process overarches the nasal groove at the point *a*, and fuses with the superior maxillary process, thus converting the nasal groove at this point into a canal. From its mode of formation it is clear that this canal at first opens just within the primitive mouth; but, by the forward growth of the parts anterior to it to form the beak, and other changes, its position becomes altered. In the fowl to some extent, but more markedly in the rabbit, the stomodæal cavity becomes divided into two chambers by the formation of a horizontal septum developed from lateral outgrowths of the superior maxillary processes (*p.*, Fig. 40, B., a diagrammatic cross section of the mouth, in which the palatal processes, *p.*, have not yet met in the middle line). Anteriorly the upper or *respiratory*

chamber is completely separated by this *palatal septum* from the
lower or mouth cavity ; but posteriorly the two chambers unite
where the palate ceases.   Finally, the upper or respiratory
chamber becomes more or less completely divided into two
chambers by a median vertical partition (*e.*, Fig. 40, B.) or *internasal septum.*

All the salivary and buccal glands of the rabbit are stomodæal
in their origin, and are therefore lined with epiblast and not
with hypoblast as in the pancreas.

The *teeth* are also stomodæal structures.   The manner of
their formation is briefly as follows :—Along the line of the jaw
the epithelium becomes thickened.   This thickened epithelial
keel, projecting downwards into the tissue of the jaw, is
known as the *primary enamel germ.*   At the special points
where teeth are to be developed, the down-growth becomes
more marked and special *enamel organs* (*en. or.*, Fig. 41) are
differentiated, one for each tooth.   Into each enamel germ

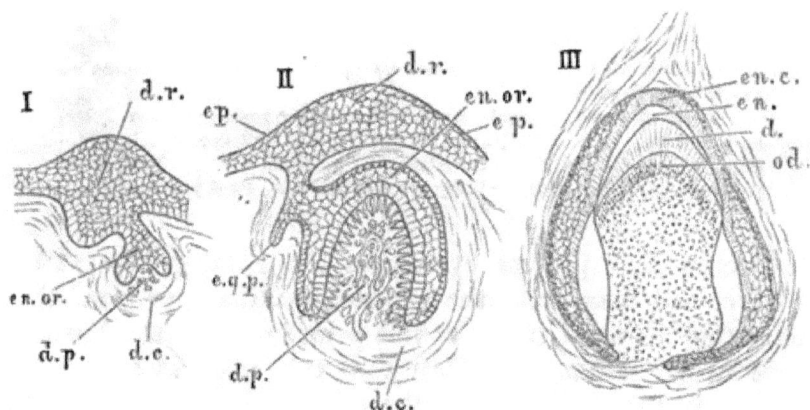

FIG. 41.—DIAGRAMS OF TOOTH DEVELOPMENT.

I. Transverse section through dental ridge soon after formation of enamel
germ. II. Transverse section of the same at a later stage. III. Transverse
section through the jaw of a kitten, in which the permanent tooth is yet
further developed.

d. Dentine. *d. c.* Dental capsule. *d. p.* Dental papilla. *e. g. p.* Enamel
germ of corresponding permanent tooth. *en.* Enamel. *en. c.* Enamel cells.
en. or. Enamel organ. *ep.* Epithelium of gum. *od.* Odontoblasts.

there projects upwards a vascular *dental papilla* (*d. p.*), and
round the enamel germ and its dental papilla the tissues become

somewhat differentiated, **thus giving** rise to the so-called *dental capsule* (*d. c.*). **(1) The** external layer of the papilla is composed of columnar nucleated **cells**, the *odontoblasts.* These give rise to long external processes, which lie in the tubules of the dentine (*d.*), **and by which that substance is formed.** (2) The internal layer of the enamel germ (*enamel cells*) **sends processes** inwards which are believed to give rise to **the enamel prisms of** the enamel (*en.*). (3) The internal cells **of the dental** capsule surrounding **the root of the** tooth form **an osteogenetic layer,** from which the cement **is formed.**

Early in its development **the** enamel organ of the milk-tooth gives rise to an outgrowth, **into** which a dental papilla **soon** projects, **and** which is the **enamel** germ of the permanent **tooth.**

The Organs of Smell, Sight, and Hearing—(1.) Smell.— The organ of smell arises as two nasal pits lined with **thick-** ened epiblast. As development proceeds and the olfactory chambers **are** formed, this layer of epiblast spreading over their surface, and over the surface of the turbinal bones or cartilages developed within them, becomes of considerable extent. Very early in development (in the chick on the third day) the olfactory nerve-fibres, outgrowths from the **anterior** end of the brain, come into relation with this olfactory **membrane.**

(2.) **Sight.**—Examine **the fresh eye of an ox, sheep, or failing** these, rabbit. The anterior **transparent** *cornea* (Fig. 42, D., *cor.*) is continuous with the opaque, **fibrous** *sclerotic* (*scl.*). **These** form the capsule of the eye. Make an incision in the **cornea** with the point of the scalpel; a lymph-like transparent **fluid** oozes out; this is the *aqueous humour* **contained** in the anterior chamber of the eye (*aq. h.*). Remove **the cornea and** examine the *iris* (*ir.*), a circular curtain with a central perforation, the *pupil* (*pu.*), through which the transparent *crystalline lens* (*l.*) is visible. The iris contains circular and radiating muscular fibres, by the contraction or relaxation of which the pupil can be enlarged or diminished according as the light is **dim or** bright. The

9

crystalline lens is attached to the sides of the eyeball by the suspensory ligament. It is contained in a delicate capsule. By the action of special muscles (ciliary) the convexity of the anterior face of the lens can be altered, so as to allow of the focussing of the image on the retina.

Cut the remainder of the capsule in half by an incision at right angles to the axis of the eye. Note the optic nerve,

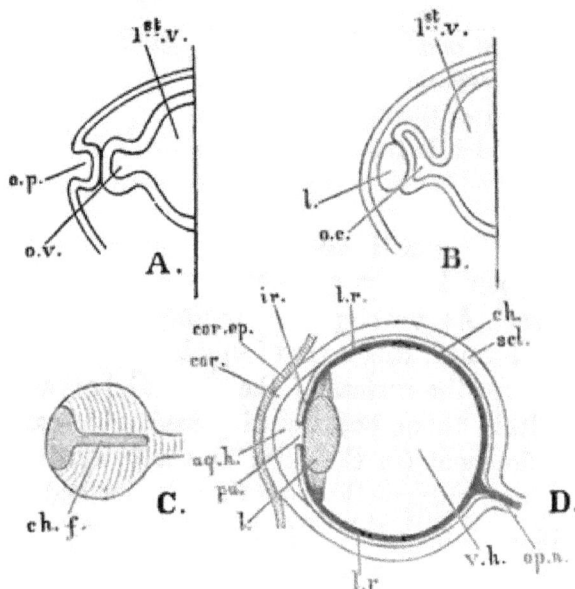

FIG. 42.—DIAGRAMS OF EYE DEVELOPMENT.

A. Transverse section, showing the **diverticulum** of the brain (*o.v.*) meeting the invagination of the surface (*o.p.*). B. Transverse section at **a later stage,** showing the lens derived from the surface invagination lying **in the pushed-**in diverticulum of the brain. C. The optic cup derived from **the diverticu-**lum of the brain seen from the ventral aspect. D. Section **of the eye.**

1st v. First vesicle of the brain (fore-brain). *aq. h.* Aqueous humour. *ch.* Choroid. *ch. f.* Choroid fissure. *cor.* Cornea. *cor. ep.* Corneal epithe-lium. *ir.* Iris. *l.* Lens. *l. r.* Limits of the retina proper. *o. c.* Optic cup. *o. p.* Optic pit. *o. v.* Optic vesicle. *op. n.* Optic nerve. *pu.* Pupil. *scl.* Sclerotic. *v. h.* Vitreous humour.

which will be seen to enter the back of the capsule; also jelly-like substance (*vitreous humour, v. h.*) with which this posterior chamber of the eye is filled. Lining the interior of this chamber will be seen the delicate transparent *retina*, ending anteriorly by a jagged border, the *ora serrata*. Between the retina and the sclerotic lies the *choroid (ch.)*, a pigmented vascular membrane, black for the most part, one portion, however, the *tapetum*, being iridescent. At the point of origin of the iris the choroid is continued into a *ciliary fold*, which bears a number of thickenings known as the *ciliary processes*. The optic nerve which enters the back of the eye-capsule is in direct continuity with the retina. In its centre runs a retinal artery and vein. In the fowl this artery does not run in the centre of the nerve; and in the fowl there is a structure which is not developed in the mammal. This is a peculiar fold projecting inwards from the choroid through the retina, and known as the *pecten*.

Let us now turn to the development of the eye. From the first vesicle of the brain there grows out on either side a hollow outgrowth known as the ***optic vesicle***. These optic vesicles come into close relation with two depressions on the exterior of the head, much resembling the olfactory pits, and known as the *optic pits*, which soon, by the growing together of their epiblastic edges, become shut sacs. Thus a hollow mass of epiblast is on each side separated from the superficial epiblastic layer; and this, by the growth of its cells, especially the more posterior, becomes a fibrous transparent mass, the crystalline lens, the cavity of which is ultimately obliterated. With this the optic vesicle comes into relation in such a way as to give rise to invagination in the manner shown diagrammatically in Fig. 42, where A. shows the optic vesicle (*o. v.*) growing out from the hollow first vesicle of the brain (*1st v.*) to meet the optic pit (*o. p.*), while B. shows the embryonic lens (*l.*) lying in the optic cup (*o. c.*) formed by the invagination of the optic vesicle.

The optic cup is not symmetrical, and its edge is indented from below in such a way that a slit, the *choroidal fissure* (C., *ch. f.*) breaks the continuity of the cup. Through this cleft mesoblastic tissue grows inwards and gives rise to the vitreous humour (*v. h.*), while in the fowl the pecten passes inwards by the same fissure. After this the fissure becomes more or less completely obliterated.

Meanwhile the cavity of the original optic vesicle (seen in 42, A. and B.) has become completely obliterated, while the stalk has become the solid optic nerve which, in the mammal, is at first somewhat flattened, and is then wrapped round the central artery. Of the two layers of which the optic cup is composed, the posterior (outer) becomes the pigmented layer of the retina, while the rest of the retina is formed out of the anterior (inner) layer. In front of the region of the retina, the limits of which are marked *l. r.* in 42, D., the two layers become fused together, and with the mesoblastic tissue of the ciliary region; and beyond this the anterior part of the optic cup forms the posterior portion (or uvea) of the iris. So that the pupil marks the limit of the optic cup, of which only the posterior portion is converted into the retina.

The remaining parts of the eye are for the most part mesoblastic in origin. The mesoblastic investment of the optic cup differentiates into the choroid internally, and the sclerotic externally. From the mesoblastic choroid the greater part of the iris is derived. The cornea is also for the most part of mesoblastic origin. But the corneal epithelium is derived from the external epiblast which covers the eye after the invagination of the lens, which structure is also, as we have seen, epiblastic in origin.

In Fig. 42, D., the dark line represents the part derived from the optic cup, which originates from the epiblast of the central nervous system. The lens and corneal epithelium, which are also epiblastic in origin, are cross-shaded. The mesoblastic parts are left unshaded.

(3.) **Hearing.**—In the frog there is no external ear; but above, and slightly behind the angle of the mouth, the *tympanic membrane* is stretched over a hard ring, and covered by pigmented skin. Within this membrane is the *tympanum*, a funnel-shaped cavity, widening outwards. This communicates with the mouth by the Eustachian passage, and is lined with mucous membrane continuous with that of the mouth. In the tympanum lies a cartilaginous rod, bony in the middle, the *columella* *auris*, attached externally to the tympanic membrane, and inserted internally into an aperture in the periotic capsule called the *fenestra ovalis.*   This aperture leads into a cavity which contains the essential part of the organ of hearing, which is known as the *membranous labyrinth*, and lies in a cavity moulded to its form in the periotic capsule.   Around the membranous labyrinth, between it and the walls of the chamber, is a fluid, *perilymph;* within the membranous labyrinth itself is a similar fluid, the *endolymph.*   The membranous labyrinth

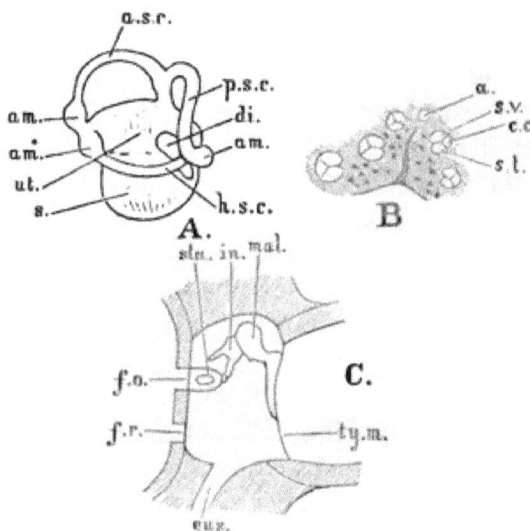

Fig. 43.—Diagrams of Auditory Organ.

A. Labyrinth of frog.   B. Mammalian cochlea in diagrammatic section. C. Mid-chamber of mammalian ear.

a. Summit of spire of cochlea. *am.* Ampulla. *a. s. c.* Anterior semicircular canal. *c. c.* Cochlear canal. *di.* Diverticulum. *eus.* Eustachian canal. *f. o.* Fenestra ovalis. *f. r.* Fenestra rotunda. *h. s. c.* Horizontal semicircular canal. *in.* Incus. *mal.* Malleus. *p. s. c.* Posterior semicircular canal. *s.* Sacculus. *s. t.* Scala tympani. *s. v.* Scala vestibuli. *sta.* Stapes. *ty. m.* Tympanic membrane. *ut.* Utriculus.

(Fig. 43, A.) consists of : (1) a vestibule constricted into two divisions, the *utriculus* (*u.*) and the *sacculus* (*s.*), on the outer side of which latter is a small dilatation ; and (2) three *semicircular canals*, of which two are nearly vertical, and the third nearly horizontal in position, each of which, near the vestibule, presents a bulging enlargement, the *ampulla*. The utriculus gives off a diverticulum (*di.*). The sacculus contains a great quantity of white crystalline calcareous *otoliths*. In the sacculus, utriculus, and ampullae of the semicircular canals there are special areas of ramification of the fibres of the auditory nerve. In the sacculus and utriculus these take the form of thickenings known as *maculæ acusticæ*, the similar thickenings in the ampulla being known as *cristæ acusticæ*. The epithelium which covers these is composed of columnar cells, between which are wedged in spindle-shaped cells prolonged into auditory hairs.

The auditory apparatus of the rabbit is more complicated. The tympanic membrane does not lie on the surface, but at the end of a short canal or *external auditory meatus*, the opening of which is partially surrounded by the large *pinna*. Between the tympanic membrane and the fenestra ovalis is a chain of three small bones (see p. 146) which take the place of the columella of the frog. The membranous labyrinth is enclosed in a bony labyrinth, which consists of the densely ossified walls of the chamber in which the membranous labyrinth lies. The sacculus and utriculus are connected by a narrow channel from which the recessus is given off. Connected with the utriculus are three semicircular canals resembling those of the frog. But connected with the sacculus is a new structure represented in the frog (if at all) by the small dilatation already mentioned. This new structure is the *cochlear canal* (*c. c.*) lying in the bony *cochlea*.

The cochlea is like a snail-shell, with a central spindle, the *modiolus*, around which winds a spiral canal with about two and a half turns. But the spiral canal is divided into three parts in the manner shown in Fig. 43, B., which represents the cochlea quite diagrammatically in cross section. *c. c.* is the cochlear canal connected with the sacculus and containing endolymph : *s. v.* is a second canal, the *scala vestibuli*, connected with the bony vestibule in which the membranous vestibule lies, and containing perilymph : *s. t.* is a third canal, the *scala tympani*, also containing perilymph. The cochlear canal gets smaller as we ascend the spire, which is supposed in the figure to be seen in cross section, and ends blindly near the top. The scala tympani and scala vestibuli also get smaller, and eventually communicate one with the other at the top of the spire at *a*. Thus it would be possible for a minute organism to swim out of the perilymph surrounding the vestibule into the scala vestibuli, and so up to the top of the spire, and then down again by the scala tympani. At the foot of that canal his way would be blocked by a membrane filling an aperture in the bone of the osseous labyrinth, and called the *fenestra rotunda*. If he were to swim back into the perilymph round the vestibule he would there find a similar membrane filling the aperture of the *fenestra ovalis*. And this membrane is connected by means of the chain of auditory ossicles (as shown in Fig. 43, C.) with the tympanic membrane. The vibrations of the tympanic membrane are therefore passed on by the ossicles (see p. 147) to the membrane of the fenestra ovalis, which sets all the perilymph in vibration, and with it the membrane of the fenestra rotunda. The partition between *s. t.* and *s. v.* in 43, B., contains a bony shelf, the *lamina spiralis*, but those between these canals and the cochlear canal are membranous, while that between *s. t.* and *c. c.*, called the basilar membrane, contains radial fibres over-arched by bowshaped fibres of Corti, and arranged somewhat as in the key-board of a piano. It has been suggested that they may have the function of vibrating to sounds of a certain definite pitch; but they cannot be described here.

In the fowl there is no external ear; the tympanic membrane is placed in connection with that of the fenestra ovalis by a columella; there is a fenestra rotunda; the sacculus and utriculus are less completely constricted off; the cochlear canal is not spiral, as in the rabbit, but is a curved process with a dilated extremity. The osseous cochlea is, as compared with the rabbit, rudimentary.

The auditory organ, like the olfactory organ and the eye, originates as a depression of the surface of the head. This *auditory pit* becomes converted into a closed epiblastic vesicle by the closure of its edges and the overgrowth of the superficial epiblast. From this *otic vesicle* and its mesoblastic investment

are developed the **membranous** and osseous labyrinths **by pro·**
cesses of differentiation.

The Urinary Organs and Generative Ducts.—The first trace
of the excretory system in the **embryo** is the *segmental duct*, which
is stated to take its origin from the epiblast, but soon appears as
a tube in the peritoneal epithelium of the somatic layer, near the
dorsal border of the body-cavity.   This tube opens posteriorly
into the cloaca, and anteriorly into the peritoneal cavity by
apertures far forward, just behind the branchial region.   Just
behind the openings the duct becomes coiled upon itself, and
sends out a number of blind diverticula.   The glandular body
thus constituted is called the *head-kidney* or *pronephros*.   Opposite the openings of the duct, at the root of the mesentery, a
highly vascular ridge is formed, which constitutes a *glomerulus.*
The portion of the body-cavity which contains this glomerulus
and the openings of the segmental duct, is temporarily shut off
from the rest of the body-cavity.   Thus a provisional and
temporary excretory organ is constituted.   In the frog it subsequently atrophies; but in the cod-fish it probably persists
throughout life.

Much further back in the body-cavity a number of coiled
tubes are formed.   They originate in the mesoblast as solid
cords, which are subsequently hollowed out into tubes, one end
of which opens into the segmental duct, while near the other
each is invaginated by a little tuft of vascular tissue, thus giving
rise to a glomerulus and its capsule (see p. 81).   In some forms,
but not in the frog, there is one of these for each portion of the
body corresponding to a mesoblastic somite or segment; they
are called *segmental tubes*.   Secondary and tertiary segmental
tubes are subsequently added; they become much convoluted
and come into such close proximity that their primitive distinctness is completely lost.   The large glandular body thus
constituted is called the *Wolffian Body* or *Mesonephros*, and
persists as the permanent kidney of the frog.

In the meanwhile important changes go on in the segmental
duct of the frog.   A cord of cells separates from its ventral side,

and this cord becomes hollowed out so as to form a tube which acquires an opening at its anterior end into the peritoneal cavity, and in the female opens posteriorly into the cloaca, but in the male ends blindly. This is known as the *Müllerian duct.* From it the *oviduct* of the female is developed. In the male it atrophies. The remainder of the segmental duct, from which this new duct was separated off, is known as the *Wolffian duct,* since it carries off the products of the kidney or Wolffian body.

A further change has to be noted. In the anterior part of the Wolffian body outgrowths from the primary segmental tubes proceed to, and eventually open into, the *testis* of the male, the products of which are thus carried by these *vasa efferentia* into the Wolffian duct, which, therefore, is functional as the *vas deferens.* As it conveys the urinary product it is also termed the ureter; but it is not strictly homologous with the ureter of the rabbit. It is best termed the urino-genital duct.

In the rabbit and the fowl there are notable differences in the mode of development of the segmental, Wolffian, and Müllerian ducts. There seems, at least in the fowl, to be a rudimentary pronephros; and in both a mesonephros is formed. The latter organ, however, though it functions as a kidney during embryonic life, does not persist as the permanent kidney in either of these types.

From the dorsal aspect of the hinder part of the Wolffian duct there is said to proceed an outgrowth which extends forwards and sends out collecting tubes into a specially differentiated portion of that mass of mesoblastic tissue from which the mesonephros was previously formed. The outgrowth is the *ureter*; the glandular mass with which it enters into relation is the *metanephros,* or permanent kidney of the bird and the mammal. The ureter does not long remain attached to the Wolffian duct, but, in the fowl, acquires a separate opening into the cloaca, and, in the rabbit, opens into that portion of the allantoic duct which is persistent as the urinary bladder.

In the fowl and the rabbit, as in the frog, a connection is established in the male between the testis and the Wolffian duct through the Wolffian body. The *vasa efferentia* thus represent a remnant of the mesonephros, and the Wolffian duct becomes the vas deferens of the male, while it atrophies more or less completely in the female. The rest of the Wolffian body is generally held to atrophy more or less completely in both sexes.

Thus in the male the Müllerian ducts atrophy, but the Wolffian ducts become the vasa deferentia, just as they do in the frog, only in that type they have also the functions of ureters. In the female it is the Wolffian ducts which atrophy, while the Müllerian ducts become the oviducts. In the fowl one—usually the right—atrophies. In the rabbit the oviducts of the two sides unite in the median line to form the *vagina*.

The embryo rabbit has, like the adult frog and fowl, a cloaca which receives the excreta from the alimentary canal, and the products of the kidneys and genital organs. This, however, becomes modified in such a way as to separate the opening of the rectum from that part (the *urino-genital sinus*) which receives the ureters and the genital ducts. Further changes occur which differ in the two sexes : one of the most remarkable of which is the passage of the testes from the peritoneal cavity into two special pouches of skin, the scrotal sacs.

The Heart.—This organ is formed very early by the coalescence of the layers of splanchnic mesoblast beneath the throat. The cleavage of the mesoblast on each side of the body into a splanchnic and a somatic layer is shown in Fig. 38A, i. This section is through the mid region of the body. Further forward the mesenteron is closed in below and converted into a tube, beneath which the splanchnic layers come together. But they do so in such a way as to leave a cavity or space between them. This space is the cavity of the primitive heart, the muscular walls of which are derived from the splanchnic mesoblast, as shown in Fig. 38A, ii.

The heart is thus in the first instance a simple tube. This soon becomes further differentiated, and bent on itself like an S (see Fig. 37, ii. and iii.). The upper limb of the S forms the atrium, which in the frog is early divided into two auricles by a septum, the ventricle remaining undivided. In the fowl and rabbit the ventricle is the first to be divided.

CHAPTER VIII.

THE SKELETON.

THE word skeleton is used in a broader and in a **narrower** sense. In the broader sense it comprises all those tissues **which** are developed for the support and protection of the rest, and which form the framework of the essential parts of the various organs. These are, in the main, connective tissue, cartilage, and **bone**. Were it possible by **some** subtle solvent to dissolve **out all the tissues** but connective tissue, we should still have a **perfect** model of the vertebrate body in this material. Bone **and** cartilage are more restricted in position. They may be said **to** constitute the skeleton in the narrower sense **in** which it is used in **this** chapter; while the dried skeleton **is composed** of bone alone.

The skeleton may be divided into the following parts :—

I. Axial Skeleton—
    1. The skull and branchial arches.
    2. The vertebral column, ribs, and sternum.
II. Appendicular Skeleton—
    1. The pectoral arch and fore-limb.
    2. The pelvic arch and hind-limb.

The Skull.—The skull, into the composition of which there enter **certain** elements derived from the branchial arches, is developed : (1) for the protection **of the brain**; (2) for the lodgment and protection of the special sense-capsules; and (3) for the support and attachment of the jaw-apparatus. The olfactory (*ol.*), optic (*op.*), and auditory (*au.*) capsules always

139

bear the same relation to the brain-case, shown in Fig. 44, B. The jaw-apparatus bounds the mouth in the antero-lateral region.

We will consider the skull under the following heads :—(1) brain-case, (2) olfactory chamber, (3) auditory chamber, (4) optic chamber, (5) maxillary region, (6) mandible, (7) hyoid.[1]

(A.) The Rabbit's Skull.—(1.) *Brain-case* (Figs. 44 A. and 45). —The posterior wall of the brain-case is formed, in the adult

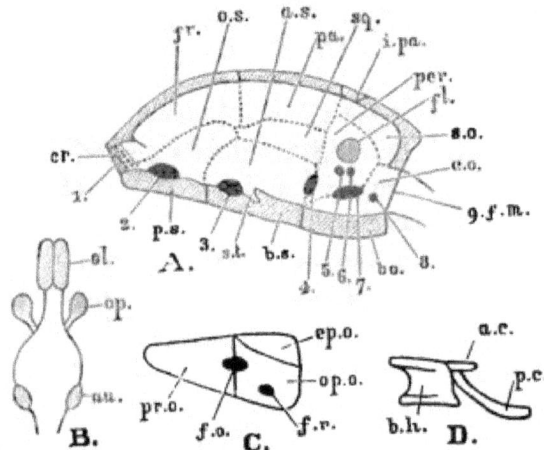

FIG. 44.—SKULL OF RABBIT.

A. Diagrammatic section. B. Diagram to show relation of sense capsules to brain-case. C. Periotic mass. D. Hyoid.

In A.—*a. s.* Ali-sphenoid. *b. o.* Basi-occipital. *b. s.* Basi-sphenoid. *cr.* Cribriform plate. *e. o.* Ex-occipital. *f. m.* Foramen magnum. *fl.* Floccular fossa. *fr.* Frontal. *i. pa.* Inter-parietal. *o. s.* Orbito-sphenoid. *pa.* Parietal. *p. s.* Pre-sphenoid. *per.* Periotic. *s. o.* Supra-occipital. *s. t.* Sella turcica. *sq.* Squamosal. 1-9. Foramina.

In B.—*au.* Auditory capsule. *ol.* Olfactory capsule. *op.* Optic capsule.

In C.—*f. o.* Fenestra ovalis. *f. r.* Fenestra rotunda. *ep.o.* Epiotic. *op.o.* Opisthotic. *pr.o.* Prootic.

In D.—*a. c.* Anterior cornu. *p. c.* Posterior cornu. *bh.* Basi-hyal.

rabbit, of a single bone (the *occipital*) pierced by a large hole, the *foramen magnum* (9. *f. m.*), through which passes the spinal cord.

---

[1] The student should prepare a young and an adult skull, with all the bones in their natural connection. This is best done by dissecting away the muscle, and carefully scooping out the brain, dipping the skull occasionally in boiling water for a *few* minutes. A young skull should also be left in water to macerate for a week or two. The several bones will then readily separate. They should

At the sides of this great foramen are two rounded eminences (the *occipital condyles, oc. c.,* Fig. 45) by which the skull articulates with the first vertebra. In the young rabbit this occipital segment of the brain-case is not a single bone, but is composed of a single median *basi-occipital* (*b. o.*) below, a single median *supra-occipital* (*s. o.*) above, and two *ex-occipitals* (*e. o.*), one on

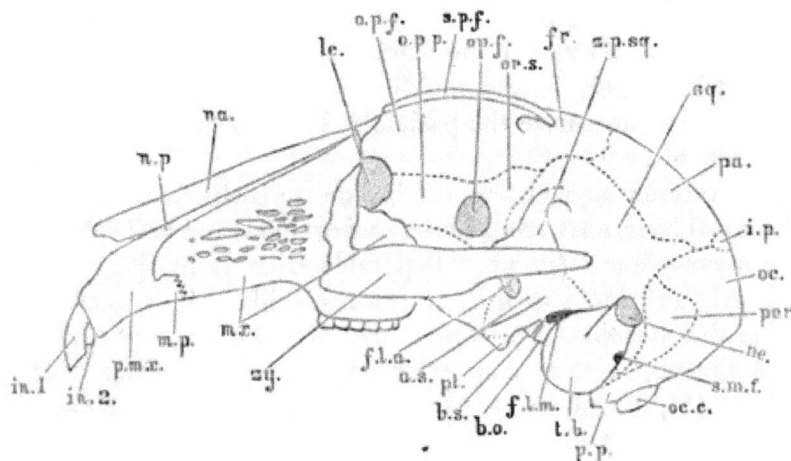

FIG. 45.—SKULL OF RABBIT (Side View).

a.s. Ali-sphenoid. *b.o.* Basi-occipital. *b.s.* Basi-sphenoid. *f. l. a.* Foramen lacerum anterius. *f. l. m.* Foramen lacerum medium. *fr.* Frontal. *i. p.* Interparietal. *in.* 1 and 2. Incisor teeth. *lc.* Lacrymal. *m. p.* Maxillary process of pre-maxilla. *mx.* Maxilla. *na.* Nasal. *ne.* Neck of tympanic. *n. p.* Nasal process of pre-maxilla. *o. p. f.* Orbital process of frontal. *o. p. p.* Orbital process of pre-sphenoid. *oc.* Occipital. *oc. c.* Occipital condyle. *op. f.* Optic foramen. *or. s.* Orbito-sphenoid. *p. p.* Par-occipital process. *pa.* Parietal. *per.* Periotic. *p. mx.* Pre-maxilla. *pt.* Pterygoid. *s. m. f.* Stylo-mastoid foramen. *s. p. f.* Supra-orbital process of frontal. *sq.* Squamosal. *t. b.* Tympanic bulla. *z. p. sq.* Zygomatic process of squamosal. *zy.* Jugal.

either side. These lateral elements are produced downwards into two *par-occipital processes* (45, *p. p.*). Such is the occipital segment, the hindermost of the three segments of the brain-case.

not be separated, however, until the student has some knowledge of their relative positions from a study of the complete skull. An adult skull should be sawn in two along the mid longitudinal line. Fig. 44, A., is a very diagrammatic view of the bisected brain-case. The parts cross-shaded are where the bones have been cut. The dotted lines show the boundaries of the bones in the side walls. The reference numbers are to the foramina.

The median segment is the *parietal segment.* Here the basal bone answering to the basi-occipital is the *basi-sphenoid* (*b. s.*). At its sides are two winglike bones, separable in the embryo, but early ankylosed (united by bony union) the *ali-sphenoids* (*a. s.*). Besides its ascending process which enters into the composition of the brain-case, each ali-sphenoid sends downwards two flattened laminæ forming together a V with its point directed forwards (see Fig. 46, *pt. pa.*). The median segment is completed above by the *parietals* (*pa.*), two large roofing bones which meet in the middle line. It is, however, incomplete at the sides, since the parietals do not reach down to the ali-sphenoids.

The anterior segment of the brain-case is the *frontal segment.* The basal bone, answering to the basi-occipital and basi-sphenoid, is the *pre-sphenoid* (*p. s.*). Separable from it in the embryo, but early ankylosed with it, are two winglike bones, answering to the ali-sphenoids. These *orbito-sphenoids* (44, *o. s.*, and 45, *or. s.*) form the posterior and superior boundaries of the large *optic foramen* (44, 2, and 45, *op. f.*) which places the two orbits (containing in the fresh state the eye-capsules) in communication, and also opens into the brain-case. The pre-sphenoid forms its anterior and inferior boundary, and then sends outwards ascending orbital processes (45, *o. p. p.*) which meet descending orbital processes (*o. p. f.*), of the *frontals* (*fr.*) large roofing bones which complete this anterior segment of the brain-case. Each sends out a supra-orbital process (*s. p. f.*) over the orbit.

The anterior end of the brain-case is bounded by a bone (cribriform plate, 44 A., *c. r.*) which is perforated like a sieve for the olfactory nerve-threads (1).

These **three** segments, occipital, parietal, and frontal, form the greater part of the walls of the brain-case. Between the occipital and parietal segments, however, there would be, were no other bones developed, a great gap at the sides, and a small gap in the roof. The posterior part of the gap on each side is filled in with the *periotic* bone, which results from the ossification of the auditory capsule. The gap in the roof is filled in by a small oval *interparietal* (45, *i. p.*). Furthermore, at the sides of **the** brain-case two other bones are affixed **to the** exterior. These are (1) the *squamosal (sq.)*, which fills the anterior part of the lateral gap, and separates the ali-sphenoid from the parietal; and (2) the *tympanic*, which forms a swollen rounded mass (tympanic bulla, 45, *t. b.*) **stuck** on to the outer side of the periotic.

The brain-case thus constituted has the following foramina (see Fig. 44, A.) :—

1. *Olfactory ;* numerous perforations in the cribriform plate for the fibres of the first nerve.

2. *Optic;* for the second nerve, just within the orbito-sphenoid. The confluence of the optic foramina places the orbits in communication. (This is not so in the dog or cat.)

3. *Sphenoidal* (or foramen lacerum anterius); between the basi-sphenoid and ali-sphenoid. It transmits the third, fourth, sixth, and first and second divisions of the fifth nerve. (In the dog or cat the second division of the fifth passes through a separate orifice, the *foramen rotundum.*)

4. *Foramen lacerum* **medium ; for the third division of the** fifth nerve **and** the internal **carotid artery. It lies between the** periotic **and** the ali-sphenoid. (**In the dog or cat the third** division of the fifth nerve passes through a separate orifice in the ali-sphenoid, the *foramen ovale.*)

5. *Facial ;* this passes through the periotic bone, and emerges by a foramen (*stylo-mastoid*), behind and above the swollen portion of the tympanic. It transmits the seventh nerve.

6. *Auditory*; the eighth nerve enters **the** periotic close to the seventh, the shallow depression common to **the two being the** *internal auditory meatus.* **Near it** is a deeper depression, **the** *floccular fossa.*

7. *Foramen lacerum posterius ;* between the periotic and ex-occipital, for the transmission of the ninth, tenth, and eleventh nerves.

8. *Condylar ;* in the occipital bone ; two small apertures for the twelfth nerve (one in dog or cat).

9. *Magnum,* within the occipital ring, for the spinal cord.

It is a fact worth remembering that the three main segments of the brain-case always have a definite relation to the nerves which pass out. The orbito-sphenoid lies between the olfactory and sphenoidal, and contains the optic foramina; the ali-sphenoid between the sphenoidal and f. l. medium, being perforated by the rotundum and ovale where they exist; the periotic between the f. l. medium and the f. l. posterius; the ex-occipital behind the f. l. posterius, and perforated by the condylar.

The inner surface of the brain-case is divided into three cranial fossæ: (1) the *olfactory fossa*, from the cribriform to an ill-marked ridge on the frontal; (2) the *cerebral fossa*, from this ridge to the tentorial plane, which is marked by a ridge along the periotic, and by the line of junction of the parietal and supra-occipital (the tentorium, which is membranous in the rabbit, is ossified in the dog or cat); (3) the *cerebellar fossa*, behind the tentorial plane. The plane of the cribriform is called the ethmoidal plane, that of the occipital bone the occipital plane. In comparative mammalian anatomy, the angles made with each other by the ethmoidal, tentorial, and occipital planes are of importance. The student should, if possible, compare them in the hedgehog, the rabbit, the dog, and man.

In the basi-sphenoid is a depression, the *sella turcica* (44, A., *s.t.*), for the lodgment of the pituitary body. It is bounded before and behind by anterior and posterior clinoid processes.

On the external surface there are sutures marking the contact of separable bones. (1) The *frontal suture*, between the two frontals; (2) the inter-parietal or *sagittal suture*, between the two parietals; (3) the fronto-parietal or *coronal suture*, between the frontals and parietals; (4) the occipito-parietal or *lambdoidal suture*, between the occipital and parietals.

(2.) *The Olfactory Chamber.*—This lies in front of the brain-case. It is bounded posteriorly by the cribriform plate, and is divided into two compartments by the median *lamina perpendicularis*, which is continued forwards by the cartilaginous *septum nasi*. The lamina perpendicularis and the cribriform plate form parts of one bone, the *mesethmoid*. Beneath the cartilaginous septum lie the coalesced *vomers*, forming a trough. The roof of the chamber is formed by the frontals posteriorly, and the *nasals* (45, *na.*) anteriorly. In front are the openings of the external nares, bounded below by the *pre-maxillæ* (*p. mx.*), which send up long ascending processes (*n. p.*) to form part of the outer wall of the nasal chamber. This outer wall is completed by the *maxillæ* (*mx.*), which are spongy in structure, and by the *lachrymals* (*lc.*), which are apt to drop out in a dry skull, as in Fig. 45. Each is perforated for the lachrymal duct. The incomplete floor is formed by the horizontal wings of the vomers and by the upper horizontal laminæ of the *palatines.*

The olfactory chamber is filled with a spongy mass of bone of threefold origin, and composed of (1) a superior pouchlike process of the nasal, the *naso-turbinal*; (2) a median mass ankylosed with the mesethmoid, and in the adult with the vomerine trough, the *ethmo-turbinals* or *parethmoids*; and (3) an anterior scroll-like mass in close connection with the maxillæ—the *maxillo-turbinals.* All these turbinal bones are covered over in the living rabbit with mucous membrane. That covering the ethmo-turbinals is supplied by the olfactory nerve, while that investing the maxillo-turbinals is innervated by a branch of the fifth.

Connected with the olfactory chambers are the narial passages (Fig. 46), of which the vomers, pre-sphenoid (*p. s.*), and part of the basi-sphenoid (*b. s.*), form the mid-line of the incomplete roof. The maxillæ (*mx.*), palatines (*pa.*), and the *pterygoids* (*pt.*), which are in relation with the

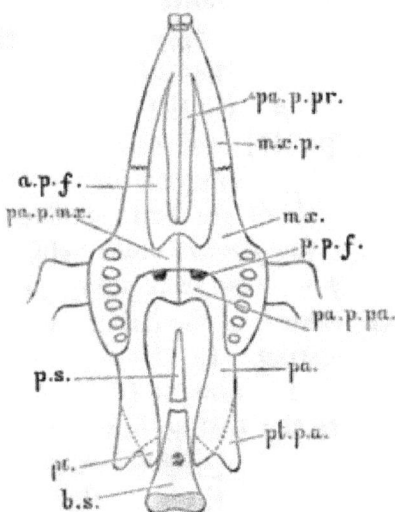

FIG. 46.—PALATE OF RABBIT (viewed from under side).

a. p. f. Anterior palatine foramen. *b. s.* Basi-sphenoid. *mx.* Maxilla. *mx. p.* Maxillary process of pre-maxilla. *p. p. f.* Posterior palatal foramen. *pa. p. mx.* Palatal process of maxilla. *pa. p. pa.* Palatal process of palatine. *pa. p. pr.* Palatal process of pre-maxilla. *pa.* Palatine. *pt.* Pterygoid. *pt. p. a.* Pterygoid process of ali-sphenoid. *p. s.* Pre-sphenoid.

10

inner limb of the ᴠ-shaped pterygoid processes of the ali-sphenoids, form the side-walls and lateral portions of the roof. A horizontal shelf formed by palatine processes of maxillæ (*pa. p. mx.*) and palatines (*pa. p. pa.*) compose the very incomplete bony floor. [This bony palate is much more complete in the cat or dog.] In it are the very large *anterior palatine foramina* (*a. p. f.*), and the small *posterior palatine foramina* (*p. p. f.*).

(3.) *The Auditory Chamber.*—The *periotic* fits into the side-walls of the brain-case just in front of the ex-occipital. It is very irregular in shape, and differs in bone-density in its anterior (*petrous*) and posterior (*mastoid*) portions. The latter is seen on the exterior (45, *per.*) behind the tympanic bulla. On its inner side are two depressions, one larger and blind, the floccular fossa (44, A., *fl.*), for the lodgment of the flocculus of the cerebellum (see p. 217); the other smaller, the internal auditory meatus, with nerve-foramina for the seventh and eighth nerves. Exter-nally there are two openings leading into the cavity of the bone, the *fenestra ovalis* before and the *fenestra rotunda* behind. They are best seen on removing the bulla. Within the bone are lodged, in the living rabbit, the essential parts of the organ of hearing. In the embryo of mammals the periotic ossifies from three centres (Fig. 44, C.), and may thus be regarded as due to the coalescence of three bones, the *prootic* (*pr. o.*), *epiotic* (*ep. o.*), and *opisthotic* (*op. o.*), which have the relative positions and the relations to the fenestra ovalis (*f. o.*) and fenestra rotunda (*f. r.*) diagrammatically shown in Fig. 44, C. (p. 140), the prootic being anterior.

External to the periotic, covering all its petrous portion, is the tympanic, of which there are two well-marked divisions, a swollen ventral portion, the bulla, and a spout-like dorsal pro-longation, the neck (Fig. 45, *ne.*).

Within the cavity of the bulla are several small bones, the *auditory ossicles.* In the fresh state membranes are stretched across the fenestra ovalis and rotunda, and a large tympanic membrane is attached to a raised ridge which marks the junction of the swollen and tubular portions of the tympanic. Attached to the membrane of the fenestra ovalis is a little stirrup-shaped bone, the *stapes* (Fig. 47, *sta.*), bearing a little knob at the summit of its arch, to which there articulates, by means of a minute intermediate *orbicular* bone, a little bone, the *incus* (*in.*), shaped somewhat like a two-fanged tooth. The fangs are not of the same length; that articulated with the stapes is the longer (*long crus*), while the shorter (*short crus*) articulates with the overhanging wall of the periotic. The crown of the tooth-

like incus is fashioned into the shape of a saddle, with which articulates the head of another small bone, the *malleus* (*mal.*), the shaft of which is affixed to the tympanic membrane. Thus the auditory ossicles form a chain of bones between the tympanic membrane on the one hand, and the fenestra ovalis on the other.

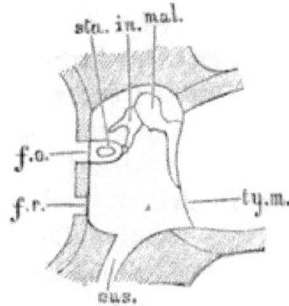

FIG. 47.—AUDITORY OSSICLES.

(4.) *The Optic Chamber.*—The capsule of the eye is not ossified, nor is it fused with the bones of the skull by cartilaginous union. It is enclosed in the *orbit*.

In the rabbit's skull there will be seen, on each side, a single large cavity, the two communicating by means of the optic foramen. [But in the cat's skull, or the dog's, this large cavity on each side will be seen to be imperfectly divided into two, of which the anterior is for the lodgment of the eye-capsule, and is the true orbit, while through the posterior, the temporal fossa, pass the great muscles for the lower jaw. In the human skull these are quite distinct.] The orbital recess of the rabbit is thus a common orbit and temporal fossa.

Its inner boundary is the wall of the brain-case, the frontal sending out a supra-orbital process (see Fig. 45, *s. p. f.*, p. 141). Posteriorly the squamosal (*sq.*) sends out a curved and distally expanded zygomatic process (*z. p. sq.*), running at first outwards and then downwards, to be applied to and form part of a lateral bar or *zygoma* (*zy.*). This is composed of a single bone forming a process of the large maxilla. In the new-born rabbit, however, the hinder part of this bar is a separate bone, the *malar* or *jugal*. [The incomplete partition in the dog or cat is formed by an ascending process of the malar, which nearly meets a descending process of the frontal.]

(5.) *The Maxillary Region.*—The tooth-bearing bones in the upper jaw are the maxillæ (*mx.*) and pre-maxillæ (*p.mx.*). The maxillæ are large and irregular. Their palatal processes have already been considered. Each bears three pre-molar and three molar teeth in the adult. The pre-maxillæ are long, and articulate with each other closely, though they do not ankylose. Each has three large processes : (1) a nasal process (*n. p.*) passing along the outer edge of the nasal bones ; (2) a maxillary process (*m. p.*) coming into relation with the maxilla ; and (3) a palatine process (Fig. 46, *pa. p. pr.*) passing backwards alongside its fellow, and partially dividing from each other the anterior palatine foramina (*a. p. f.*). Each bears two incisor teeth, one behind the other ; the anterior large, the posterior small. A long space, or *diastema*, separates the incisors from the pre-molars. There are no canine teeth.

(6.) *The Mandible.*—The lower jaw is composed of two symmetrical *rami* which meet, but do not ankylose, at the symphysis. It articulates directly with the skull by a longitudinally elongated condyle, which fits in beneath the zygomatic process of the squamosal. Running up towards the condyle, from the tooth-bearing or horizontal portion, is an ascending portion which bears an incurved coronoid process. At the junction of the two portions is an inferior dental foramen for a branch of the fifth nerve. The angle is rounded, and has a slight shelf projecting inwards.

The lower jaw in the adult bears two incisor teeth, one in each ramus ; no canines ; two pre-molars and three molars on each side. The following dental formula will now be understood. The numbers above and below the line represent the teeth in the upper and lower jaw respectively—those of the opposite sides being separated by the —— :—

Rabbit, i. $\frac{2-2}{1-1}$, c. $\frac{0-0}{0-0}$, p. m. $\frac{3-3}{2-2}$, m. $\frac{3-3}{3-3}$, =28.

This may be compared with the dental formulæ of the cat, the dog, and man, all of which have canines :—

$$\text{Cat,} \quad \text{i. } \frac{3-3}{3-3}, \text{ c. } \frac{1-1}{1-1}, \text{ p. m. } \frac{3-3}{2-2}, \text{ m. } \frac{1-1}{1-1}, = 30.$$

$$\text{Dog,} \quad \text{i. } \frac{3-3}{3-3}, \text{ c. } \frac{1-1}{1-1}, \text{ p. m. } \frac{4-4}{4-4}, \text{ m. } \frac{2-2}{3-3}, = 42.$$

$$\text{Man,} \quad \text{i. } \frac{2-2}{2-2}, \text{ c. } \frac{1-1}{1-1}, \text{ p. m. } \frac{2-2}{2-2}, \text{ m. } \frac{3-3}{3-3}, = 32.$$

(7.) *The Hyoid.*—This bone, developed from one of the visceral arches, consists of a stout body or *basi-hyal* (Fig. 44, D., *b. h.*), small *anterior cornua* (*a. c.*), and larger *posterior cornua* (*p. c.*). It lies in the root of the tongue. In the dog the anterior cornua are longer, and consist of three elements.

(B.) The Fowl's Skull.—In the skull of an adult fowl, several years old, the bones of the brain-case, and some other parts, become so closely ankylosed, and completely fused, that it is impossible to trace the sutures. In the skull of the young chick, under a month old, some of the bones have not begun to ossify. In the skull of a fowl some six months old, while some of the bones are already ankylosed, the ossification of the pre-sphenoid has not commenced. It will be convenient to take for description the skull of a recently hatched chick, and the skull of an adult bird. Fig. 48 gives a side view and ventral view of the adult skull, the positions of some of the obliterated sutures being dotted in.

(1.) *The Brain-case.*—This is rounded above and posteriorly, but much encroached upon by the large orbits. In the chick the basi-occipital, ex-occipitals, and supra-occipital are distinct; but in the adult they fuse with each other, and with the other bones of the brain-case (*b.o., e.o., s.o.*). There is a single occipital condyle, hemispherical and grooved dorsally. Anterior to the basi-occipital is the basi-sphenoid. This bone is not, however, visible in B., since it is hidden by a large expanded *basi-temporal* bone (*b.t.*), at the anterior ends of which are the openings of the Eustachian canals. The ali-sphenoids (48, A., *a.s.*) form part of the anterior wall of the brain-case. Each sends out a strong sphenoidal process which meets a process of the squamosal. In the chick there is a membranous fenestra in each. The parietals (*pa.*), on the other hand, form part of the roof of the

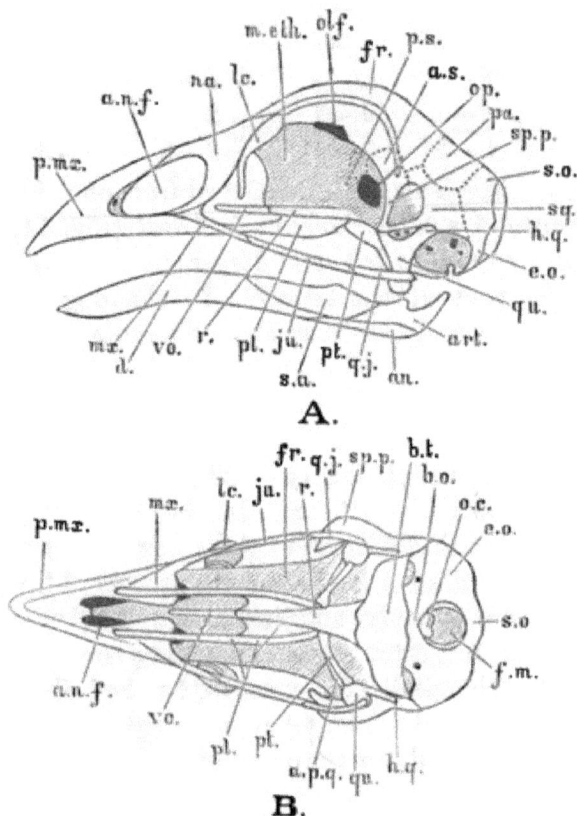

FIG. 48.—SKULL OF FOWL.

A. From the side.   B. From below.

a. n. f. Anterior nasal fossa. *a. p. q.* Anterior process of quadrate. *a. s.* Ali-sphenoid. *an.* Angular. *art.* Articular. *b. o.* Basi-occipital. *b. t.* Basi-temporal. *d.* Dentary. *e. o.* Ex-occipital. *f. m.* Foramen magnum. *fr.* Frontal. *h. q.* Head of quadrate. *ju.* Jugal. *lc.* Lachrymal. *m. eth.* Meseth-moid. *mx.* Maxilla. *na.* Nasal. *olf.* Olfactory foramen. *op.* Optic foramen. *pa.* Parietal. *pl.* Palatine. *pt.* Pterygoid. *p. mx.* Pre-maxilla. *p. s.* Pre-sphenoid. *qu.* Quadrate. *q. j.* Quadrato-jugal. *r.* Rostrum. *s. a.* Supra-angular. *s. o.* Supra-occipital. *sp. p.* Sphenoidal process. *sq.* Squamosal. *vo.* Vomer. *o. c.* Occipital condyle.

posterior region of the brain-case.   Between these two elements of the parietal segment lies the large squamosal (*sq.*), behind and below which is the auditory capsule.

Of the anterior segment of the brain-case the frontal bones are the most conspicuous and the earliest formed (*fr.*). They constitute the whole of the central arch of the skull, and give rise to a prominent supra-orbital ridge, of which, however, the postero-inferior portion is formed by the ali-sphenoid. From the ridge each is continued sharply inward, within the orbits, as the orbital plate of the frontal. These orbital plates, and the ali-sphenoids which lie below them, are in the chick separated from the inter-orbital septum by a membranous tract, below which is the optic foramen (*op.*). In this tract the orbito-sphenoids make their appearance between the second and third months of chick life. Anterior to them is the cartilaginous inter-orbital septum which contains a large membranous inter-orbital fenestra. Behind this fenestra—between it and the optic foramen—the pre-sphenoid (*ps.*, the reference line passes through the small orbito-sphenoid) begins to ossify at about the ninth month, thus completing the frontal segment. In front of the fenestra is the mesethmoid ossification. Beneath the mesethmoid and the inter-orbital septum, stretching back along the mid-line of the base of the skull as far as the basi-temporal, is a grooved splint of bone, the *rostrum* (*r.*). Anterior to this is the grooved vomer (*vo.*).

In the adult fowl all these various skull-elements, except the vomer, run together, and fuse into a continuous bony mass in which the sutures are completely obliterated. The bones of the three segments, the basi-temporal and rostrum below, the squamosal and bones of the auditory capsules at the sides, the mesethmoid in front, all unite firmly together in the skull of the adult fowl. The inter-orbital septum becomes a continuous plate of bone, except where the olfactory and optic foramina place the two orbits in communication. And the frontals fusing with the nasals anteriorly, the cranial and facial portions of the skull become continuous.

The foramina for the exit of the nerves have the same essential relations as in the rabbit. The student should compare the two.

There is no olfactory fossa. The cerebral fossa is large, and on each side of it there is an optic fossa for the large optic lobe. The cerebellar fossa is well marked, and there is a strong tentorial ridge. The floor is hollowed out for the reception of the medulla oblongata.

(2.) *The Olfactory Chamber.*—On either side of the beak are large *anterior nasal fossæ* (*a. n. f.*), so that the nasal chambers are less completely enclosed than in the rabbit. Each fossa is bounded by the nasals behind (*na.*) and the pre-maxillæ (*p. mx.*) in front. The mesethmoid is continued forwards in the fresh skull by a cartilaginous plate, the septum nasi, which forms a median vertical partition between the nasal cavities. Beneath it lies the vomer. It is produced on each side into scroll-like masses of cartilage, the turbinals. The anterior nares are far back, just in front of the eyes; the posterior nares open almost directly beneath them in a longitudinal slit in the palate. There are thus no narial passages like those in the rabbit.

(3.) *The Auditory Chamber.*—The auditory capsule is seen on the inner side of the brain-case, lying just external to the foramen magnum and adjoining the optic and cerebellar fossæ. It is limited above by a well-marked semi-circular ridge indicating the position of the anterior semi-circular canal. In the chick there are three distinct ossifications: (1) the prootic, which forms a distinct bone for some time and constitutes the main part of the capsule; (2) the epiotic, which fuses early with the supra-occipital; and (3) the opisthotic, which fuses with the ex-occipital.

No tympanic bulla covers over the tympanic chamber, which is an irregular hemispherical depression. In the inner wall of this cavity is a largish aperture, the fenestra ovalis, and somewhat behind and below this a smaller one, the fenestra rotunda. In place of the chain of auditory ossicles there is a partly cartilaginous, partly osseous rod, the *columella auris.* This is expanded at its proximal end into a plate-like stapes which fits into the fenestra ovalis. The cartilaginous distal end is affixed to the tympanic membrane, which is stretched across the aperture of the cavity.

(4.) *The Optic Chamber.*—The capsule of the eye—within which a circle of small imbricating bones, the *sclerotic plates*, is developed —is enclosed in the large rounded orbit. This is bounded behind by the anterior wall of the brain-case; on its inner side by the inter-orbital septum; and anteriorly by the nasals and the lachrymals (*lc.*), small bones with a sharp downward-directed process, and not ankylosed to the skull even in old birds. (The sharp process internal to that of the lachrymal is part of the ethmoid bone.) Below, the orbit is bounded internally by the palatines and pterygoids (*pl. pt.*), and externally by a slender bar composed of the maxillæ (*mx.*), the jugal (*ju.*), and the *quadrato-jugal* (*q. j.*), which is attached by ligament to the *quadrate* bone (*qu.*).

(5.) *The Maxillary and Palatal Region.*—There are no tooth-bearing bones. The pre-maxillæ (*p. mx.*), which form the greater part of the upper mandible, are, in the fresh skull, en-sheathed in horn. Each has long slender nasal processes running up to the frontals internal to the nasals, and stout maxillary processes, each of which sends inwards a palatal shelf, to which the palatine is attached in the adult. The maxillæ (*mx.*) at their anterior ends send inwards flattened maxillo-palatine processes which nearly meet in the middle line; for the rest of their length they are slender. The palatines (*pl.*) are also slender for three-fourths of their length; but in the posterior fourth they send inwards curved scroll-like processes which meet in the middle line beneath the rostrum. Their hinder extremities articulate with the pterygoids (*pt.*), short, stout, mallet-shaped bones, the handles of which pass outwards and backwards to articulate with the quadrate. The quadrate (*qu.*) is a large irregular somewhat Y-shaped bone. By one limb of the Y it articulates with the dorsal part of the tympanic cavity (*h. q.*). The other limb of the Y is seen at *a. p. q.* (Fig. 48, B.), partially hidden by the pterygoid. At the foot of the stem there is a condyle for the articulation of the lower jaw, near which it articulates with the quadrato-jugal (*q. j.*).

(6.) *The Mandible.*—The two rami ankylose at the symphysis, and the component elements are more or less completely fused together in the adult. In the chick, however, each ramus is

distinctly composite.   There is a central core of (Meckel's) cartilage which is ensheathed anteriorly by the *dentary* (d.).   In
the hinder third of the jaw are three splints of bone, the *angular*
below (*an.*), the *supra-angular* (*s. a.*) and the *splenial* on the
inner side.   And within the cartilage, at the hinder end, is a
further ossification, the *articular* (*art.*).   All these tend to run
together in the mandible of the adult skull, in which posterior
and internal articular processes are well-marked.

It is especially to be noted that the lower jaw does not articulate directly with the squamosal as in the rabbit, but is connected with the brain-case by the intervention of a quadrate bone.

(7.) *The Hyoid.*—The hyo-branchial apparatus is composed of a
series of slender bones, the relations of which will be seen from
Fig. 49.   The short anterior cornua (cerato-hyals) coalesce to

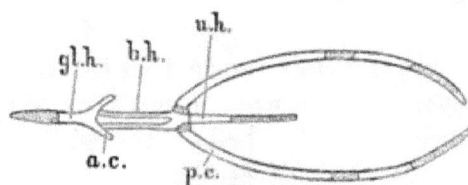

Fig. 49.—Hyoid of Fowl.

a. c. Anterior cornu. *b. h.* Basi-hyal. *gl. h.* Glosso-hyal. *p. c.* Posterior
cornu. *u. h.* Uro-hyal.

form a Y-shaped glosso-hyal.   Lying behind this are two slender
median bones, one behind the other, the basi-hyal (*b. h.*) and the
uro-hyal (*u. h.*).   From their point of junction spring the long
posterior cornua, in each of which there are two ossifications, a
proximal and a distal.

(C.) **The Frog's Skull.**—In the skull of even an adult frog
there is much more unossified cartilage than there is in the fowl
or the rabbit.   There is, in fact, a chondrocranium or cartilaginous skull, which, partly invaded by bone, may be obtained by
maceration and stripping off the splints of bone which overlie it.
Fig. 50, A., shows a dorsal view of half this chondrocranium;
B. and C. are a ventral and dorsal aspect of the right side of
the bony skull; E. shows the lower jaw, and D. the hyoid.

(1.) *The Brain-Case.*—The three typical segments are not represented in bone. In the hinder segment there is no basi-occipital, and the supra-occipital is not represented. On the other hand, the ex-occipitals (*e. o.*) are large and well developed, and may comprise also the epiotic and opisthotic elements. There

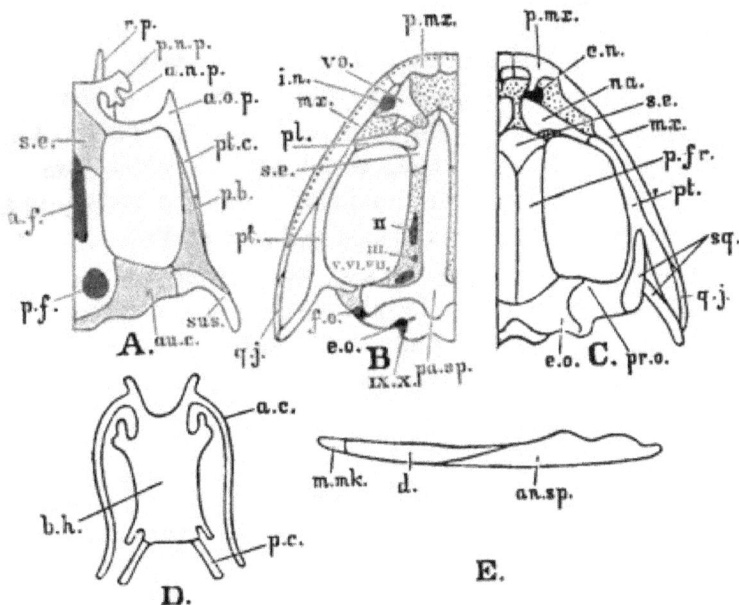

FIG. 50.—SKULL OF FROG.

A. Chondrocranium, with invading bone, from above.   B. Skull from below.
C. Skull from above.   D. Hyoid.   E. Lower jaw.

a. c. Anterior cornu of hyoid. *a. f.* Anterior fontanelle. *a. n. p.* Ali-nasal process. *a.o. p.* Antorbital process. *an. sp.* Angulo-splenial. *au. c.* Auditory capsule. *b. h.* Basi-hyal. *d.* Dentary. *e. n.* External nares. *e. o.* Ex-occipital. *f. o.* Fenestra ovalis. *i. n.* Internal nares. *m. mk.* Mento-meckelian (probably from lower labial cartilage). *mx.* Maxilla. *na.* Nasal. *p. b.* Pterygoid bone. *p. c.* Posterior cornu of hyoid. *p. f.* Posterior fontanelle. *p. fr.* Parieto-frontal. *p. mx.* Pre-maxilla. *p. n. p.* Pre-nasal process. *pa. sp.* Para-sphenoid. *pl.* Palatine. *pr. o.* Prootic. *pt.* Pterygoid. *pt. c.* Pterygoid cartilage. *q. j.* Quadrato-jugal. *r. p.* Rhinal process. *s. e.* Sphen-ethmoid. *sq.* Squamosal. *sus.* Suspensorium. *vo.* Vomer.

are two occipital condyles. In the parietal segment there is no basi-sphenoid and no ali-sphenoid. The parietals are, however, fully developed, and in comparatively young frogs completely fuse with the frontals to form the parieto-frontals (*p. fr.*). In the frontal segment there are no separate sphenoid or ethmoid bones; the curious girdle-bone *sphenethmoid* (*s. e.*) represents these skull-elements. The frontals are represented by the anterior portions of the large **parieto-frontals.** The front wall of the brain-case is formed by the sphenethmoid, a dice-box-shaped bone pierced by the olfactory nerves. In the side walls the prootics (*pr. o.*) are large well-ossified bones; but the squamosal takes no share in closing in the brain-case. In the absence of other basal elements, the *para-sphenoid* (*pa. sp.*) underlies and strengthens the cartilaginous floor of the skull. This is a conspicuous dagger-shaped bone, the blade of which probably answers to the **rostrum** in the fowl, while the crossguard is perhaps homologous with the large basi-temporal mass.

Thus in the brain-case of the frog we have the paired parieto-frontals above, the single para-sphenoid below, the sphenethmoid in front, the prootics at the sides, and the ex-occipitals behind.

The positions of some of the **nerve exits** is seen in **50,** B. The foramen for the ninth and tenth passes through the ex-occipital mass. That for the fourth or pathetic is very minute above and slightly in front of the optic foramen. The brain lies loosely in the brain-case, which has no fossæ.

(2.) *The Olfactory Chamber.*—The sphenethmoid appears in the roof of the skull, between and anterior to the two **parieto-frontals.** Separated by maceration, it is seen to be perforated by two foramina for the olfactory nerves. Posteriorly these foramina open into a common cavity, incomplete above, in which are lodged in the living frog the olfactory lobes of the brain. Anteriorly there are two cavities separated by a mesethmoidal partition. This is carried forwards in the fresh skull by a cartilaginous septum nasi, which forms the partition between the two nasal chambers. These are roofed over by the nasals (*na.*), somewhat kite-shaped bones, with the thinner tail end directed outwards and backwards. Between these and the pre-maxillæ (*p. mx.*) are the external nares (*e. n.*). The internal nares (*i. n.*) are below and somewhat behind them, immediately in front of the palatines. There is no prolonged narial passage as in the rabbit. On the inner side of the internal nares are two somewhat irregular tri-radiate bones, the *vomers* (*vo.*), which send their longest ray backwards and inwards towards the point of the para-sphenoidal dagger. Each vomer bears a little patch of teeth. There are no turbinal bones or distinct cartilages within the nasal chambers.

(3.) *The Auditory Chamber.*—The anterior part of the ex-occipital, that in front of the foramina for the ninth and tenth nerves may represent epiotic and opisthotic elements. The prootic is well developed as a separate bone, though it too eventually fuses with the ex-occipitals. It sends outwards a shelf (tegmen tympani) over the irregular and ill-defined tympanic cavity. In the fresh skull there is a ring of cartilage over which the tympanic membrane is stretched. At the bottom of the cavity is the fenestra ovalis (*f. o.*), a large hole in the prootic in the dried skull, but partly filled in with fibrous tissue in the fresh state. A stapes, to which the columella is attached, fits into the fenestra ovalis.

(4.) *The Optic Chamber.*—The large orbits have the following boundaries :—*Anterior*, nasals above and palatines (*pl.*) below, the latter bones being slender transverse curved rods. *Posterior*, prootic above and guard of the para-sphenoid below. *External*, the pterygoids (*pt.*), large tri-radiate bones with the longest ray abutting against, and supported by, the maxillæ (*mx.*), and the shortest bending round towards the para-sphenoid. The upper and hinder part of the external boundary is formed by the hammer-like head of the squamosal. *Internal*, parieto-frontals above, para-sphenoidal blade below, and sphenethmoid anteriorly.

(5.) *The Maxillary Region.*—Both pre-maxillæ and maxillæ bear teeth. The former are small bones (*p.mx.*), with well-marked ascending processes. The maxillæ are long, and they too have ascending nasal processes. They are supported by the out-wardly directed palatines (*pl.*), and the longer limbs of the tri-radiate pterygoids (*pt.*). A slender quadrato-jugal connects the maxilla with the outer limb of the pterygoid and the handle end of the squamosal.

(6.) *The Mandible.*—The lower jaw articulates with a cartila-

ginous *suspensorium* directed outwards, and so far backwards
that the articular surface is nearly on a level with the occipital
condyles. At the end of this suspensorium a quadrate ossification is formed in the cartilage, and constitutes the posterior end
of the quadrato-jugal. Dorsally the cartilage is strengthened
and supported by the overlying handle of the squamosal.
Ventrally it is supported by the outer limb of the pterygoid,
the inner limb of which is moveably articulated to the para-
sphenoid, while the anterior limb runs alongside the maxilla.

By means of this suspensorial apparatus the articulation of
the lower jaw, which differs from that of both rabbit and fowl,
is carried away from the brain-case outwards and backwards.
This gives the long broad gape of the adult frog. In young
frogs this swinging outwards and backwards of the suspensorium
is not carried nearly so far. In a minute frog, the tail of which
has only just been absorbed, the suspensorium is directed as
much forwards as, in the adult frog, it is directed backwards, so
that the gape is very much shorter.

The lower jaw itself contains a rod of cartilage (Meckel's).
There are three ossifications in each ramus: (1) a small mento-
meckelian (*m. mk.*) anteriorly; (2) a dentary (*d.*) forming the
upper part of the anterior half of the remainder of the jaw;
and (3) an angulo-splenial, by some regarded as articulare,
ensheathing the rest of Meckel's cartilage.

(7.) *The Hyoid.*—The form of the hyo-branchial plate (*b. h.*),
with its posterior and anterior cornua (*p. c., a. c.*), will be seen
by reference to Fig. 50, D. The posterior cornua pass one on
each side of the glottis. The anterior cornua are confluent
with the periotic capsule close to the fenestra ovalis.

The Chondrocranium.—The form of the chondrocranium will
be made out by reference to Fig. 50, A. In it we find:—(1) A
cartilaginous brain-case interrupted by an anterior and two
posterior fontanelles (*a. f., p. f.*), which lie beneath the parieto-
frontal bones. The cartilage is to a large extent invaded by
bone forming posteriorly the ex-occipitals, and anteriorly the
sphenethmoid (*s. e.*). (2) The auditory capsules (*au. c.*), largely
invaded by the prootics and, according to some observers, the

anterior part of the ex-occipitals, and fused into the general
mass of the brain-case. (3) The nasal chamber, with the median
septum nasi carrying forward the mid-plate of the sphenethmoid. The septum gives off above and below horizontal wings
of cartilage forming the roof and floor of the cartilaginous nasal
chamber. The roof gives off a *pre-nasal* (*p. n. p.*) and an *ali-
nasal process* (*a. n. p.*), the floor a slender *rhinal process* (*r. p.*).
Anteriorly the septum gives off on each side a flattened vertical
plate meeting the horizontal wings above and below, and forming the front wall of the chamber. (4) A bow-shaped orbital
bar on each side, constituted anteriorly by an *antorbital* or *palatine bar* with a forward projecting process (*a. o. p.*), and laterally
a *pterygoid bar* which bifurcates posteriorly, the dorsal limb
passing into the ventral crus of the suspensorium, and the
ventral articulating with the periotic capsule, and forming the
pedicle of the suspensorium. The pterygoid cartilage (*pt. c.*) is
partly invaded by the pterygoid bone (*p. b.*). (5) The suspensorium lying between the squamosal and pterygoid bones, and
dividing dorsally into two crura, of which the ventral crus is
continuous with the pterygoid cartilage, and the dorsal crus is
attached to the periotic capsule (*p. c.*).

Development and Homologies of Parts.—A generalised description
of the development of the skull must suffice. The main features
of development are the same in all *craniata* (vertebrates with
skulls), but the details differ considerably, and for them the
student is referred to larger treatises.

The parts of the skull are first indicated in mesoblastic tissue,
then developed in cartilage, and finally invaded or overlaid by
bone. Some parts are developed in relation to the brain-case,
others in relation to the sense-capsules, and others in relation
to the visceral arches.

The first indication of the skull is the formation of two
parachordal plates underlying the mid and hind brain on either
side of the notochord (Fig. 51, A., *ch.*), with which they combine to form a continuous *basilar plate* (*b. p.*). In front of this
plate two bars are carried forwards from the parachordals, with
which they are from the first continuous in the fowl but not in

the frog or rabbit. These *trabeculæ* (*tr.*) are united behind, but diverge forwards, and then curve in so as to enclose a pituitary space (*pit.*). Anteriorly they unite again, and thus are some-

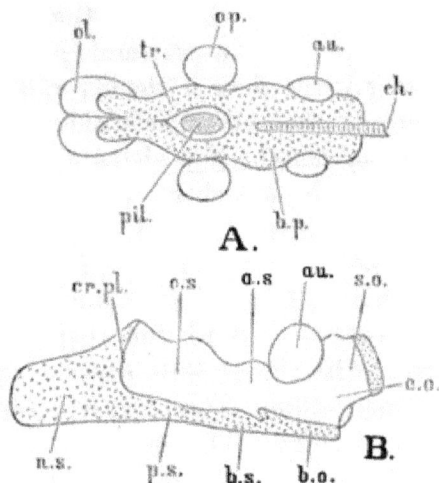

FIG. 51.—DIAGRAMS OF SKULL DEVELOPMENT.

A. Plan of cartilaginous embryonic skull. B. Section of a somewhat older skull.

a. s. Ali-sphenoid region. *au.* Auditory capsule. *b. o.* Basi-occipital region. *b. p.* Basilar plate. *b. s.* Basi-sphenoid region. *ch.* Notochord. *cr. pl.* Cribriform plate. *e. o.* Ex-occipital region. *n. s.* Nasal septum. *ol.* Olfactory capsule. *op.* Optic capsule. *o. s.* Orbito-sphenoid region. *p. s.* Pre-sphenoid region. *pit.* Pituitary space. *s. o.* Supra-occipital region. *tr.* Trabeculæ.

what lyre-shaped. They lie beneath the fore-brain, the base of which, to begin with, projects downwards into the pituitary space.

The basilar plate represents the future basi-occipital region of the skull. Its sides grow up fusing with the periotic capsules of the auditory sacs, and eventually meet above. Thus the whole occipital segment is preformed in cartilage, and in this cartilage are subsequently formed ossifications starting from basi-occipital, ex-occipital, and supra-occipital centres (*b.o.*, *e.o.*, *s.o.*), while in the periotic capsules the otic bones are subsequently formed.

The trabeculæ grow together so as to almost completely close the pituitary space. In the region of the fore-brain they grow up at the sides as two pairs of wings, an anterior and a posterior, more or less continuous; but the roof of the cartilaginous brain-case in this region is incomplete. In this region the basi-sphenoid and pre-sphenoid (*b. s., p. s.*) subsequently ossify in the mid-ventral line, and the ali-sphenoids and orbito-sphenoids (*a. s., o. s.*) are formed in the side wings. Anterior to this the trabecular plate narrows and forms a median vertical lamina, which further forward comes into relation with the nasal sacs, curving over on either side to form their roof, or sending out wings beneath to form their floor, or providing turbinal ingrowths. In this region the ethmoidal and turbinal bones are subsequently developed.

We have now to consider the part played by the visceral arches. It will be remembered that in the tadpole there are gill-slits leading outwards from the pharynx to the exterior. Homologous gill-slits are formed in the fowl and the rabbit; but in these organisms they never have a branchial function. There are four such visceral clefts. In front of each, and behind the last, is a visceral fold. Of the five visceral folds or arches, the first lies just behind the mouth, and nearly meets its fellow in the mid-line. This is the mandibular arch. It does not remain simple, but bifurcates into two processes, a superior maxillary process (Fig. 52, *s. m. p.*) in front of the mouth, and an inferior maxillary process (*i. m. p.*) behind the mouth. The second or hyoid arch is the strongest in the embryo. The third, fourth, and fifth are progressively shorter and weaker.

The first post-oral visceral cleft persists in the adult as the Eustachian tube and tympanic cavity. The other three function as branchial clefts in the frog, but are early obliterated in the fowl and rabbit.

The further history of the first two arches is remarkable and important. In the superior maxillary process of the first a cartilaginous palato-pterygoid bar is developed, in relation to which the palatine and pterygoid bones are formed. In the inferior maxillary process of the mandibular arch a cartilage (Meckel's) is formed which constitutes the basis of the lower jaw. In the frog its proximal portion forms the suspen-

11

sorium. In the fowl the proximal portion forms the quadrate bone, the succeeding portion in the lower jaw ossifying as the articular. In the rabbit it seems probable that the

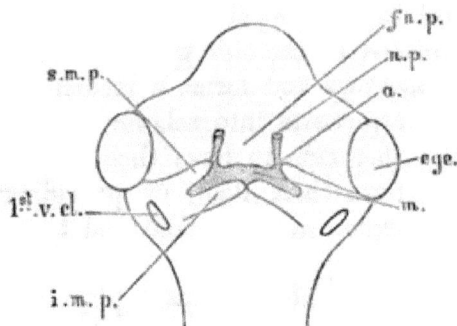

FIG. 52.—THE MANDIBULAR ARCH.

a. Point where fronto-nasal process overarches the nasal groove. *fn. p.* Fronto-nasal process. *i. m. p.* Inferior maxillary process. *m.* Mouth. *n. p.* Nasal pit. *s. m. p.* Superior maxillary process. *1st v. cl.* First visceral cleft.

proximal portion, answering to the quadrate of the fowl, separates early and ossifies as the incus, while the succeeding portion, answering to the articular of the fowl, is segmented off from the rest of Meckel's cartilage, and ossifies as the malleus ; while the main mass of the lower jaw is ossified (perhaps in connection with lower labial cartilages) around, but not within, the remainder of Meckel's cartilage. It may be, however, that ossification of the most distal portion of Meckel's cartilage takes place. Concerning the homologies of the quadrate, articular, malleus, and incus, more than one interpretation has been offered.

The uppermost end of the hyoid arch would seem to be segmented off and to ossify as the stapes, while the columella of the frog and fowl are also developed from this arch. The remainder of the arch forms the anterior cornua of the hyoid, the basi-hyal giving rise to the body. The posterior cornua of the rabbit and the fowl are remnants of the first branchial arch, but in the frog of the fourth branchial arch, the third and first two being represented by processes of the hyoid plate, the

hinder portion of which is basi-branchial in origin, and represents the fused basal elements of these branchial arches.

So far all the bones of the skull, with the exception of the
palatines and pterygoids in some forms (*e.g.* fowl), are cartilage
bones, ossification taking place within a preformed mass of
cartilage. All the other bones are splint bones, ossification
taking place in membrane without the intervention of cartilage.
Such are the parietals, frontals, lachrymals, nasals, pre-maxillæ,
maxillæ, vomers, squamosals, jugals, quadrato-jugals (in the fowl,
and their anterior portion in the frog); dentary, angular, supra-
angular, and splenial, in the lower jaw; and para-sphenoid at
the base of the skull. This is a single bone in the frog with
rostral and basi-temporal divisions. In the fowl the rostrum
and basi-temporal form distinct ossifications, and in the rabbit
the para-sphenoidal rostrum is entirely suppressed, and only
minute rudiments [larger in the guinea pig] of the basi-temporal
guard have been discovered. The question of the homology of
the basi-temporal, however, is one that has received more than
one answer.

The Vertebral Column.—(1.) *In the Rabbit.*—The vertebral
column of the rabbit has some 45 vertebræ. They are, reckoning
from before backwards, atlas, axis, 5 more cervical; 12 thoracic,
with free ribs; 7 lumbar; 4 sacral; about 15 or 16 caudal.

We may take the fourth thoracic vertebra for illustration
(Fig. 53.) Below there is a solid mass of bone, the body or
centrum (*c.*), the ends of which are flattened and vertical. From
this springs the *neural arch* (*n.*) through which runs the spinal
cord. Above the neural arch rises the *neural spine* (*n. s.*). At
its sides project *transverse processes* (*tr.*) which bear at their ends
tubercular facets (*t. f.*), with which the tubercular processes of the
ribs articulate. If the third and fifth be placed in position
before and behind it, it will be found to articulate with them by
means of little bony shelves in front and behind. Those in front
(*prezygapophyses, pr. z.*) face upwards and outwards; those behind
(*postzygapophyses, pt. z.*) face downwards and inwards. When the
vertebræ are thus placed together a lateral space is seen between

the peduncles of the neural arches. It is the *intervertebral foramen*, by means of which the spinal nerves make their exit. And when the centra are in contact a *capitular facet* for the articulation of the head of the rib is formed by the juxtaposition of two demi-facets (*c. s.*), one on each centrum.

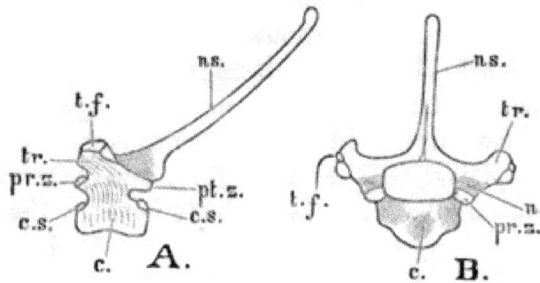

FIG. 53.—FOURTH THORACIC VERTEBRA—RABBIT.
A. From the side. B. From the anterior end.

c. Centrum. *c. s.* Capitular semi-facet. *n.* Neurapophysis of neural arch. *n. s.* Neural spine. *pr. z.* Prezygapophysis. *pt. z.* Postzygapophysis. *t. f.* Tubercular facet. *tr.* Transverse process.

The following points may be noted in other parts of the column:—The transverse processes of the cervical vertebræ are pierced by a *vertebrarterial canal*. The second cervical (axis) has a peg-like *odontoid* process; the first (atlas) is a mere ring of bone with facets for the condyles of the skull, and flattened transverse processes. In the eleventh thoracic vertebra the neural spine becomes vertical, sloping forward in the twelfth. In the vertebræ of this region there is a large process (the *metapophysis*) just behind the prezygapophysis. In the first three lumbar vertebræ there is a flattened process (*hypapophysis*) beneath the centrum, and in the first five lumbar small pointed processes (*anapophyses*) looking backwards, beneath the postzygapophyses. The four sacral vertebræ are fused into one mass; but only the first, or at most, the first two, come into relation with the ilia by means of their expanded transverse processes or ankylosed ribs. The metapophyses are here smaller, and there are no ana- nor hypapophyses. The caudal vertebræ gradually diminish in size and in complexity until they are reduced to mere centra.

In the young rabbit there are cartilaginous intervals in each centrum near the ends ; these separate the main mass (*diaphysis*) of the centrum from its end plates (*epiphyses*) which are characteristic of mammalian vertebræ.

It will be remembered that the first indication of the axial skeleton in the body is the notochord which underlies the neural axis. This becomes invested by a cartilaginous sheath, which comes into relation with a number of cartilaginous arches derived from the mesoblastic investment of the spinal cord. As growth proceeds, the notochord is encroached upon by its cartilaginous investment, which becomes differentiated into vertebral and intervertebral regions. In the vertebral region the notochord is finally obliterated ; but in the intervertebral region it persists throughout life as part of the *nucleus pulposus* in the axis of the intervertebral ligament. Before the differentiation of the vertebræ, there is split off from the upper and outer part of each somite a muscle-plate (see p. 120), from which is formed the voluntary muscle of that part of the trunk. It is to be noted, however, that the vertebral segmentation takes place in such a way that the centrum of the newly-formed vertebra does not correspond in position to the middle of the muscle-plate, but to the line separating two muscle-plates ; the muscles formed from which, therefore, act upon two adjoining vertebræ.

The odontoid process of the axis ossifies from a distinct centre, and is at first a distinct bone. It is indeed a portion of the centrum of the atlas, which has lost its primitive connection and become conjoined to the axis. The so-called transverse processes of some (cervical, sacral) vertebræ are to be regarded as ankylosed ribs.

(2.) *In the Fowl* there are sixteen cervical vertebræ, of which the last is ankylosed to the first three thoracic. The fourth thoracic is free ; but the fifth or last coalesces with a number of succeeding vertebræ to form the compound sacrum. In this sacral mass there are ankylosed : one (the fifth) thoracic ; three lumbar ; five (ribless) sacral ; six (with ankylosed ribs) urosacral ; fifteen in all. Following on the sacrum are six free caudal ;

and the vertebral column is terminated by a *pygostyle* or plough-share bone, composed of several completely ankylosed vertebræ.

Note the extreme coalescence of parts in the column. The sacrum forms a compact mass, which in the adult is firmly anky-losed to the iliac bones (see Fig. 56, A) by means of the trans-verse processes, coalesced ribs, and anteriorly by the expanded crest of the coalesced neural spines.

The form of the centrum in birds is noteworthy. The anterior face is convex in vertical section, and concave in hori-zontal section ; while the posterior face is concave in vertical section and convex in horizontal section. In other words, both faces are saddle-shaped, but the anterior saddle is horizontal and the posterior saddle vertical.

(3.) *In the Frog.*—The vertebral column consists of nine vertebræ and a posterior elongated *urostyle*. There is no distinction into cervical, dorsal, and lumbar ; but the ninth, the expanded trans-verse processes of which articulate with the ilia, may be regarded as the sacral vertebra. The urostyle articulates with the last vertebra by two concave facets. Along its dorsal surface runs a prominent ridge, in which are two coccygeal foramina for the tenth spinal nerves. The faces of the centra of the second to the seventh vertebræ are concave anteriorly and convex pos-teriorly. Both faces of the eighth are concave. The ninth is convex anteriorly and has two convex tubercles for articulation with the urostyle posteriorly. The atlas has facets for the con-dyles of the skull. The centra contain vestiges of the notochord.

General Considerations.—In studying the successive vertebræ in one organism, or comparing the vertebræ in different organisms, we are struck by the general similarity of plan in the midst of differences of detail. We see specialisation, and yet adherence to type. After considering all the forms presented to us by the vertebræ, we may construct an ideal vertebra, which we may call a typical vertebra. And then we may trace the modifica-tions which have led to greater or less divergence from the typical form. These modifications fall under three heads (Huxley's law) : (1) *Coalescence*, where parts typically distinct run together and become fused ; (2) *Suppression*, where parts typically present are undeveloped; and (3) *Metamorphosis*, where there is a notable departure from the typical form. The student is, however, to bear in mind that the typical form has not necessarily any existence as such in nature, but is a conception of the human mind. To explain: photographs are now taken which represent, *e.g.* the typical criminal. A number of photographs of representative criminals are separately taken, and then the images of all are combined on one plate. The result is a typical photograph which resembles no one individual, but represents all. So too the ideal vertebra resembles no one individual vertebra, but represents all. If, therefore, we speak of a particular vertebra as typical, we mean that its divergence from the ideal type is comparatively slight.

The **Ribs and Sternum.**—(1.) *In the Rabbit.*—Free ribs are connected with and characterise the thoracic vertebræ. There are seven true ribs which, with the sternebræ below, constitute inferior or hæmal arches. Behind these are five false ribs, incomplete below. Each true rib consists of a dorsal osseous portion, the *vertebral rib*, and a ventral cartilaginous portion, the *sternal rib*. The vertebral rib has a head (*capitulum*), which articulates with the capitular facets of the vertebræ, and a *tubercle* on the dorsal side, which articulates with the tubercular facet on the transverse process of the vertebra. Just external to the tubercle is a *dorsal process*.

The sternum consists of six segments or *sternebræ*, of which the first, *manubrium*, is large and keeled ventrally. The last has a rounded plate of cartilage (the *xiphoid process*). Seven pairs of ribs articulate with the sternum, the first two pairs with the manubrium, the others with the interspaces between the sternebræ.

(2.) *In the Fowl* there are five complete or true ribs. Each has a capitulum and tuberculum, and is divided by a transverse articulation into a dorsal (vertebral) and a ventral (sternal) rib, both of which are osseous. At their junction they form an open angle, which is widened during the process of respiration. Each vertebral rib has a backward and upward process (the *uncinate pr.*).

Anterior to the true ribs are three free incomplete false ribs, the form of which resembles the vertebral portion of the true ribs.

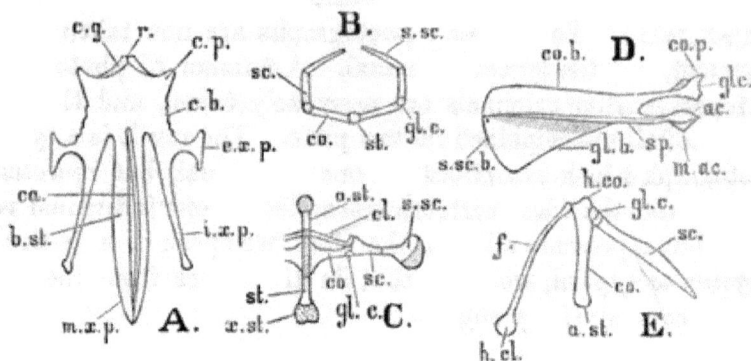

FIG. 54.—STERNUM AND SHOULDER-GIRDLE.

A. Sternum of fowl from below. B. Sternum and shoulder-girdle of frog in diagrammatic section. C. The same from below, with upper part straightened out. D. Scapula of rabbit. E. Shoulder-girdle of fowl.

a. st. Part of coracoid which articulates with sternum. *ac.* Acromion. *b. st.* Body of sternum. *c. b.* Costal border. *c. g.* Coracoid groove. *c. p.* Coracoid process in fowl. *ca.* Carina or keel. *cl.* Clavicle. *co.* Coracoid. *co. b.* Coracoid border. *co. p.* Coracoid process in rabbit. *e. x. p.* External xiphoid process. *f.* Furcula. *gl. b.* Glenoidal border. *gl. c.* Glenoidal cavity. *h. cl.* Hypocleidium. *h. co.* Head of coracoid. *i. x. p.* Internal xiphoid process. *m. ac.* Metacromion. *m. x. p.* Middle xiphoid process. *o. st.* Omosternum. *r.* Rostrum. *sc.* Scapula. *s. sc.* Supra-scapula. *s. sc. b.* Suprascapular border. *sp.* Spine. *st.* Sternum. *x. st.* Xiphi-sternum.

The sternum of the fowl is a large and well-ossified structure. The central portion or *body* of the sternum (Fig. 54, *b. st.*) is boat-shaped and concave dorsally, and is prolonged backwards as an elongated *middle xiphoid process* (*m. x. p.*). Ventrally there is a well-marked keel or *carina* (*ca.*). Anteriorly the sternum is produced into a vertically-flattened *rostrum* or manubrium (*r.*), on either side of which, running backwards and outwards along the front edge of the body, is a *coracoid groove* (*c. g.*). On either side of the sternum there are three large processes. The most anterior is the *costal process* (*c. p.*); behind which, at the side, is the *costal border* for the articulation of the ribs (*c. b.*). Posterior to this are two processes having a common

stem.   The more anterior is the *external xiphoid process* (*e. x. p.*),
the more posterior the *internal xiphoid process* (*i. x. p.*).

The plate of cartilage from which the sternum is formed arises
as two bands, which are segmented off from the ventral ends of
the ribs, and unite to form a common plate.

(3.) *In the Frog* there are no ribs.   The sternum is well ossified, and is in close connection with the pectoral arch.   Behind
it is an expanded plate of cartilage, the *xiphi-sternum* (Fig. 54,
x. st.).   Anterior to the pectoral arch is a median bony process,
the *omosternum* (*o. st.*), carrying a rounded cartilaginous plate.

The Pectoral Arch.—The shoulder-girdle of the frog comes
into close relation with the sternum and omo-sternum.   Between
these sternal elements two bones pass outwards from each side
of the mid-ventral line.   Of these the more posterior and stouter
is the *coracoid* (54, B. and C., *o.*) the more anterior and slenderer
is the *clavicle* (*cl.*).   Outside these, articulating with both, and
passing nearly vertically on each side, is a somewhat hourglass-
shaped bone, the *scapula* (*sc.*).   Dorsal to this, and lying over the
transverse processes of the second, third, and fourth vertebræ, is
a flattened plate of bone and cartilage, the *supra-scapula* (*s. sc.*).
At the point of junction of the coracoid and scapula is the
glenoidal cavity for the articulation of the fore-limb (*gl. c.*).   In
Fig. 54, C., the parts are flattened out.   In B. they have their
natural position.   The inner edges of the coracoids remain
cartilaginous.   Posterior to the clavicle (mainly a membrane
bone) lies a *precoracoid* bar of cartilage.

In the Rabbit, while the ribs form an attachment with the
sternum which is absent in the frog, the attachment of the
shoulder-girdle is much less perfect.   The main element is the
scapula or shoulder-blade (Fig. 54, D.).   The coracoid is held
to be represented by a process (*co. p.*) of the scapula.   Doubts
have, however, recently been thrown upon the validity of this
view.   The small clavicle is the only part which has attachment to the sternum, being connected by ligament on the one
hand with the *acromion* (*ac.*), and on the other with the sternum.
The form and relations of the scapula will be readily made out

with the aid of the figure, in which *sp.* is the crest or spine which forms the **acromion** border ; *gl. b.* is the glenoidal border ; *gl. c.* the glenoidal cavity ; *co. b.* the coracoid border ; *s. sc. b.* the supra-scapular border, which is postero-dorsal in position.

In the Fowl the coracoid is a separate bone (54, E., *co.*). The glenoidal cavity (*gl. c.*) is at the junction of coracoid and scapula (*sc*) ; *h. co.* is the head of the coracoid which articulates with the **sternum** by *a. st.* The clavicles are represented by the **merry-thought** or *furcula* (*f.*), of which the flattened median portion (*h. cl.*) is called the *hypocleidium.*

The Fore Limb.—The typical fore-limb consists of the following parts: (1) a single proximal bone in the arm, the *humerus*

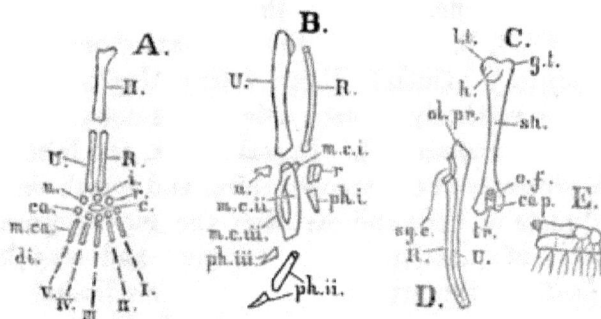

FIG. 55.—RIGHT FORE-LIMBS.

A. Typical. B. Fore-limb of fowl. C. **Humerus of rabbit.** D. Radius and ulna of rabbit. E. Carpus of rabbit.

H. Humerus. *R.* Radius. *U.* Ulna. *c.* Centrale. *ca.* Carpus. *cap.* Capitellum. *di.* Digits. *g. t.* Greater tuberosity. *h.* Head of humerus. *i.* Intermedium. *l. t.* Lesser tuberosity. *m. ca.* Metacarpals. *mc.* i. ii. iii. Metacarpels of fowl. *o. f.* Olecranon fossa. *ol. pr.* Olecranon process. *ph.* i. ii. iii. Phalanges of fowl. *r.* Radiale. *sh.* Shaft. *sy. c.* Sigmoidal cavity. *tr.* Trochlea. *u.* Ulnare.

(Fig. 55, A., *II.*) ; (2) two bones side by side in the fore-arm, one pre-axial the *radius* (*R.*), and one post-axial, the *ulna* (*U.*) ; (3) the *carpus*, composed of nine bones ; (*a.*) a proximal row of three bones, *radiale* (*r.*), *intermedium* (*i.*), and *ulnare* (*u.*) ; (*b.*) a single (sometimes double) bone, the *centrale*, between the proximal and distal row ; (c.) a distal row of five *carpalia*, numbered from the radial to the ulnar side ; (4) the *metacarpus*, composed of five *metacarpalia* articulated to the five carpalia ; and (5), the five digits with articulating phalanges.

In the rabbit the humerus has the form shown in dorsal view in Fig. 55, C., where *h.* is the *head*, *l. t.* and *g. t.* the *lesser* and *greater* tuberosity, *sh.* the *shaft*, *o. f.* the *olecranon fossa*, *tr.* the *trochlea*, and *cap.* the *capitellum.* On the ventral aspect there is a well-marked *deltoid ridge.* The radius and ulna have the form shown in D. where *ol. pr.* is the *olecranon*, and *si. c.* is the *sigmoid cavity* for articulation with the humerus. The two bones are closely applied, and do not admit of that change of relative position which gives us the power of rotating our hand and wrist. In the carpus (E.) carpalia iv. and v. coalesce to form a single bone (unciform), while the centrale is wedged in between carpalia ii. and iii. To the left of the ulnare in E. is seen a little bone (shaded) ; this is the pisiform, hitherto regarded as an extraneous bone developed in tendon. But the homologies of these parts are being re-investigated, and some change of view (*e.g.* in the nature of the pisiform) seems probable. Special names are given to the bones of the mammalian carpus. In the proximal row of the carpus the bones are (in E., reading from left to right) *pisiform, cuneiform, lunar,* and *scaphoid ;* in the distal row they are *unciform, magnum, centrale* (wedged in from above), *trapezoid,* and *trapezium.*

In the fowl the humerus is a strong light bone. In the ventral aspect the greater tuberosity is strongly marked ; in the dorsal aspect, close to the lesser tuberosity, is an excavation leading to the pneumatic foramen, by means of which air from the interclavicular air-sac finds access to the shaft. The radius and ulna (B.) do not call for especial notice. The manus is, however, curiously modified in relation to flight. Of the carpus only the ulnare and radiale are developed. The metacarpals i., ii., iii., ankylose, and are peculiar in form. The others are suppressed. Of the digits, one phalange of i., two of ii., and one of iii., are developed. The rest are wanting in the adult.

In the frog the humerus is short and curved, with a deltoid

crest more pronounced in male than female. The radius and ulna coalesce to form a *radi-ulna*, which shows, however, at the distal end signs of its composite nature. The carpus consists of six bones, two articulating with the radi-ulna, three with the metacarpals (of which the largest articulates with iii., iv., and v.) and one having an intermediate position on the radial side. There are five metacarpals, the first very small; and four complete digits, the pollex (thumb) being present only in rudiment.

For the changes of position which the fore-limbs have undergone see Chapter II., pp. 7, 14, and 19.

The Pelvic Arch.—Three elements enter into the composition of the haunch-bone of either side. And these three meet at the *acetabulum* or articular cavity for the hind-limb. These three are the *ilium* (anterior); the *ischium*, postero-dorsal; and the *pubis*, postero-ventral. Internal to the pubis there is, however, in some forms a fourth bone, the *acetabular*, which may in some cases (*e.g.* rabbit) shut out the pubis from any share in the formation of the acetabular cavity.

In the frog these bones (Fig. 56, C. D.) are tolerably easily separable, though the elements of the two sides fuse somewhat closely in the mid-line at the symphysis. The part marked *pu.* is regarded as pubis. The feature to be specially noticed is the great lengthening of the ilia (*il.*), and their articulation with a single sacral vertebra, through the agency of supra-iliac cartilages. The swinging back of the point of articulation of the hind-limb, brought about by the lengthening of the ilia, is more marked in the adult than in the young frog. The student should compare the pelvic arch of the frog with that of the newt, where the acetabulum is not far removed from the sacrum.

In the rabbit (56, B.) the ilium (*il.*), ischium (*isc.*), and pubis (*pu.*) (together with a small acetabular, which, in the very young rabbit, shuts out the pubis from the acetabulum), completely fuse into a single *innominate bone*. The two innominates unite by synchondrosis, or cartilaginous union, at the symphysis (*sy.*). They articulate with the expanded lateral plates (ankylosed ribs) of the first two sacral vertebræ. In B. *t. isc.* is the tuber ischii, *isc. b.* the ischial border, *s. il. b.* the supra-iliac border, *pu. b.* the pubic border, *ac.* the acetabulum, *ac. b.* the acetabular border, and *ob. f.* the obturator foramen.

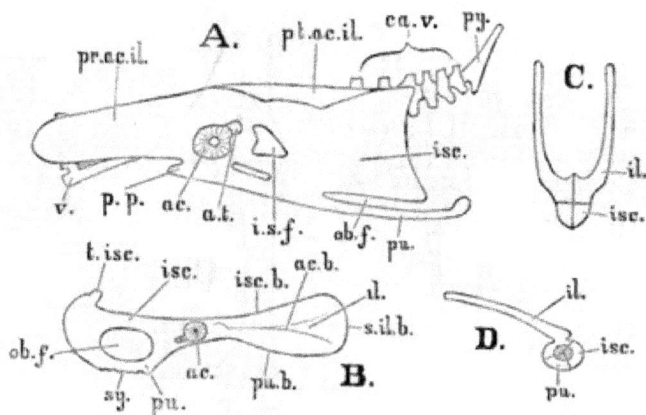

FIG. 56.—HIP-GIRDLE.

A. Innominate bones and sacrum of fowl. B. Innominate of rabbit. C. Innominate of frog, from above. D. The same from the side.

ac. Acetabulum. *ac. b.* Acetabular border. *a. t.* Anti-trochanter. *ca. v.* Caudal vertebræ. *il.* Ilium. *i. s. f.* Ilio-sacral foramen. *isc.* Ischium. *isc. b.* Ischial border. *ob. f.* Obturator foramen in rabbit, obturator fissure in fowl. *p. p.* Pectineal process. *pu.* Pubes. *pu. b.* Pubic border. *pr. ac. il.* Pre-acetabular part of ilium. *pt. ac. il.* Post-acetabular part. *py.* Pygostyle. *s. il. b.* Supra-iliac border. *sy.* Symphysis. *t. isc.* Tuber ischii. *v.* Last thoracic vertebra.

In the adult fowl the ilium, ischium, and pubes, not only fuse into a single innominate (56, A.), but this becomes ankylosed with the compound sacrum. On the other hand, there is no symphysis as in the frog and the rabbit. The student should endeavour to obtain a just-hatched chick or a nestling pigeon, in which the bones are still separate. In Fig. 56, A., *pr. ac. il.* is the pre-acetabular part of the ilium, *pt. ac. il.* the post-acetabular part, *ca. v.* the caudal vertebræ, *py.* the pygostyle, *isc.* the ischial portion, *pu.* the pubis, *ob. f.* the obturator fissure, *i. s. f.* the ilio-sciatic foramen, *a. t.* the anti-trochanter, *ac.* the acetabulum, *p. p.* the *pectineal process*, perhaps derived from an acetabular, and *v.* the last thoracic vertebra.

The Hind Limb.—The typical hind limb (Fig. 57, A.) consists
of the following parts : (1) a single proximal bone articulating at
the acetabulum, the *femur* (F.) ; (2) two bones side by side in the
crus, one pre-axial, the *tibia* (*Ti.*), and one post-axial, the *fibula*

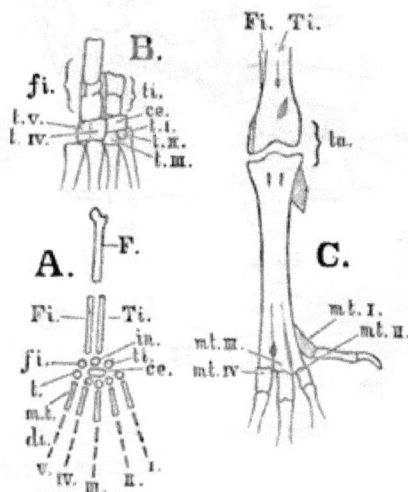

FIG. 57.—RIGHT HIND LIMB.

A. Typical.   B. Tarsus of rabbit.   C. Distal part of hind limb of fowl.

F. Femur.   Fi. Fibula.   Ti. Tibia.

ce. Centrale. di. Digits. fi. Fibulare. in. Intermedium. m.t. i.-iv. Metatarsals. t. i.-v. Tarsals. ti. Tibiale.

(*Fi.*) ; (3) the *tarsus,* composed of nine bones—(*a.*) a proximal row
of three bones, *tibiale* (*ti.*), *inter-medium* (*i.*), and *fibulare* (*fi.*) ; (*b.*) a
single bone, the *centrale* (*ce.*), between the proximal and distal
rows ; (*c.*) a distal row of five *tarsalia,* numbered from the tibial to
the fibular side ; (4) the *metatarsus* (*mt.*), composed of five metatarsals ; and (5) the five digits with their phalanges.   The
student cannot fail to notice how marked is the apparent
homology of the parts of the fore and hind limbs, femur answering to humerus, tibia to radius, and so on.

In the rabbit the femur is a strong bone, with rounded
head, borne upon a neck, below which is the lesser trochanter,
and above which is the strongly-marked greater trochanter,
overhanging a trochanteric fossa.   A third trochanter is on

the opposite side of the bone to the lesser trochanter. At
the distal end of the bone there is an internal and external
condyle (the former on the same side as the head), separated
by the inter-condylar notch, on the pre-axial side of which
the small *patella* of the knee-cap fits. The tibia is a large
strong bone, with which the small fibula fuses distally on the
outer side. The tarsus (Fig. 57, B.) consists of six bones : a
large *calcaneum* or fibulare (*fi.*), and a smaller *astragalus* or tibiale
(*ti.*), in which the inter-medium is also probably merged; a centrale, or navicular (*ce.*), a bone (the cuboid) on the fibular side,
which represents tarsale iv. and v. coalesced, tarsale iii., known as
the ecto-cuneiform, and tarsale ii. or meso-cuneiform. Tarsale i.
(ento-cuneiform) may be represented by a small process of meta-
tarsale ii. The homologies of the tarsus are, however, being
re-investigated. There are only four metatarsals, i. being suppressed ; and four digits, the hallux (great toe) being absent.

In the fowl the femur is a stout somewhat curved bone with a
strong great trochanter. In front of its distal end is a small
patellar nodule. On the femur follows a compound bone known
as the *tibio-tarsus*, the distal end of which (Fig. 57, C.) represents
one or more of the proximal row of tarsal bones. The fibula
(*Fi.*) forms a thin bone lying as a splint on the tibio-tarsus, but
with an expanded head. Upon the tibio-tarsus follows another
compound bone, the *tarso-metatarsus.* Its proximal end represents
the distal row of tarsal bones which becomes fused on to the ends
of metatarsalia ii., iii., iv., ankylosed side by side into a single
bone. The first metatarsal is found at the distal end of the
tarso-metatarsus united to it only by ligament. The fifth metatarsal is absent. There are four digits, of which three are
directed forwards, and rest with their whole ventral surface on
the ground. The other digit, the hallux, is directed backwards,
and raised off the ground at the proximal end. The fifth digit is
suppressed.

Thus in the fowl the heel-joint is not, as in the rabbit,
between the tibia and the tarsus, but between the proximal and
distal portions of the tarsus, which are respectively ankylosed
with the tibia and the fused metatarsus. It is therefore a *meso-*

tarsal articulation. The hallux has two phalanges; the second, third, and fourth digits have three, four, and five phalanges respectively. Both the hind and fore-limb of the bird are thus seen to be remarkably modified.

In the frog the femur is a thin elongated bone. The tibia and fibula fuse together into a tibio-fibula, which shows signs, even in the adult, of its compound origin. The astragalus and calcaneum are remarkably elongated, fusing to a considerable extent at their ends. They aid in giving the extreme length to the pes. The distal row of the tarsals is represented by two partly ossified cartilages. There are five metatarsals and five well-developed digits, and, in addition to these, the *calcar*, which is situated on the radial side; it is composed of two or three pieces, and may represent a sixth digit.

For the changes of position which the hind-limbs have undergone see Chapter II., pp. 7, 14, and 19.

CHAPTER IX.

NUTRITION AND METABOLISM.

No steam-engine can continue to work without fresh supplies of fuel; no organism can maintain its vital activity without fresh supplies of food. It is the combustion of the fuel which gives to the engine its energy or power of doing work; and in the animal it is to a kind of slow combustion that the maintenance of the vital energy is due. But the chemical processes in the latter case are vastly more complex, and are not confined, as in the engine, to a special part of the machine; for the animal is a cunningly wrought piece of mechanism, and the most efficient machine we know. But we must not forget that the processes which go on in the animal, though they are chemical, physical, and mechanical, are at the same time something more. They are vital or organic; and this, not through any mysterious addition of something from without, but through more complex combination of the self-same materials and energies.

In this chapter we must consider shortly some of the *metabolic processes* (as the chemical processes within the organism are termed) upon which the continued vitality of the organism depends.

The Alimentary Canal.—The alimentary canal is a continuous more or less coiled tube running through the body from the mouth to the vent, and for a considerable part of its course suspended in the body-cavity. It does not in any way communicate directly, that is, by any opening or canal, with the blood-vessels. And when we remember that its lining membrane is continuous at the lips and vent with the epidermis of the skin, we shall see that the contents of the canal, although

12

within the body, are still in a sense outside it. They are in the same position as the fuel in a Cornish or Lancashire boiler which is placed in the tube or tubes that run through the midst of the boiler, but which can hardly be said to be inside it. Hence the elaborated food-stuff has to be absorbed through the walls of the canal before it enters the body.

But though there are no tubes which carry material from the canal into the system, there are several large, and a multitude of minute tubes which convey material into the canal. These are the ducts of the various glands which minister to digestion. They are divided into : (1) *salivary glands* in the region of the mouth; (2) *gastric glands* in the stomach; and (3) *intestinal glands*.

The digestive tube itself has the following structure (for details see Chapter V. p. 78, Fig. 28) : Externally there is a peritoneal investment; then follow the muscular walls with longitudinal and circular fibres; within this is the mucous membrane. The epithelium that lines the tube passes up into and lines the ducts of the glands, large and small. At the mouth and vent it is continuous with the epidermis.

This epithelium, and its supporting mucosa and sub-mucosa, does not form a smooth and even lining to the canal. In the stomach of frog and rabbit it is thrown into folds or *rugæ*. In the small intestine of the rabbit it is doubled inwards so as to form crescentic or nearly circular folds (*valvulæ conniventes*). And if a small portion of the small intestine of a rabbit or fowl be examined under water with a lens it will be seen to have a velvety appearance, due to great numbers of closely-set minute processes, the *villi* (see Fig. 28). In the small intestine of the frog there are minute dependent processes; but they do not form so close-set a velvety pile as the villi of fowl or rabbit. In the large cæcum of the rabbit there is a spiral infolding of the tube readily seen when this curious appendage is removed, cleaned out by directing a stream of water through it, distended with air, and dried. In the colon of the rabbit the whole wall is characteristically puckered.

In the fowl, answering to the stomach of frog or rabbit, there is anteriorly a proventriculus, the walls of which are richly supplied with glands, the orifices of which may be seen by careful examination with a lens ; and posteriorly the gizzard, the walls of which are exceedingly muscular, the fibres of either side radiating from a central tendinous aponeurosis. The epithelium here develops a dense horny coat. It is here in fact, and not in the mouth as in the rabbit, that the food is triturated ; and to aid in this grinding process the fowl swallows small stones.

The Glands.—We must now briefly consider the glands that minister to digestion ; first their structure, and then their products. It will be necessary to confine our attention to the rabbit.

The *salivary glands* of the rabbit are : (1) the *parotid*, lying between the base of the ear and the angle of the lower jaw, and opening forwards into the side of the mouth by the parotid duct (Stenson's) ; (2) the *sub-maxillary*, lying between the angles of the lower jaw, and opening forwards into the floor of the mouth by the sub-maxillary duct (Wharton's); (3) the *sub-lingual*, small, and lying along the inner side of the mandible, with many ducts opening separately into the mouth; (4) the *infra-orbital*, lying in the antero-inferior region of the orbit, the duct of which passes downwards to the mouth ; and (5) the small buccal and other glands set in the mucous membrane of the lips, palate, tongue, and pharynx.

The buccal glands are simple or branched tubes lined with epithelial cells, which are flattened in the duct, but columnar in the deeper secreting part or body of the gland. The other salivary glands are all of the racemose or branched tubular type. The duct is elongated into a long tube, which divides and sub-divides a great number of times, and so gives rise to a multitude of branched tubes, each of which ends in a slight dilatation (*alveolus*). The tubes and their alveoli are imbedded in connective tissue, and aggregated into lobules separated by connective tissue septa.

The cells which line the alveoli of the so-called serous salivary glands, differ curiously according to the state of rest or activity of the gland. After a period of rest they are large, and have a markedly granular appearance, the protoplasm of the

cell being much reduced in size. After a period of secretion
the cells decrease in size, so that the lumen (or cavity), which
was before almost or quite obliterated, becomes obvious. The
cell-contents are much less granular, and the granules are restricted to that part which lies nearest the lumen. The protoplasm is relatively increased in amount, and the nucleus,
before disc-shaped and seen with difficulty, becomes more
clearly visible, and spherical.

The *gastric glands* are of two kinds, cardiac and pyloric. The
cardiac glands, which are developed in all parts of the stomach
except the pylorus, are simple tubes, of which two or three may
open into the same duct. They are lined with epithelial cells,
columnar and transparent in the body of the gland, more
cubical in the neck. Between these epithelial cells and the
supporting membrane is a discontinuous layer of large granular
oval cells (parietal cells). The *pyloric* glands have longer ducts
and are more curved. The epithelial cells of the body of the
gland are cubical, and resemble those in the neck of the cardiac
gland. There are no parietal cells.

Of the *intestinal glands, Brunner's* are practically continuous
with the pyloric glands, but are more decidedly branched, almost
racemose, in character. They are confined to the duodenum.
But throughout the whole of both small and large intestine there
are great numbers of the simple tubular glands (or crypts) of
Lieberkuhn which are lined with columnar epithelium.

There remain the pancreas and liver. Of these the *pancreas*
is a racemose gland resembling the salivary glands in structure,
the lobules containing the dilated ends of the branched tubes.
The *liver* has a different structure, each lobule containing a
reticulum of bile canaliculi. Each lobule has, moreover, a double
blood-supply. It receives arterial blood from the aorta by the
hepatic artery : it receives venous blood from the alimentary
canal by the portal vein. The blood is carried off by the
hepatic veins to the post-caval. Traced backwards each hepatic
vein is found to result from the union of a great number of
factors, the small hepatic veins, and these again receive yet
smaller factors, the sub-lobular veins. Around these sub-lobular

veins the lobules cluster, being aggregated round, and, to some extent, supported on minute intra-lobular venules. The lobules are divided from each other by connective tissue, in which there run: (1) the inter-lobular venules, which are the ultimate branches of the portal system; and (2) the inter-lobular arterioles, the ultimate branches of the hepatic artery. Thus each lobule has a double blood-supply, the blood from each source being collected by the intra-lobular venule. In addition to the blood-capillaries within the lobule there are also bile canaliculi. The hepatic cells, of which the lobule is composed, are polygonal in form, and are so arranged as to leave minute channels (the bile-canaliculi) between them. At the margins of the lobules these canaliculi are connected with minute tubes, the inter-lobular bile ducts, which form networks in the inter-lobular connective tissue, and eventually, by continued fusion one with another, pass into the larger bile ducts, by which the product may either be delivered directly to the alimentary canal, or may pass into the storage reservoir, the gall-bladder. Thus the duct of this great gland does not break up into a number of blind and dilated tubes, but into a network of canaliculi. It is not a racemose, but a reticulated gland.

The Gland Products.—The product of the salivary glands is saliva. It is a thin, watery, slightly alkaline fluid, which is seen under the microscope to contain, besides flattened epithelial cells from the mucous membrane of the mouth, a number of smaller rounded corpuscles which may show amœboid movements. The salivary fluid is especially secreted by the parotid and the sub-maxillary, the sub-lingual secreting a mucous fluid. An important constituent of the salivary fluid is a ferment called *ptyalin*. It is one of those curious chemical bodies, a very small quantity of which may induce extensive chemical changes in other substances. The chemical change induced by ptyalin is the conversion of starch $((C_6H_{10}O_5)_n)$ first into its isomer dextrin (a body of similar chemical composition, but with widely different properties), and then into sugar (glucose, $C_6H_{12}O_6 + H_2O$, or in some cases maltose, $C_{12}H_{22}O_{11} + H_2O$). When we learn or remember that starch mucilage will not diffuse through a membranous partition, while with a **solution** of sugar this diffusion (or dialysis) takes place with extreme readiness, we shall see the physiological importance of this change set up by ptyalin.

A little thin starch mucilage may be prepared by rubbing a few grains of starch into a paste with cold water, and then boiling with more water. In such a solution the presence of the starch **may be** shown by the deep blue coloration with a **solution of iodine.** If now a little saliva be added to some of **the mucilage in a test** tube, and the mixture be kept for some **time at a temperature of** about 38°C., **the** starch will be converted into sugar. Iodine will no longer give a blue coloration. At an earlier stage it may give a violet coloration, due to the **mixture of the** starch-blue with a claret colour, due to the action of iodine on dextrin. The presence of the sugar may be **shown** by adding a few drops of a solution of copper sulphate, and then a little potassic hydrate, when, on boiling, an orange-red pre**cipitate of cuprous oxide** will be formed. If, **however,** the solution containing starch and saliva be acidified **so as to** contain about ·1 per cent. HCl, the starch will remain unchanged. Acid **prevents** the action **of the** ferment if it does not destroy it.

Thus we see that **saliva** contains ptyalin and mucin, and the question arises, how are **these** secretions formed in the salivary and mucous glands? The gland-cells are bathed on the one hand in the plasma of the blood, and **on** the other hand their secretion is **poured** forth into the lumen of the tube which leads to the duct. **Is the secreted ptyalin or mucin** simply extracted **from the plasma, or is it manufactured out of** the materials of **the** plasma? **The latter seems** the more probable **view.** They seem to be **formed by a** process of chemical change or metabolism in the gland-cell. Moreover, it **is** probable that they are **not** formed directly, but **that,** during the **resting** condition, the protoplasm of the cell, **in and** through its special vitality, elaborates the plasma into an intermediate substance, from which the mucin or ptyalin will be elaborated in the active stage. We may call the special intermediate substance mother of mucin, or mother of ptyalin ; but to **such** intermediate substances in general the term *mesostates* **has been** given.

There are two kinds of metabolism. In one the new molecules formed are more complex and more unstable than those out of which they are formed. This is termed *anabolism*, and the intermediate substances which may be formed during the process are called *anastates*. The process involves the storing up of energy. In the other the new molecules are less complex and less unstable than those out of which they are elaborated. It is called *katabolism*, the intermediates being termed *katastates*. The process involves the setting free of energy in the form of heat or visible motion. It is under this latter category that the processes of secretion fall. They are katabolic processes, and the mother of mucin and mother of ptyalin are katastates. On the other hand the conversion of dextrin into glucose is an anabolic process, the molecule of sugar being more complex than that of dextrin, while maltose, as an intermediate substance, is an anastate.

The product of the cardiac and pyloric glands of the stomach is *gastric juice*. The fluid is strongly acid (HCl), and contains the ferment *pepsin*, with small quantities of a *rennet* ferment. The ferments may be extracted from the minced mucous membrane of the stomach with four or five times its bulk of glycerine, which should be allowed to act for some days. The fluid may then be strained off through muslin, and added to from ten to twenty times its volume of dilute (·2 per cent.) hydrochloric acid. Such artificial gastric juice will be found to have no action on starch.

If, however, some chopped white of egg or meat be placed in the solution, and be kept in the warm (38°C.), the white of egg will be dissolved, and of the meat only a pulp of connective tissue and fatty matter will remain. And if some of the solution be carefully neutralised and added to fresh milk, the milk will clot, owing to the coagulation by the rennet ferment of casein, which will dissolve through the action of the pepsin if the solution be again acidified and placed in the warm.

Whereas saliva, therefore, has the property of converting starch into sugar, gastric juice has the property of converting proteids (*e.g.* albumin, fibrin, myosin, casein), and such nitrogenous materials as gelatin and chondrin, into bodies called

peptones, which are not only very soluble, but very readily
diffuse through such a moist membrane as lines the stomach and
intestinal canal. The process is one of anabolism or building up
of more complex molecules out of less complex, and in the course
of the conversion of proteid into peptone there are probably
several mesostates, quite possibly of both orders, katastates and
anastates. That is to say, the original proteid molecules may
very possibly be first broken down into less and less complex
molecules, and then built up anew into more and more complex
molecules.

Of the constituents of gastric juice the hydrochloric acid is
probably secreted by the parietal cells of the cardiac glands,
which are said to select sodium phosphate and sodium chloride
from the blood by the interaction of which the HCl results.
It is questionable, however, whether the change is so simple as
this, and, in any case, the reaction is brought about by the
inherent vitality of the cells. The pepsin is probably the product of katabolism (with pepsinogen, mother of pepsin, as a
katastate) in the lining cells of both cardiac and pyloric glands.

The product of the small intestinal glands is termed *succus
entericus*, which is said to be a yellowish alkaline fluid. It is
also said that the secretion of Brunner's glands converts proteids
into peptones, and that of Lieberkuhn's glands is said, like
pancreatic juice, to act upon all the constituents of food.

The product of the pancreas is called *pancreatic juice*, a colourless, transparent, viscid, and alkaline fluid. A pancreatic extract
may be made by soaking the gland in water, mincing, and treat-
ing with glycerine. The strained extract should be mixed with
ten to twenty volumes of a dilute (1·5 per cent.) solution of sodium
carbonate and filtered. Such artificial pancreatic juice, best
prepared from the pancreas of an animal killed soon after a meal,
will be found (1) to act on starch in a similar way to saliva,
converting it into sugar. This it does by means of a ferment
similar to, or indistinguishable from, ptyalin. The extract
will also be found (2) to act on proteids in a similar way to
gastric juice, converting them into peptones, or even altering
them still further. This it does through the special ferment

trypsin. It will also (3) curdle milk through a rennet ferment. And lastly (4) it has an important **action** on fatty and oily substances. Not only does it (*a.*) convert them into an emulsion, that is, cause them to be suspended in **very** minute globules, like the butter globules in milk, as may be seen by shaking up some oil with the extract, but it also, (*b.*) not improbably through the action of a fourth ferment (*steapsin*), splits up the fats into their fatty acids and glycerine. These fatty acids, with an alkaline carbonate, of which there is about ·75 per cent. in pancreatic juice, form soluble soaps, and these soaps further aid in emulsifying the fats. Pancreatic juice, therefore, (1) converts starch into sugar; (2) converts proteids into peptones; (3) coagulates casein from milk; and (4) emulsifies and saponifies fats.

The product, or rather a product, of the liver is bile. This is a viscid, slightly alkaline fluid, green in the rabbit and most herbivorous animals, reddish in the carnivora. It contains (1) certain organic bile salts (sodium glycocholate and taurocholate); (2) certain organic colouring matters; and (3) fatty matter and cholesterin (an univalent alcohol), together with **mucous and** inorganic salts. Its main functions in digestion would seem to be (1) emulsifying fatty matters; **and** (2) precipitating the gastric peptones, together with pepsin and bile acids. In addition to these functions it assists digestion by moistening the mucous membrane and **so facilitating** absorption, and by acting as a natural purgative.

Like the other secretions before considered, the bile is not derived as such from the blood, but is elaborated out of **the** materials of the blood by a process of katabolism, and **it is** possible that there may be hitherto undiscovered **katastates.** The process of secretion goes on most rapidly during digestion; **but it does** not cease at other times. At such times the bile, instead of passing down the common bile-duct into the duodenum, regurgitates through the cystic duct into the **gall-bladder, where** it is stored up for future use.

Having thus considered the general structure of the glands, the nature of their products, and the part they play in digestion, we may now pass on to consider the process of digestion as a whole, and the manner in which the elaborated products are absorbed.

Digestion and Absorption.—The organic food-stuffs, exclusive, that is, of water and saline matters, are of two kinds, *nitrogenous*, such as the proteids and gelatins ; and *non-nitrogenous*, such as the carbo-hydrates, starch and sugar, on the one hand, in which the oxygen and hydrogen are in due proportion to form water, and on the other hand oil and fat, in which there is a relative deficiency of oxygen. These nitrogenous and non-nitrogenous food-stuffs have to be consumed in such proportion as to make good the material lost to the body in nitrogenous excreta, in water, and in carbonic acid gas.

In the mammalia they are prepared for further digestion by mastication. In the mouth the food is mixed with saliva, which has no action on the fats or proteids, but converts the starch into glucose. Besides having this direct chemical action, the saliva serves to moisten the food and make it more readily swallowed. This process of deglutition is a somewhat complicated one, bringing many muscles into play. The morsel of food is pushed back into the pharynx, where it is grasped and pushed onwards, the circular muscles of the part of the tube in front of it relaxing while the muscles behind it are successively contracted. Thus it passes down the œsophagus into the stomach.

There it is for a while imprisoned by the firm closure of the cardiac and pyloric apertures, which only momentarily open for the admission of food or the emission of its contents. By the action of the strong gastric muscles the food is rolled about and thoroughly mixed with gastric juice, the acidity of which at once stops the further action of ptyalin on starch. Gastric digestion thus succeeds salivary digestion. The initiation of the conversion of starch into glucose ceases, but the proteids are converted into peptones, and the fats are, by the breaking down of their proteid framework, set free, and incorporated with the food, though they are not further acted upon. Thus the food is converted into *chyme,* a fluid of the consistency of pea-soup, which is allowed to pass through the pylorus into the duodenum.   At the same time, however, the process of absorption begins.   Probably some little soluble matter is absorbed in the mouth.   But in the stomach a little of the peptones there prepared, together with the glucose prepared by the saliva, and other diffusible fluids, may pass by osmosis through the membranous walls of the stomach, and so entering the blood of the capillaries, are carried to the portal vein, and thus to the liver.

When it passes into the duodenum the acid chyme is subject to the action of the alkaline fluids, bile, pancreatic juice, and succus entericus.   The fats are emulsified and partly saponified. Under the influence of the pancreatic fluid the conversion of starch into sugar is renewed.   The bile causes the precipitation of the peptones, while the gastric pepsin is also precipitated and rendered inert, and thus the field is left open for the digestive influence of the trypsin of the pancreatic fluid.   Thus, as it passes onwards through the small intestine by the wave-like peristaltic contractions of its muscular walls, the three chief constituents of the food, the proteid, the starch, and the fat, are so acted upon as to be converted, the first two into diffusible fluids, the fat partly into a soluble soap, but chiefly into an emulsion.

Absorption of these materials, which are still, it must be remembered, strictly speaking outside the body, is in the small intestine effected by the *villi*.   Each of these minute processes has in its centre the termination of a lymphatic canal, which is here called a *lacteal* ; and surrounding this is a close plexus of blood-capillaries.   The diffusible fluids pass through the epithelial layer, and are caught up by the blood-stream in the capillaries and hurried away to the liver.   The emulsified fats, and such soluble matters as have escaped absorption into the blood-stream, pass into the lacteal rootlet, and so into the lymphatic system.   This passage is aided by the alternate shortening and elongation of the villus.   By the former process the contained fluid is forced through the lacteals, whence its return is prevented by valves : by the latter the lacteal rootlet in the villus tends to suck in the surrounding material.   But still the question *how* the fats pass into the lacteals, through the cells or between the cells, is not well understood.   It certainly is largely through the agency of the cells themselves that the process is effected; and it may be through the intervention of certain amœboid cells, which lie at the bases of the columnar epithelial cells, and send up long processes between the cells. In any case it is eminently probable that neither the absorption of emulsified fats, nor the diffusion of soluble peptones and sugar is merely a mechanical process. Both are probably processes effected through the vitality of the tissues; and there are good reasons for believing that the epithelial cells themselves exercise a sort of digestive property upon the ingested fat.

The absorption of fatty matters is probably completed in the small intestine where alone the villi are developed; but the absorption of diffusible matters may probably continue in the large intestine where water is largely absorbed. And one may well suppose that in the large cæcum of the rabbit some further process of digestion, not at present well understood, is effected. As they pass onwards the contents of the large intestine become more solid and acquire the character of fæces, and are finally expelled from the body. Notwithstanding the very large quantity of bile poured forth into the small intestine, the amount of bile salts to be found in the fæces is quite insignificant. Whence it has been inferred that the biliary salts are decomposed and the taurine and glycine thus set free are reabsorbed.

Before leaving the subject of absorption attention may be drawn to the fact that, in the lower vertebrates, the epithelial cells would seem to be capable of pushing forth amœboid absorptive processes, and that leucocytes have been shown in some cases to migrate through the epithelium into the lumen of the alimentary canal, and thus come in contact with the prepared food. They have therefore been called *phagocytes*. Other phagocytes may take up food particles which have passed through the epithelium into the connective tissue. Phagocytes are also said to absorb, and thus render innocuous, noxious substances, or portions of tissues which have undergone retrogressive metamorphosis within the body.

Preliminary Metabolism of Absorbed Products.—We have thus seen how the products of digestion are absorbed, partly into the

blood and partly into the lymphatic **system.** We have now to
note that these products do not pass unaltered into the general
system of the circulation. For the blood which receives the
diffusible products has to pass through the liver before it reaches
the heart; and the *chyle*, as the milky fluid in the lacteals is
termed, has to pass through the lymphatic glands before it
passes by the thoracic duct into the blood-stream.

The exact changes which **go** on in the liver have still to be
fully worked out. **It is** certainly the seat of most important
chemical changes, and the high temperature of the blood which
leaves this organ seems to indicate that katabolic changes are in
excess. Of these chemical changes perhaps the most important
is the metabolism of nitrogenous matters. From this metabolism, effected by the vital activity of the hepatic cells, three
products result: (1) a non-nitrogenous starchy substance,
readily convertible into glucose and termed *glycogen*; (2) **the**
slightly nitrogenous *bile*; and (3) the highly nitrogenous *urea*,
or a mesostate thereof. Of these the bile, as we have already
seen, is poured out into the alimentary canal; the urea passes
away in the blood to be eliminated by the kidneys; the destina-
tion of the glycogen is still a matter of uncertainty. By some it
is held to be at once converted into glucose, which, by oxidation in the lungs, is a source of heat to the body. By others it
is regarded as a store of carbo-hydrate ready to be converted
into sugar and drawn upon by the organism when need arises.
By others it is regarded as a mesostate in the formation **of fat.**

In any case we see that the series of metabolic changes **which**
the absorbed material has to undergo are initiated in the liver
before the portal blood reaches the heart. So, too, do important
metabolic changes take place in the chyle before it passes into
the blood-stream by the thoracic duct.

The lymph may be regarded **as** to a large extent the overflow
of the blood, the plasma of which has exuded into the tissues,
and has been drained off by the lymphatic vessels. In the
region of the alimentary canal, the lymph, during digestion,
becomes chyle, a milky, slightly alkaline fluid, the whiteness
and opacity of which is due to the emulsion absorbed by the

lacteal rootlets.   In the course of the lacteals there are numerous lymphatic glands, many of which lie in the mesentery.
Glandular tissue of similar character is developed in the walls
of the intestines, and forms in parts oval white patches (Peyer's),
and is well seen in the appendix and sacculus rotundus.   In
them there are rounded masses of lymphatic tissue (*follicles*), or
elongated strands (*medullary cords*) of similar tissue, crowded
with leucocytes or pale blood corpuscles.   Around these
follicles and cords there are lymph-spaces and lymph-channels.
The metabolism that goes on in these lymphatic glands is
not well understood.   But there seems no doubt that the
chyle from the thoracic duct (that is, after it has passed
through the glands), readily coagulates, owing to the formation
of fibrin, contains an immense number of white corpuscles
(leucocytes), and some red corpuscles, which may not improbably have escaped from the vascular system in the spleen or
elsewhere.   The chyle from the lacteals before passing through
the lymphatic glands does not readily coagulate, has much
fewer pale corpuscles, and no red corpuscles.

Further Metabolism.—In the liver and in the lymphatic glands,
however, the metabolism of the products of digestion is scarcely
more than begun.   By the general blood-system it is carried in
a more or less altered condition throughout the body, and has
to make good the tissue wasted by its vital activity.   Of the
complex metabolism that takes place in the tissues we at
present know but little.   We may, however, indicate in a few
words the direction in which the experiment, observation, and
inference of modern physiology would seem to be tending.

We have already seen that in the secreting organs the protoplasm of the secreting cells elaborates from the plasma, during
the resting stage, a material, the mother of ferment, from which
during the period of activity the ferment itself is produced by
further katabolic changes.   In a somewhat similar way it would
seem that, in a muscle, the protoplasm of the muscle fibre is
occupied during the resting stage in elaborating from the plasma,
with the aid of the oxygen brought by the red corpuscles, a
material called *inogen*, which it incorporates with itself.   During

the period of activity this inogen **undergoes** katabolism with the formation of carbonic acid and sarcolactic acid, and this process is accompanied by the shortening of the muscle **fibre.   In** any case it seems clear that the **katabolism of the muscle fibre** at the moment of contraction, **by which carbonic acid and sarco**lactic acid are produced, is not the result **of the direct** oxidation of the substance of the fibre at the expense of the oxygen of the blood.   A muscle containing no free oxygen will contract in an atmosphere perfectly free of oxygen.   The change is a true katabolism or breaking down of more complex into less complex molecules.   **The** katabolism of the inogen may be likened to an explosion, where the explosive material, **more or less** suddenly, by **a** rearrangement of molecular groupings, falls from a state of unstable to one of stable equilibrium.

Of the metabolism of **nervous tissue we know still less than of that which** occurs in muscular **tissue.   But it is probable** that here too, by the vital activity of the protoplasmic element in each cell, **there** is produced a mesostate which, at the moment of activity, passes **by katabolism into waste products, so called,** of less complex composition.

A metabolism similar in principle goes **on in fat cells.   Here** the inherent **vital activity of** the protoplasm **of certain** connective tissue cells, causes the production **of** this peculiar storage tissue.   It is exceedingly improbable that there is anything like **a** direct deposition **in** the fat cells **of fat** absorbed into the lacteals.   Each individual **has, by its** vital activity, to **build up its own** tissues for itself.

So far, however, **we** have assumed in each **cell a protoplasmic** element which initiates the temporary or more enduring **storage** of a mesostate.   How **is this** protoplasm itself elaborated ?   At present we **do not know.   We may well suppose that it results** from a process **of** metabolism, **and is probably preceded by one** or more anabolic mesostates.   **We know that the protoplasmic** pale corpuscles **or leucocytes are largely produced in the lym**phatic glands.   But how ?

In any case **we** may fairly regard the **complex** chemical changes which **go on within the body as a series of** ascending and descending steps, which together form a pyramid. The summit of that pyramid is protoplasm, which the student must not regard as a body of unchanging chemical composition, but rather as a group of bodies, each the terminal product of a series of anabolic processes. It is scarcely probable that the protoplasm of muscle, nerve, and glandular tissue is identical. But in each case it is the highest product of a special line of anabolism. And in general we may say that protoplasm is the summit of a step-faced metabolic pyramid. The ascending steps **are anastates;** the descending steps katastates. The ascent of the pyramid involves the absorption of energy or work done on the matter involved. The descent of the pyramid involves the setting free of energy, in the form of visible motion or of **heat,** or work done by the matter **involved.** But the pyramid

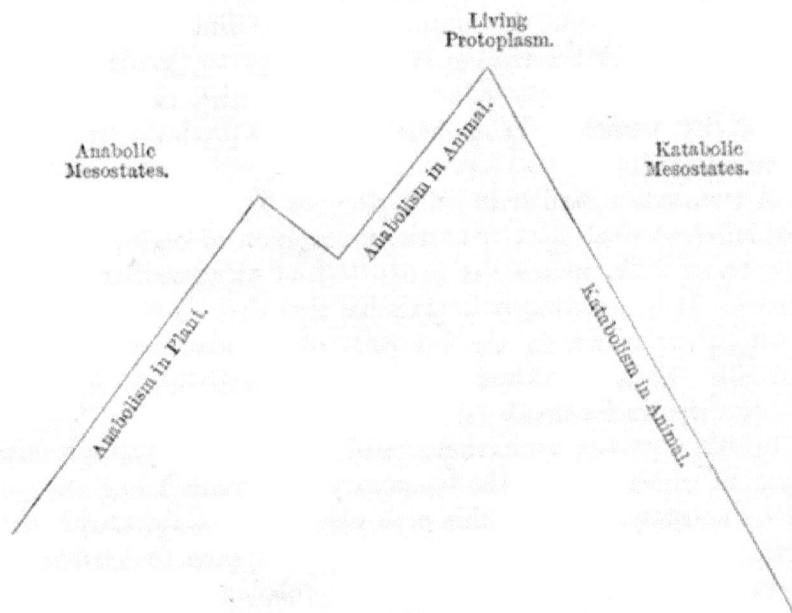

FIG. 58.—METABOLIC LIFE-PYRAMID.

is not symmetrical. Its katastatic **side,** with descending steps, is longer than its anastatic side **with** ascending steps; for the food starts at a higher level of energy and complexity than that

of the **ultimate products** of waste.   The katastatic surplus marks
the energetic capital with which the animal is credited.   In a
complete life-pyramid, however, this will not be so; for the
lower anastatic stages by which the **food** is raised to its high
level of energy and complexity falls within the life-work of the
plant.   The accompanying figure (58) will roughly symbolise
this conception, the notch between plant protoplasm and animal
protoplasm being intended to convey the fact that in the animal
the food prepared by the plant is partly broken down into lower
products before it is finally built up into higher.

The following nitrogenous substances are among those found
in animal tissue.   They are arranged in order of chemical
degradation :—

Cerebrin,	$C_{17}H_{33}NO_3$.
Tyrosin,	$C_9H_{11}NO_3$.
Hippuric Acid,	$C_9H_9NO_3$.
Leucin,	$C_6H_{13}NO_2$
Guanin,	$C_5H_5N_5O$.
Hypoxanthin,	$C_5H_4N_4O$.
Kreatin,	$C_4H_9N_3O_2$
Kreatinin,	$C_4H_7N_3O$.
Xanthin,	$C_5H_4N_4O_2$
Uric Acid,	$C_5H_4N_4O_3$.
Allantoin,	$C_4H_6N_4O_3$.
Urea,	CH_4N_2O.

13

CHAPTER X.

THE HEART AND CIRCULATION.

Preliminary Considerations.—The main organ of circulation is the heart ; the circulating fluids are the blood and the lymph ; the channels of circulation are the blood-vessels and lymphatics. The principal vessels which convey blood from the heart are the arteries ; when these are much reduced in size, through oft-repeated branching, they are called arterioles ; the arterioles end in an anastomosing network of very fine tubes, which ramify throughout the organs of the body—these are the capillaries. The blood returning from the capillaries collects into venules ; by oft-repeated unions with each other the venules form veins, which convey the blood back to the heart from which it started.

The tissues may be said to be irrigated by the blood which flows within the capillaries, through the delicate walls of which some of the plasma of the blood exudes and collects in the crevices and in the spaces between the connective tissue strands. The fluid that thus exudes, which contains less albumen than blood plasma and some products of the waste of the tissues, forms the lymph, and the crevices are the so-called lymph-rootlets. From the lymph-rootlets the lymph passes into fine tubes with endothelial (epithelioid) lining called lymph capillaries ; the larger vessels which receive the lymph from these capillaries are the lymphatics. In the course of the lymphatics are the lymphatic glands. The lymphatics in the lower vertebrates are not infrequently in connection with large lymph-spaces,

194

such as those beneath the skin and that at the back of the pleuro-peritoneal cavity in the frog—the latter communicati· ; by means of stomata with the cœlom.

In the frog, but not in the fowl or rabbit, there are four lymph-hearts, whose function it is to pump the lymph back from the lymphatic system into the veins. The anterior pair lie close to and beneath the transverse processes of the third vertebra. They communicate with the innominate veins. The posterior pair lie one on each side of the urostyle about one-third of an inch from its termination. They communicate with the renal-portal veins. In each case valves prevent the passage of blood from the veins into the rhythmically contractile lymph-hearts. In the bird and the mammal the lymph returns to the veins through the thoracic ducts or duct without the intervention of lymph-hearts.

Thus the course of the blood is, disregarding the lymph overflow, heart, arteries, arterioles, capillaries, venules, veins, heart.

From what has been said in the last chapter, however, it will be understood that the blood is no mere inert fluid which returns to the heart in much the same state in which it left it, but is throughout its course undergoing continuous meta-bolism. Nor are the blood-vessels simple inelastic tubes. Each artery and vein has, besides its smooth lining of endothelial (epithelioid) cells, a triple coat of elastic tissue, in the mid layer of which unstriped muscular fibres may be developed. In the arterioles these muscular fibres are numerous. In the venules the elastic coat is much reduced ; and in the capillaries both it and the muscular coat are absent. So that we may say that the arteries are very thick-walled and highly elastic ; the arterioles are moderately thick-walled, elastic, and contractile ; the capil-laries have only an epithelioid (endothelial) wall ; the venules are thin-walled and slightly elastic ; the veins are relatively thin-walled and slightly elastic. The veins (of the rabbit, *e.g.*) are, moreover, in some parts of the body, where they are liable to pressure, provided with flap-like valves which offer no resistance to blood flowing towards the heart, but close and block the way to blood attempting to flow in the opposite direction.

Into this system of **tubes the** blood is forced at each contraction (systole) of the muscular walls of the ventricles of the heart. The arteries are swollen with blood, the heart-pressure being to some extent stored up in the stretched fibres of the **elastic coats of** these vessels. After the systole of the ventricles is over, the blood is passed on through the arteries and arterioles **to the** capillaries. And now the muscular contractility of **the walls** of the arterioles comes into play. Nerves from the sympathetic system are distributed to their muscular walls. Suppose, then, that an organ is quiescent and in need of but little nourishment. Through the influence of the nerve-fibres the muscular walls of the arterioles contract ; less blood is supplied, and **the** blood-pressure in the capillaries falls. But now suppose that the organ is active and in need of much blood and nourish- **ment.** The walls of the arterioles are caused to relax ; a full **tide of blood flows** to the part ; blood-pressure **increases** in the capillaries ; **there is** more exusion of plasma through the epithelioid walls, and nutrition increases.

When the blood reaches the capillaries, unless the arterioles be very widely distended, the wave-like flow, noticeable when a great artery is cut, has ceased and given place to a continuous flow. The elasticity of the walls of the elongated tubes, and the distribution of the pressure over a large area of capillaries, has effected this. **The** pulse that we feel in an artery of the human wrist is the shock of systole propagated through the vessel, and must not be confounded with the onward flow of the blood.

It only remains to be noticed, in these preliminary remarks, that the partial vacuum in the thorax of the rabbit, and sundry other minor causes, may slightly add to or slightly impede the flow of the blood.

The Cod Fish.—The heart of the cod (Fig. 59) consists of a thin-walled *sinus venosus* (*s. v.*), which receives blood from the *hepatic veins* (*he. v.*) and the *ductus Cuvieri* (*d. c.*), of which only the right vein receives an *inferior jugular* factor (*i. j. v.*). Below and slightly in front of the sinus venosus is the *auricle* (*au.*), through which all the blood passes on its way to the thick-walled *ventricle* (*v.*). From the ventricle it passes into the thick-walled *bulbus arteriosus* (*b. a.*), and thence to the *ventral aorta* (*v. ao.*). Note that there is one undivided auricle and one undivided ventricle.

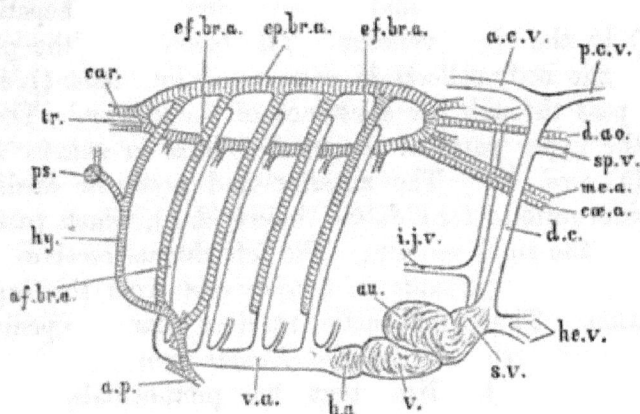

FIG. 59.—DIAGRAM OF HEART AND GREAT VESSELS OF COD-FISH.

a. c. v. Anterior cardinal vein. *a. p.* Artery supplying pelvic fins. *af. br. a.* Afferent branchial arteries. *au,* Atrium or auricle. *b. a.* Bulbus arteriosus. *car.* Carotid artery. *cœ. a.* Cœliac artery. *d. ao.* Dorsal aorta. *d. c.* Ductus Cuvieri. *ef. br. a.* Efferent branchial arteries. *ep. br. a.* Epibranchial artery. *he. v.* Hepatic vein. *hy.* Hyoidean artery. *i. j. v.* Inferior jugular vein. *me. a.* Mesenteric artery. *p. c. v.* Posterior cardinal vein. *ps.* Pseudobranchia on inner surface of operculum. *s. a.* Subclavian artery. *s. v.* Sinus venosus. *sp. v.* Spermatic vein. *tr.* Transverse vessel completing circulus cephalicus. *v. a.* Ventral aorta. *v.* Ventricle.

The arrangement of the great vessels is shown diagrammatically in Fig. 59. The ventral aorta (*v. ao.*) gives off on either side four *afferent branchial arteries* (*af. br. a.*) to the four complete gills. From the gills there pass on each side four *efferent branchial arteries* (*ef. br. a.*), which open into a vessel (*epibranchial artery, ep. br. a.*), running along the dorsal ends of the gill arches. Traced forwards the epibranchials give off the *carotid arteries,* and curving round unite to complete the *circulus cephalicus.* Traced backwards the epibranchials unite posteriorly to form the *dorsal aorta* (*d. ao.*), which runs backwards beneath the spine, and becomes the caudal artery. But before thus uniting, each epibranchial gives off a *subclavian artery* (*s. a.*) to the pectoral fin, while the right epibranchial gives off also a *cœliac* (*cœ. a.*) and a *mesenteric artery* (*me. a.*) to supply the stomach and intestines.

From the stomach and intestines the blood is delivered by a *portal vein* to the liver, and thence passes by the hepatic veins (*he. v.*) to the sinus venosus. The blood from the posterior parts of the body collects in *posterior cardinal veins* (*p. card. v.*), which pass through the substance of the kidney.[1] The blood from the upper parts of the head collects in *anterior cardinal veins* (*a. card. v.*). The anterior and posterior cardinals of each side unite to form *ductus Cuvieri* (*d. c.*), which pass downwards to the sinus venosus. The left ductus receives near its dorsal aspect a *spermatic* or *ovarian vein* from the organs of generation. The right ductus receives near its opening into the sinus venosus an *inferior jugular vein* from the lower parts of the head. Note that the epibranchials, by curving round and uniting anteriorly, form a closed circle (*circulus cephalicus*), which lies beneath the skull, and may be seen by removing the lower jaw and the mucous membrane of the roof of the mouth. In the figure the afferent and efferent branchial arteries and the ductus Cuvieri are placed side by side. They are connected by the capillaries of the gills, and curve upwards from the ventral aspect, where the heart lies, to the dorsal aspect, along which run the epibranchials and cardinals. The vessels containing arterial blood are shaded.

The Frog.—The heart of the frog consists of four divisions :—

1. *Sinus venosus:* dorsal, thin-walled, receiving venous blood by the precavals and postcaval.

2. *Atrium :* anterior, thin-walled, divided into two chambers :

(*a*) right auricle, receiving venous blood from the sinus venosus;

(*b*) left auricle, receiving arterial blood from the pulmonary vein.

[1] That of the left side is aborted for some distance in the kidney.

3. *Ventricle:* posterior, muscular, undivided, receiving venous and arterial blood *simultaneously* from the right and left auricles.

4. *Truncus arteriosus:* ventral, anterior, and to the right; thick-walled and muscular, receiving *successively* venous and arterial blood from the ventricle.

To ensure the passage of blood in the right direction there are the following valves :—

(a) *Sinu-auricular valve :* between the sinus venosus and right auricle.

(b) *Auriculo-ventricular valve :* between the atrium and the ventricle—prevented from flapping back into the auricles by fine tendinous cords.

(c) *Semi-lunar valves* in the truncus arteriosus (two sets).

(d) A median *longitudinal valve*, attached to the dorsal wall of the truncus arteriosus, and partially dividing it into two passages.

The arrangement of the great vessels which enter and leave the heart will be seen from the diagram Fig. 60, which shows the ventral aspect. The *truncus arteriosus (tr. a.)* bifurcates, and each branch gives rise to three arteries. The *carotid (ca.)* anteriorly, the *systemic aorta (sy. ao.)* in the middle, and the *pulmo-cutaneous (p. cu.)* posteriorly. The carotid supplies the head and brain. On it is developed a *carotid gland (ca. gl.)*, a muscular vesicle with septa in its interior. Close to the carotid gland the artery gives off a *lingual* branch *(li.)* to the tongue. The systemic aorta of each side curves round to meet its fellow in the middle line. It gives off the *vertebral (ve.)* and the *subclavian arteries (s. cl. a.)* to the vertebral column and fore-limb. At or near the point where the two systemic arches meet to produce the dorsal aorta which runs down the roof of the pleuro-peritoneal cavity, a *cœliaco-mesenteric (c. m.)* is given off to supply the stomach and intestines. It soon branches into a *cœliac (c.)* to stomach and liver, and a *mesenteric (m.)*. A small *inferior mesenteric artery* supplies the base of the large intestine.

The blood is returned from the stomach and intestines by the *portal vein (por.)*, which breaks up into a capillary plexus in the liver. It is formed by the union of two factors; the *gastric* (*g.*) from the stomach, and the *lieno-intestinal* (*l. i.*) from the intestines. Blood from the hinder part of the body also reaches the

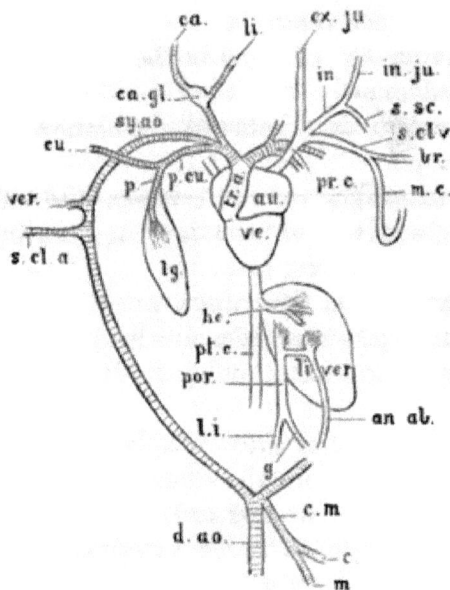

FIG. 60.—HEART AND GREAT VESSELS OF FROG.

an. ab. Anterior abdominal. *au.* Auricle. *br.* Brachial vein. *c.* Cœliac artery. *c. m.* Cœliaco-mesenteric artery. *ca.* Carotid artery. *ca. gl.* Carotid gland. *cu.* Cutaneous artery. *d. ao.* Dorsal aorta. *ex. ju.* External jugular. *g.* Gastric vein. *he.* Hepatic vein. *in.* Innominate vein. *in. ju.* Internal jugular vein. *li.* Lingual artery. *l. i.* Lieno-intestinal vein. *lg.* Lung. *m.* Mesenteric artery. *m. c.* Great cutaneous vein. *p.* Pulmonary artery. *p. cu.* Pulmo-cutaneous artery. *por.* Portal vein. *pr. c.* Precaval vein. *pt. c.* Postcaval vein. *s. cl. a.* Subclavian artery. *s. cl. v.* Subclavian vein. *s. sc.* Subscapular vein. *sy. ao.* Systemic aorta. *tr. a.* Truncus arteriosus. *ve.* Ventricle. *ver.* Vertebral artery.

liver by the *anterior abdominal vein* (*an. ab.*). A connecting branch unites the anterior abdominal and the portal, and a small factor comes from the truncus arteriosus. From the liver the blood passes into the *postcaval vein* (*pt. c.*) by the *hepatic veins* (*he.*).

The postcaval delivers to the sinus venosus blood from the large kidneys; the blood from the head and fore-limbs reaches the sinus venosus by two *precavals* (*pr. c.*), one on either side.

Each is formed by the union of three factors, *external jugular* (*ex. ju.*) from the jaw and tongue, *innominate* (*in.*), and *subclavian* (*s. cl. v.*). The innominate has two factors, *internal jugular* (*in. ju.*) from the brain, etc., and *subscapular* (*s. sc.*) from the region of the shoulder. The subclavian has also two factors, the *brachial* (*br.*) from the fore-arm and manus, and the *great cutaneous* (*m. c.*) chiefly from the skin of the back.

Fig. 61 shows diagrammatically the arrangement of the vessels in the pelvic region, seen from the ventral aspect. The *dorsal aorta* (*d. ao.*) gives off vessels to the kidneys (one side shown), and then bifurcates into two *iliac* arteries (*il.*), which, after giving off a *hypogastric* branch (*hy.*) to the bladder and abdominal walls, continues on to supply the leg. From the leg the blood is returned by two veins, the *femoral* running down the front of the thigh, and the *sciatic* running alongside the artery in the back of the thigh. These two are united by an anastomosing branch (*a. b.*). The femoral then branches into two, the *pelvic vein* (*pe.*), and the *renal portal* (*r. p.*), into which the sciatic also falls. The pelvic veins, on either side, pass towards the ventral aspect, and unite to form the anterior abdominal vein, which, as we have seen, passes to the liver. The renal-portal passes to the outer edge of the

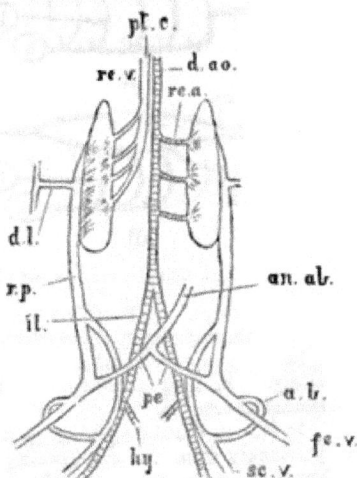

FIG. 61.—RENAL-PORTAL AND PELVIC VESSELS : FROG.

a. b. Anastomosing branch between sciatic and femoral veins. *an. ab.* Anterior abdominal vein. *d. ao.* Dorsal aorta. *d. l.* Dorso-lumbar vein. *fe. v.* Femoral vein. *hy.* Hypogastric artery. *il.* Iliac artery. *pe.* Pelvic vein. *pt. c.* Postcaval vein. *re. a.* Renal artery. *re. v.* Renal vein. *r. p.* Renal-portal vein. *sc. v.* Sciatic vein.

kidney, and breaks up into a capillary plexus in that organ. It receives a *dorso-lumbar vein* from the spinal canal and the body-wall.

Let us now see, with the aid of Fig. 62, I. II. III., how the three great arterial trunks are developed, and how a passage is made from the fish condition to the frog condition through the tadpole condition.   Fig. 62, I., shows the four branchial arches

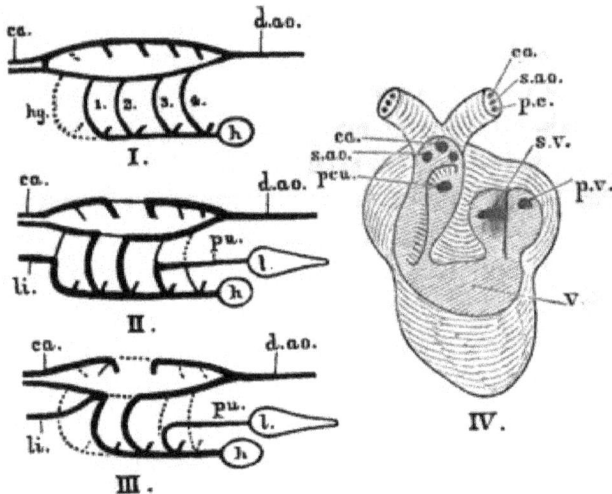

Fig. 62.—Development of Aortic Arches—Heart (Frog).

i. Fish stage.   ii. Tadpole stage.   iii. Frog stage.   iv. Heart of frog dissected so as to show the internal structure, all detail being omitted.

In i., ii., iii.—*ca.* Carotid artery.   *d. ao.* Dorsal aorta.   *h.* Heart.   *hy.* Hyoidean artery.   *li.* Lingual artery.   *l.* Lung.   *pu.* Pulmonary artery.   1, 2, 3, 4. Branchial arteries.

In iv.—*ca.* Carotid opening.   *p. c.* Pulmo-cutaneous opening.   *p. v.* Opening of pulmonary vein.   *s. ao.* Opening of systemic aorta.   *s. v.* Opening of sinus venosus.   *v.* Cavity of ventricle.

of the fish, represented here as continuous, the afferent and efferent branchial vessels being, for the sake of simplicity, merged into one (*cp.* Fig. 59).   The dotted arch in front is the hyoid arch, not found in the adult fish.   Fig. 62, II., shows the state of things in the tadpole.   The hyoid arch is developed and gives origin to a lingual artery, the dorsal connection with the carotid part of the epibranchial being slender.   The first branchial arch is well developed, and gives rise to the carotid. The second branchial arch is also well developed, and gives rise to the systemic aorta.   A slender dorsal connecting branch represents this part of the epibranchial.   The third branchial arch is well developed below, but gives rise to a large posterior pulmonary branch running to the lung. The dorsal connection with the epibranchial (systemic arch) is slender. The fourth arch is incomplete above, but joins by a slender tube (in the newt if not in the frog) the pulmonary branch, of which pulmonary branch the fourth arch is indeed most probably the true parent, the connection with the third arch being subsequent. Fig. 62, III., shows the state of things in the adult frog. The hyoid arch has atrophied, and the lingual artery derived from it has become connected with the carotid. The slender connection between the carotid and the systemic aorta has atrophied, as has also the dorsal connection of the third arch, and the whole of the fourth. Thus in the adult frog—

The lingual artery is derived from the hyoidean arch. The carotid artery is derived from the first branchial arch. The systemic aorta is derived from the second branchial arch. The pulmo-cutaneous artery is believed to be derived partly from the third branchial arch, and partly from the fourth branchial arch.

We may now consider, with the aid of Fig. 62, IV., how the heart works. The truncus arteriosus, ventricle, and auricles are partially laid open. By the contraction of the atrium venous blood from the sinus venosus (*s. v.*) and right auricle, and arterial blood from the pulmonary vein (*p. v.*) and left auricle, are received simultaneously into the cavity of the ventricle (*v.*). By the contraction of the ventricle venous and arterial blood pass successively into the truncus arteriosus. The longitudinal swing-valve in the truncus arteriosus is so arranged that blood passing to the pulmo-cutaneous orifice (*p. c.*), passes to the left, while blood passing to the carotid (*ca.*) and systemic (*s. ao.*) orifices passes to the right.

Now since—

1. The opening into the truncus arteriosus is well over to the right side of the ventricle ;
2. The blood in the ventricle is arterial to the left and venous to the right, the two not completely mixing owing to the spongy texture ;

3. (*a*) The resistance to the blood-flow is *maximum in the carotid* trunk owing to the small size of the vessels and the carotid gland ;

(*b*) The resistance to the blood-flow is *minimum in the pulmo-cutaneous* trunk owing to the short course and large capillaries of the lungs ;

(*c*) The resistance in the *systemic aorta being medium* ;

therefore it follows that—

1. Venous blood passes first along the line of minimum resistance, to the left of the swing-valve, to the lungs ; **then the** capillaries of the lungs being gorged with blood,

2. Mixed blood flows next along the line of medium **resistance** to the system ;

finally, as these capillaries also become gorged,

3. Arterial blood flows last along the line of maximum resistance to the brain and head.

It will be understood that, as the capillaries of the lungs become gorged, the resistance in the pulmo-cutaneous artery rises so as to exceed that in the systemic aorta ; wherefore the blood passes along that branch. And as it flows along the right of the longitudinal swing-valve, it tends to fold it down to the left over the pulmo-cutaneous orifice. **Then as** the systemic capillaries in turn become gorged, the resistance along that line **becomes** greater than that which was at first the maximum in the carotid ; which thus becomes the channel of the most arterial blood.

The Rabbit.—The student should lay open the thoracic cavity of the rabbit and dissect away the thymous gland and other tissues which hide the origin of the great vessels, so as to display the heart and vessels as shown diagrammatically in Fig. 63.

The heart has two auricles and two ventricles. From the right ventricle (*r. v.*) there arises the *pulmonary artery* (*pul. a.*), the part of the ventricle leading up into it being known as the conus arteriosus. The pulmonary artery passes over towards the left side, and represents the pulmonary arch of that side only. It soon bifurcates to supply the two lungs of which the right is removed in the figure. From the lungs the blood is returned by the *pulmonary veins* (*pul. v.*) to the left auricle (*l. au.*). Thence it passes into the left ventricle (*l. v.*). From

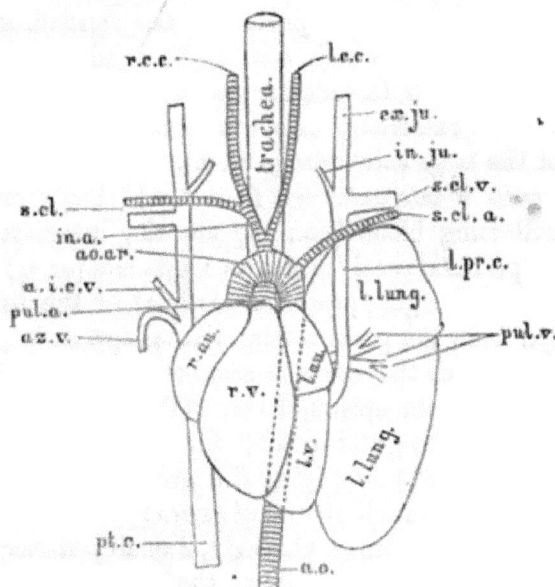

FIG. 63.—HEART AND GREAT VESSELS—RABBIT.

a. i. c. v. Anterior intercostal vein. *ao.* Aorta. *ao. ar.* Aortic arch. *az. v.* Azygos vein. *ex. ju.* External jugular vein. *in. a.* Innominate artery. *in. ju.* Internal jugular vein. *l. au.* Left auricle. *l. c. c.* Left common carotid artery. *l. pr. c.* Left precaval vein. *l. v.* Left ventricle. *pt. c.* Post-caval vein. *pul. a.* Pulmonary artery. *pul. v.* Pulmonary vessels. *r. au.* Right auricle. *r. c. c.* Right common carotid artery. *r. v.* Right ventricle. *s. cl. a.* Subclavian artery. *s. cl. v.* Subclavian vein. *s. cl.* Subclavian artery and vein of the right side.

this chamber it passes by the single aortic arch (*ao. ar.*) (which answers to the left systemic aorta of the frog). At the summit of the arch arises a vessel which at once bifurcates, giving rise to the *left common carotid* (*l. c. c.*) and the *innominate* (*in. a*), which latter soon again bifurcates to form the *right common carotid* (*r. c. c.*) and the *right subclavian artery*. The *left subclavian artery* (*s. cl. a.*) takes its origin from the aortic arch independently.

The aortic arch passes down behind (dorsal to) the heart, and supplies the trunk, alimentary canal, and hind limbs.

The blood thus distributed throughout the body is returned to the right auricle by three vessels, the *postcaval* (*pt. c.*) and the *right and left precavals* (*p. c.*). The postcaval passes through the substance of the diaphragm, and runs forward a little to the right of the median plane. The precavals are constituted by the *external jugular* (*ex. ju.*) (which nearer the head is seen to arise from the union of two factors, an *anterior* and a *posterior facial*, the latter being external),[1] the much smaller *internal jugular* (*in. ju.*), and the large *subclavian* (*s. cl. v.*). Nearer the heart *each* receives an *anterior intercostal vein* (*a. i. c. v.*) (shown only on the left side), collecting blood from the anterior intercostal spaces; and the *right* precaval receives also an *azygos vein* (*az. v.*) (probably answering to the right posterior cardinal of the fish), which collects blood from the posterior intercostal spaces.

Traced backwards the aorta is seen to give off a large *cœliac artery* to the stomach, spleen, liver, gall-bladder, and anterior part of the duodenum; and a little further down a still larger *anterior mesenteric artery* supplying the greater part of the intestines. Next are given off the *renal arteries* to the kidneys, and then, some way further down the body, a small *posterior mesenteric artery* (Fig. 64, *p. m. a.*), supplying the rectum. After giving off the *spermatic* or *ovarian artery* (*sp. a.*) to the testis or ovary, the aorta bifurcates to form the *common iliac arteries* (*com. il. a.*). To follow these arteries the lower part of the innominate bones should be removed, together with the rectum and genital organs. As shown in the figure, they give off *ilio-lumbar* branches (*il. l. a.*) to the abdominal walls of this region, and then bifurcate to form (1) the *external iliacs* (*ex. il. a.*), which, after giving off branches to the bladder (and uterus), become the *femoral arteries* supplying the leg; and (2) the *internal iliacs* (*in. il. a.*), which pass along the dorsal wall of the pelvic cavity. One more artery is to be noted, the *median sacral* (*m. s. a.*), which is drawn aside

---

[1] Occasionally the two external jugulars are united by a connecting trunk in the posterior region of the neck. In the higher mammalia there is but one precaval vein—the right—into which the external jugulars of both sides fall.

to the right so as to be visible near its origin passing behind the
ilio-lumbar vein. It passes backwards to the pelvis and tail,
and is the direct continuation of the dorsal aorta, **answering to**
the caudal artery in the fish.

The veins returning blood from the pelvic region and **hind-**

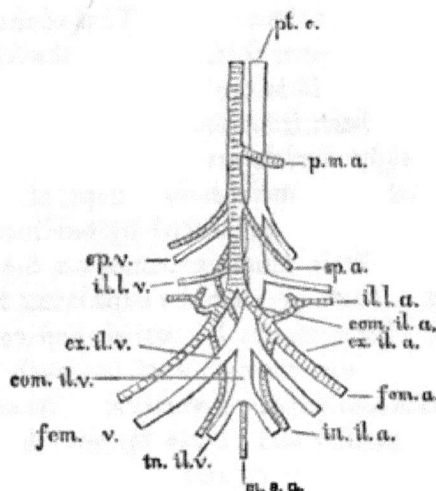

FIG. 64.—PELVIC VESSELS—RABBIT.

com. il. a., *com. il. v.* Common iliac artery and vein. *ex. il. a.*, *ex. il. v.* External iliac artery and vein. *fem. a.*, *fem. v.* Femoral artery and vein. *in. il. a.*, *in. il. v.* Internal iliac artery and vein. *il. l. a.*, *il. l. v.* Ilio-lumbar artery and vein. *m. s. a.* Median sacral artery. *p. m. a.* Posterior mesenteric artery. *pt. c.* Postcaval vein. *sp. a.*, *sp. v.* Spermatic artery and vein.

limb will be understood by a study of a careful dissection, aided by
the figure. The *internal iliacs* unite into a common trunk, which
is successively joined by the *femorals, ilio-lumbars,* and *spermatics.*

The blood is conveyed from the stomach and intestines by a
portal system to the liver. The portal is composed of (1) a large
anterior mesenteric vein, which receives a small *posterior mesenteric*
(from the rectum); and (2) a *lieno-gastric vein* from the stomach
and spleen.

Turn now to the heart itself. The auricles are thin-walled,
especially in the *atrial portion* lying dorsally. Those portions
which project over the base of the ventricles and are visible in

Fig. 63, are strengthened with a network of muscular bands, the *musculi pectinati.* In the septum between the auricles there is an oval area which is thinner than the rest. This is the *fossa ovalis,* and represents a communicating aperture, the *foramen ovale,* of the fœtus. The apertures of the pulmonary veins have no valves. That of the postcaval has a fold (*Eustachian valve*) from its posterior margin towards the septum. That of the right precaval has posteriorly a semi-lunar fold. That of the left precaval has no valve. Just within it is the small aperture of the *coronary rein,* bringing blood back from the tissues of the heart itself.

Between the right auricle and ventricle there is a *tricuspid valve,* consisting of three membranous flaps, of which the irregular posterior edges are connected by tendinous threads, the *chordæ tendineæ,* to little muscular cones on the ventricle, the *papillary muscles.* These prevent the flaps being forced upwards into the auricle. Between the left auricle and ventricle there is a *bicuspid* or *mitral valve,* consisting of two such flaps, also provided with chordæ and papillary muscles. The septum between the ventricles is muscular and convex towards the right ventricle.

Both the pulmonary artery and the aorta are guarded near their point of origin from the ventricle by a set of three *semi-lunar valves,* arranged like watch-pockets. If blood attempts to flow towards the ventricles the watch-pockets fill, and their flaps, meeting in the middle of the channel, completely block the passage.

The Pigeon (Fig. 65).—The heart has two ventricles and two auricles. The pulmonary artery bifurcates at once (*pul. ar.*), its two branches being the right and left pulmonary arches. The pulmonary veins fall into the left auricle in the V-shaped space between the two precavals. The aorta gives off two large branches, almost immediately at its origin. These are the *innominate arteries* (*in. a.*). The continuation of the *aortic arch* is relatively smaller than these two branches. It crosses over the right bronchus, and represents the right trunk in the frog. Each innominate gives off a *common carotid,* and becomes the *subclavian artery* (*s. cl. a.*), which branches into a smaller *brachial*

artery (*br. a.*) to the wing, and a larger *pectoral artery* (*pc. a.*) to
the great muscles of the breast.   The common carotids almost

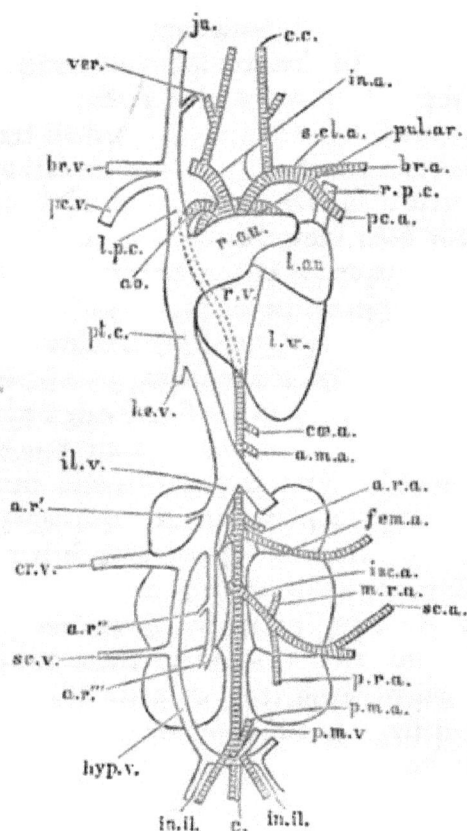

Fig. 65.—Circulation—Pigeon.

a. m. a. Anterior mesenteric artery.   *a. r. a.* Anterior renal artery.   *a. r.',*
a. r.", a. r.''' Afferent renal veins.   *a. o.* Aortic arch.   *br. a., br. v.* Brachial
artery and vein.   *c.* Caudal vessels.   *c. c.* Common carotid.   *cœ. a.* Cœliac
artery.   *cr. v.* Crural vein.   *fem. a.* Femoral artery.   *he. v.* Hepatic vein.
hyp. v. Hypogastric vein.   *il. v.* Iliac vein.   *in. a.* Innominate artery.
in. il. Internal iliac vessels.   *isc. a.* Ischiatic artery.   *ju.* Jugular vein.
l. au. Left auricle.   *l. pc.* Left precaval.   *l. v.* Left ventricle.   *m. r. a.*
Middle renal artery.   *pc. a., pc. v.* Pectoral artery and vein.   *p. m. a.* Posterior mesenteric artery.   *p. m. v.* Posterior mesenteric vein.   *p. r. a.*
Posterior renal artery.   *pt. c.* Postcaval vein.   *pul. ar.* Pulmonary artery.
r. au. Right auricle.   *r. pc.* Right precaval vein.   *r. v.* Right ventricle.
sc. a., sc. v. Sciatic artery and vein.   *s. cl. a.* Subclavian artery.   *ver.* Vertebral vessels.

14

unite in the middle line under the vertebræ. Each gives off a *vertebral artery (ver.)*.

Traced backwards the aorta gives rise to a *cœliac (cœ. a.)* and an *anterior mesenteric (a. m. a.)*, and sends an *anterior renal (a. r. a.)* branch on each side to the front lobes of the kidney. The *femoral arteries (fem. a.)* are not of great size. The *ischiatic arteries (isc. a.)* are larger, and give off **renal** branches to the second and third lobes of the kidney, and eventually become the *sciatic arteries (sc. a.)*. At the posterior end of the body the aorta divides into four small vessels, the *caudal* to the tail (*c.*), the *posterior mesenteric (p. m. a.)* to the rectum and cloaca, and the paired *internal iliacs (in. il.)* to the roof of the pelvis.

Turning now to the veins, the right auricle receives the two *precavals (r. and l. pc.)* and the *postcaval (pt. c.)*. Each precava receives as factors a large *pectoral (pc. v.)*, a *brachial (br. v.)* and a *jugular (ju.)*. Traced forwards up the neck the two jugulars curve round and unite in the middle line beneath the head. The postcaval receives several large *hepatic (he. v.)* factors from the liver. Slightly further back it is formed by the union of the two large *iliac veins (il. v.)*, which receive *afferent renal veins (a.r.', a.r.'', a.r.''')* from the three lobes of the kidney those from the more posterior lobes uniting into a common trunk Following the iliacs down, they are seen to be composed of a large *crural vein (cr. v.)*, and a *hypogastric vein (hyp. v.)* which passes through the substance of the kidney and receives a small *sciatic vein (sc. v.)*. Posteriorly the hypogastrics receive *internal iliac veins* from the roof of the pelvis. **Further** back they are seen to arise from the division of a *posterior mesenteric vein (p. m. v.)*, which receives a small *caudal vein* near its point of division. The posterior mesenteric vein returns blood from the rectum and part of the small intestine. The blood in its posterior (*coccygeo-mesenteric*) portion flows into the hypogastrics. **That** in its anterior (*inferior mesenteric*) portion joins the *portal vein*, which also receives blood (1) by a *gastro-duodenal* from the right side of the gizzard, the duodenum, and the last loop of the small intestine, and (2) by the *superior mesenteric* from the rest of the small intestine.

In the heart itself the student will make out the general resemblance to that of the rabbit. The main difference is that the tricuspid valve is replaced by a strong muscular fold, nearly as thick as the walls of the right ventricle, which applies itself over the auriculo-ventricular orifice, and converts the whole ventricle, except at the orifice of the pulmonary artery, into a muscular chamber. There are no chordæ tendineæ nor papillary muscles in this ventricle.

Notes on Embryonic and Fœtal Circulation in Fowl and Rabbit.

1. The aortic arch is developed from the second branchial arch, the right arch being suppressed in the mammal, the left in the bird.

2. The carotids are developed from the arch or arches anterior to the second branchial.

3. The pulmonary arteries are developed from the arch or arches posterior to the second branchial. In the bird the arches of both sides persist; in the mammal that of the left only.

4. In both bird and mammal, before lung-respiration sets in, the blood flows in the main from the pulmonary into the aorta by a connecting branch (*ductus arteriosus*), similar to that shown for the tadpole in Fig. 62, II.

5. From the aorta the blood is delivered in the main by the *vitelline artery* to the umbilical vesicle or yolk-sac, and by the *allantoic arteries* to the allantois.

6. From the umbilical vesicle and the allantois the blood returns by vitelline and allantoic veins, which unite and receive a small mesenteric branch from the intestines. They then pass into the liver, some of the blood being distributed to that organ, but much of it passing through it by the *ductus venosus* to the sinus venosus. (See Fig. 66.)

7. From the anterior part of the body the blood is returned by anterior cardinal veins (*a. c.*), one on each side; and from the Wolffian bodies and posterior regions by posterior cardinal veins (*p. c.*), one on each side. The anterior and posterior cardinals on each side combine and form factors of the *ductus Cuvieri* (*d. c.*) which leads the blood to the heart.

8. But when the kidneys are developing, the blood from them and from the hind limbs is led by a new median vein, the post-caval (*pt. c.*), into the sinus venosus.

9. As the postcaval increases in size the posterior cardinals diminish and tend to atrophy.

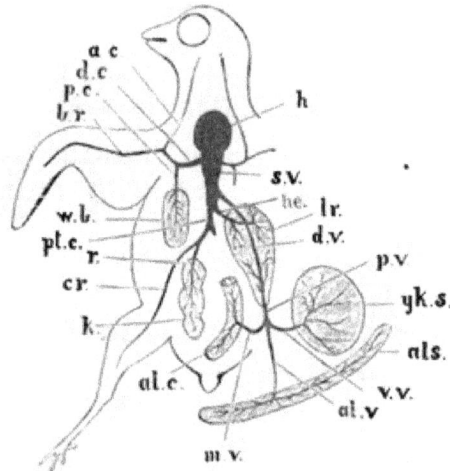

Fig. 66.—Diagram of Venous Circulation of Chick previous to Hatching.

a. c. Anterior cardinal vein, the jugular of the adult. *al. c.* Alimentary canal. *al. v.* Allantoic vein. *als.* Allantois; this and its vein are provisional embryonic structures. *br.* Brachial vein. *cr.* Crural vein. *d. c.* Ductus Cuvieri, the precaval of the adult. *d. v.* Ductus venosus, leading blood through the liver; atrophies in the adult. *h.* Heart. *he.* Hepatic vein, collecting blood from the liver. *k.* Kidney. *lr.* Liver. *m. v.* Mesenteric vein. *p. c.* Posterior cardinal vein, bringing blood from the Wolffian bodies, atrophies in the adult. *p. v.* Portal vein, delivering some blood to the substance of the liver, but delivering more through the ductus venosus to the sinus venosus. *pt. c.* Postcaval vein. *r.* Renal vein. *s. v.* Sinus venosus. *v. v.* Vitelline vein. *w. b.* Wolffian bodies, atrophy with the growth of the kidneys. *yk. s.* Yolk-sac, which, with its vitelline vein, are provisional embryonic structures.

All details diagrammatic : relative size of parts not indicated.

10. With the atrophy of the posterior cardinals the ductus Cuvieri become the precavals, which also receive factors from the wing and breast.

11. During fœtal life there is a communication between the two auricles by the foramen ovale, through which the arrow

passes in Fig. 67, A. ⁻An Eustachian valve or fold of membrane, running from the opening of the postcaval (*pt. c.*) to the foramen ovale, directs most of the blood from the posterior regions of the body through that orifice into the left auricle, while it prevents the blood passing in by the precavals (*pr. c.*) from leaving the right auricle. The blood from the posterior regions has been arterialised in the allantois.

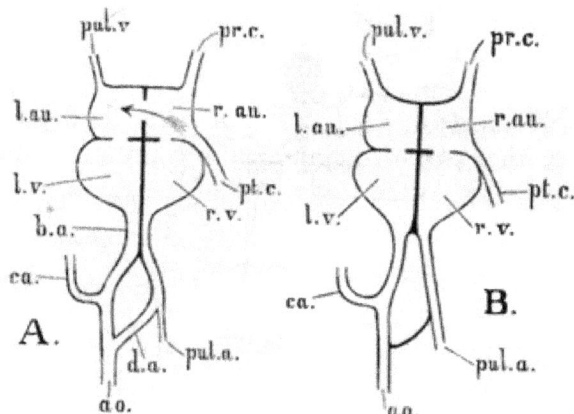

FIG. 67.—FŒTAL AND ADULT CIRCULATION IN HEART.

A. Fœtal.   B. Adult.

ao. Aorta.   *b. a.* Bulbus arteriosus.   *ca.* Carotid artery.   *d. a.* Ductus arteriosus.   *l. au.* Left auricle.   *l. v.* Left ventricle.   *pr. c.* Precaval vein.   *pt. c.* Postcaval vein.   *pul. a., pul. r.* Pulmonary artery and vein.   *r. au.* Right auricle. *r. v.* Right ventricle.

12. The blood leaves the ventricles by a bulbus arteriosus (*b. a.*), which is divided by a septum into two chambers, which will be converted into the root of the aorta and that of the pulmonary artery.

13. The blood from the left ventricle (arterialised in the allantois) flows along the aortic chamber of the bulbus arteriosus, and so into the carotid (*ca.*) and aorta (*ao.*). The blood from the right ventricle which has returned from the head passes into the pulmonary chamber, which thus delivers blood along the ductus arteriosus (*d. a.*) to the aortic arch, and in small quantities to the as yet functionless lungs (by *pul. a.*).

Thus the placenta and hinder regions receive less arterialised blood.

14. Note that during fœtal life **it is the** blood that flows along the aorta that will be arterialised. Hence the necessity of keeping open the foramen ovale, and directing by means of the Eustachian valve the blood which falls into the right auricle **across into** the left. In adult life it is the blood that flows **along the pulmonary** artery that will be arterialised. Hence **the foramen** ovale **closes,** and the Eustachian valve becomes functionless.

In **Fig.** 67 the fœtal (A.) and adult (B.) condition are compared. For the sake of diagrammatic convenience the bulbus arteriosus is placed at the lower **part** of the figure.

CHAPTER XI.

THE BRAIN AND NERVES.

THE central nervous system is differentiated very early in development. Beginning as a neural plate, it soon, by the up-growth of laminæ dorsales, which bend over and eventually meet in the mid-line, becomes converted into a neural tube (see p. 102). We have seen that the anterior part of this tube becomes differentiated into three swellings or vesicles, the fore-brain, mid-brain, and hind-brain. These in the course of development become further differentiated, so that we have eventually five regions, viz. :—

Hind-Brain, { 1. Myelencephalon, or region of the medulla oblongata.
{ 2. Metencephalon, or region of the cerebellum.

Mid-Brain, 3. Mesencephalon, or region of the optic lobes.

Fore-Brain, { 4. Thalamencephalon, or region of the thalamus.
{ 5. Prosencephalon, or region of the cerebral hemispheres, with their outgrowths, the olfactory lobes.

These regions are somewhat differently developed in Frog, Pigeon, and Rabbit. We will consider them *seriatim* from behind forwards.

(A.) Hind-Brain—1. *Myelencephalon.*—The floor and sides of the canal in this region become much thickened, while the roof becomes very thin. The central canal here expands to form the *fourth ventricle* of the brain (iv.). The pia mater covering this ventricle becomes very vascular, forming the tela vasculosa. Into this region of the brain the dorsal and ventral fissures of

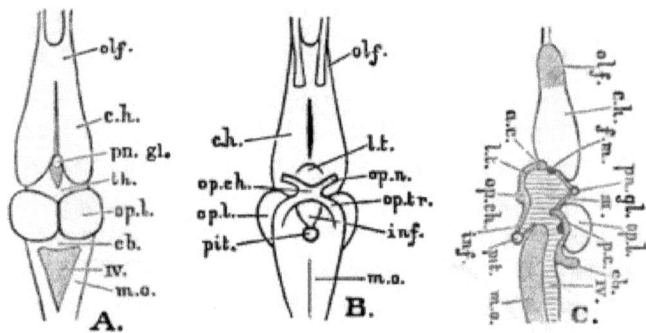

FIG. 68. BRAIN OF FROG.—A. From above. B. From below. C. In longitudinal section.

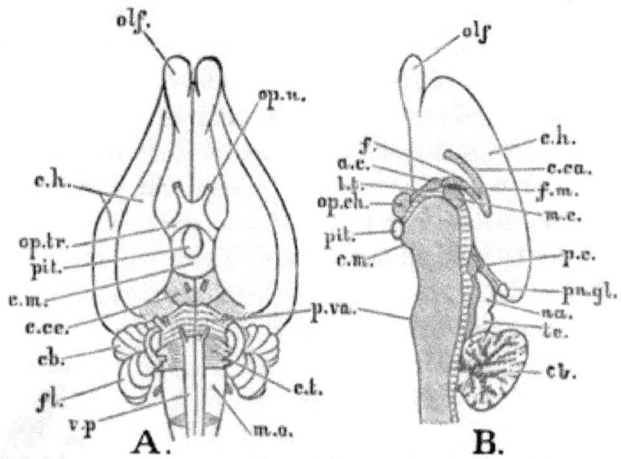

FIG. 69. BRAIN OF RABBIT.—A. From below. B. In longitudinal section.

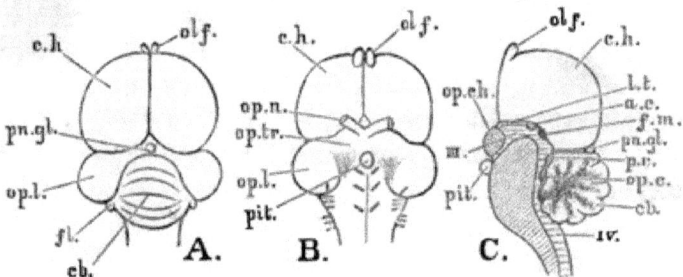

FIG. 70. BRAIN OF PIGEON.—A. From above. B. From below. C. In longitudinal section.

a. c. Anterior commissure. c. ca. Corpus callosum. c. ce. Cruri cerebri. c. h. Cerebral hemisphere. c. m. Corpus mammillare. c. t. Corpus trapezoideum. cb. Cerebellum. f. Fornix. f. m. Foramen of Monro. fl. Flocculus. inf. Infundibulum. l. t. Lamina terminalis. m. c. Middle commissure. m. o. Medulla oblongata. na. Natis. olf. Olfactory lobe. op. c. Optic commissure. op. ch. Optic chiasma. op. l. Optic lobe. op. n. Optic nerve. op. tr. Optic tracts. p. c. Posterior commissure. p. va. Pons Varollii. pn. gl. Pineal body. pit. Pituitary body. te. Testis. th. Optic thalamus. v.p. Ventral pyramid. iii. Third ventricle. iv. Fourth ventricle.

the spinal cord are continued. On either side of the dorsal fissure are the *dorsal pyramids ;* on either side of the ventral fissure are the *ventral pyramids (v. p.).* The dorsal pyramids diverge at the fourth ventricle. In the rabbit there is further differentiation. External to the ventral pyramids, is on either side, an olivary tract, and (anteriorly) the *trapezoid body (c. t.),* a band of transverse fibres. External to the dorsal pyramids, which form but narrow areas, are the *restiform bodies.* In the medulla oblongata the white matter is external and the grey internal.

2. *Metencephalon.*[1]—Here the sides and floor of the canal are thickened. In the frog and pigeon they are not well differentiated from the medulla ; but in the rabbit they are differentiated into strong bands of transverse fibres forming the *pons Varolii (p. va.).* The roof undergoes special development, giving rise to the *cerebellum (cb.),* in which the white matter is internal and the grey external. In the frog this is a small tongue-shaped band overhanging the anterior part of the fourth ventricle. In the pigeon it is large, uni-lobed, overlying the fourth ventricle behind, and abutting against the cerebral hemispheres in front. It is marked with ridges and grooves. At each side there is a small elevation, the *flocculus (fl.).* It is connected with the thickened floor of this region by strong *peduncles.* A delicate sheet of nervous tissue, the *valve of Vieussens,* connects it with the optic lobes (*op. l.*). In the rabbit the cerebellum is large and tri-lobed. Each lateral lobe has externally a large flocculus. A valve of Vieussens connects it anteriorly with the testes (*te.*), and on each side this band is thickened to form the *anterior peduncles ; middle peduncles* pass downwards to the pons Varolii ; *posterior peduncles* are continued into the restiform bodies of the medulla.

(B.) **Mid-Brain**—3. *Mesencephalon.*—In the frog the floor is thickened, but not further differentiated. The roof gives rise to large *optic lobes* (*op. l.*), which are apposed in the middle line.

---

[1] Some authors apply the term *metencephalon* to that part of the brain which has been described as *myelencephalon,* applying the term *epencephalon* to the cerebellum.

In the pigeon the thickened floor is not well differentiated from that of the hind-brain. The optic lobes are pressed outwards to the sides by the large cerebral hemispheres and cerebellum, the band of fibres uniting them being known as the *optic commissure.* In the rabbit the floor is differentiated into two strong diverging bands of white fibres, the *crura cerebri* (*c. ce.*), proceeding outwards and forwards. The optic lobes are differentiated into two pairs—*nates* (*na.*) anteriorly and *testes* (*te.*) posteriorly—forming the *corpora quadrigemina.* In this region the central canal becomes narrowed to form the *iter* connecting the fourth and third ventricles. The cavities contained within the optic lobes of the frog and pigeon communicate with the iter. In the optic lobes the grey matter **is external.**

(C.) **Fore-Brain**—4. *Thalamencephalon*, or primary fore-brain. In the frog the hinder part of the floor in this region gives rise to a funnel-shaped depression, the *infundibulum* (*inf.*), leading down to a small mass of tissue (pinched off in development from the stomodæum), known as the *pituitary body* (*pit.*). In front of this the floor gives rise to the *optic chiasma* (*op. ch.*), formed of *optic tracts* (*op. tr.*) posteriorly, and the *optic nerves* (*op. n.*) anteriorly. The sides give rise to thickened masses of mixed grey and white matter, the *optic thalami* (68, A., *th.*). The roof becomes very thin anteriorly, forming the *velum interpositum.* Posteriorly it is thickened to form a transverse band of fibres, the *posterior commissure* (*p. c.*), the anterior portion of which rises on either side to form the *peduncles of the pineal gland* (*pn. gl.*). In the pigeon the infundibulum is small, the optic chiasma well marked, and the optic tracts thick. The optic thalami are small, covered by the velum interpositum, and connected by the posterior commissure, above which on either side **are** the peduncles of the pineal gland.[1] In the rabbit the infundibulum is small, and behind it is a rounded mass, the *mammillary body* (*c. m.*). The optic chiasma is well marked. The optic thalami

[1] The function of the pineal body has long been problematical. It has recently been discovered that in the reptile (*Hatteria*) the pineal gland is the seat of an eye constructed on the invertebrate type. The saccular *brow-spot* between the eyes of the frog is probably a vestige of the **same organ.**

are large, and united (1) by a posterior commissure, on either side of which are the peduncles of the pineal gland; and (2), by a *middle or soft commissure* of grey matter (*m. c.*). The roof forms a thin velum interpositum.

In this region the central canal expands to form the *third ventricle* of the brain (iii.).

5. *Prosencephalon*, or secondary fore-brain. In the frog the *cerebral hemispheres* (*c. h.*), composed externally of grey matter and internally of white, are formed as outgrowing buds from the anterior end of the primitive fore-brain. In their adult form they are elongated anteriorly, but do not extend backwards over the succeeding parts of the brain. They contain *lateral ventricles* which communicate with the third ventricle by the *foramen of Monro* (*f. m.*). The outer wall of each lateral ventricle is somewhat thickened. The inner faces of the two hemispheres are united anteriorly in the region of the so-called olfactory lobes. The anterior wall of the brain in the mid-line is the *lamina terminalis* (*l. t.*), which contains a band of transverse fibres, the *anterior commissure* (*a. c.*). In the pigeon the cerebral hemispheres (*c. h.*) are rounded and smooth, and extend backwards, so as to separate the optic lobes and cover much of the thalamencephalon. The cavity of the lateral ventricle is much diminished, the floor and lower part of its outer wall being thickened by the development of an eminence of grey matter forming the *corpus striatum*. Above the lamina terminalis (*l. t.*) is a band of transverse fibres connecting the corpora striata, and forming the anterior commissure (*a. c.*).

In the rabbit the early development of the cerebral hemispheres (*c. h.*) resembles that in the frog. In their later development matters become much more complicated. Externally they are seen to be divided by a ventral and longitudinal groove into a larger superior and a smaller inferior lobe. A slight angulation on the ventral side of the superior lobe marks the position of that which becomes in higher types the *fissure of Sylvius*. The hemispheres grow rapidly, and extend backwards so as to cover the thalamencephalon, and almost hide the optic lobes (nates and testes). The floor becomes thickened, giving rise to the *corpus striatum* (Fig. 71, *c. st.*), which projects upwards into

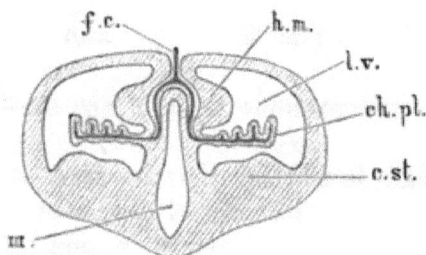

FIG. 71.—DIAGRAMMATIC SECTION OF FŒTAL
BRAIN (after Kölliker).

c. st. Corpus striatum. *ch. pl.* Choroid plexus.
f. c. Falx cerebri. *h. m.* Hippocampus major.
l. v. Lateral ventricle. III. Third ventricle.

the lateral ventricles (*l. v.*). Posteriorly the corpus striatum comes into close relation with the optic thalamus. The outer wall of the hemisphere thickens, while the inner wall becomes relatively thinner. In this inner wall two curved folds then make their appearance, and project towards the interior of the lateral ventricle. Of these the upper gives rise to a thickened mass of white nervous tissue, the *hippocampus major* (Fig. 71, *h. m.*). The wall of the lower fold becomes thinned to the merest film, and a vascular plexus (*ch. pl.*), derived from the connective tissue septum between the hemispheres (*falx cerebri, f. c.*) is formed in the fold, in the manner shown in the transverse section of the brain of a fœtal sheep (after Kölliker) Fig. 71. This constitutes the *choroid plexus* of the lateral ventricle. It is also present in the pigeon.

In the lamina terminalis (*l. t.*), besides the transverse fibres of the anterior commissures (*a. c.*), two vertical bands of fibres are developed. These constitute the *anterior pillars of the fornix* (*f.* in 69, B.). They are continued downwards along the floor of the third ventricle, to end in the mammillary body (*c. m.*). Traced upwards they meet in the middle line in the roof of the third ventricle, thus forming the *body of the fornix*. Thençe bands of fibres (the *posterior pillars of the fornix*) pass backwards, arching over the foramen of Monro, to unite with the hippocampus major.

The main commissure, uniting the cerebral hemispheres, is the *corpus callosum* (*c. ca.*). It may be seen by carefully separating the cerebral hemispheres on the dorsal side, and constitutes a flat white band. It is seen in section in Fig. 69, B. At its anterior end it bends slightly downwards (*anterior genu*), doing the same posteriorly (*posterior genu*), where it comes into relation with the **fornix.** Between the two is an area called the *septum lucidum.* It is formed of the inner walls of the lateral ventricles of this region. In higher types this area is more completely enclosed by the bending down of the more marked anterior genu of the corpus callosum. In this way a space between the two walls is more or less shut in, and constitutes the so-called fifth ventricle, which is, however, it should be noted, outside the brain. Anteriorly the cerebral hemispheres give rise, as outgrowths, to *olfactory lobes.* In the frog the so-called olfactory lobes (*olf.*) are relatively large, their bases being closely united in the middle line. They contain minute cavities, which are prolongations of the lateral ventricles. Anteriorly they become nerve-like cords. It is very probable, however, that these so-called olfactory lobes are the anterior parts of the cerebral hemispheres. In the pigeon the olfactory lobes (*olf.*) are small and apposed, but not united in the middle line. They send forward nerve-like cords. In the rabbit the olfactory lobes (*olf.*) are small (though larger than is usual in the mammalia), and are not united in the mid-line. They send out many fibres through the cribriform plate.

Cranial Nerves.—In the frog there are ten pairs of cranial nerves :—

I. *Olfactory ;* sensory nerves derived from the olfactory lobes and passing to the organ of smell.

II. *Optic ;* sensory nerves arising as outgrowths of the thalamencephalon, but acquiring a secondary connection with the optic lobes, and passing to the organ of vision. The fibres of the optic nerves interlace at the chiasma.

III. *Oculo-motor ;* motor nerves arising from the floor of the mesencephalon, and passing to four of the eye-muscles.

IV. *Pathetic* or *trochlear ;* motor nerves arising from the floor of the mesencephalon, and passing to the superior oblique eye-muscle.

V. *Trigeminal ;* arising from the myelencephalon, and, just before leaving the skull, expanding into a Gasserian ganglion. Outside the **skull each** divides into an anterior *orbito-nasal* branch, passing to the **nasal capsule** and integument in its neighbourhood, and a *maxillo-mandibular* branch which shortly divides into a *maxillary* branch which supplies the upper jaw and lower eyelid, and a *mandibular* branch which supplies the lower lip and muscles of the floor **of the mouth.**

VI. *Abducent ;* arising from the myelencephalon, entering the Gasserian ganglion, and supplying certain eye-muscles.

VII. *Facial ;* arising from the myelencephalon and entering into close connection with the Gasserion ganglion. Each leaves the skull just behind the trigeminal, and soon divides into (1) a *palatine* branch, which runs forward beneath the mucous membrane of the palate; (2) a *hyo-mandibular* branch, which **passes** back over the columella, and down to the angle of the **mouth.** It then gives rise to two branches; one, the *hyoidean*, supplying the muscles of the anterior cornu of the hyoid, the **other, the** *chorda tympani*, running along the inner edge of the lower jaw.

VIII. *Auditory ;* arising from the myelencephalon, and passing to the organ of hearing.

IX. *Glosso-pharyngeal ;* arising from the myelencephalon. Each **leaves the skull** by a foramen, just behind the auditory capsule, **gives off a** branch which joins the hyo-mandibular branch of **the** facial, and thus passes by a somewhat sinuous course to the base of the tongue.

X. *Pneumogastric* or *Vagus ;* arising from the myelencephalon, and leaving the skull in close company with the glosso-pharyn **geal.** Outside the skull each bears a ganglionic swelling. It gives **off** cutaneous branches to the dorsal integument, and then gives rise to (1) a *laryngeal* branch, forming a loop over the pulmo-cutaneous **artery ;** (2) a *splanchnic* branch ; this subsequently gives rise to (*a*) a *pulmonary* branch, passing to the lung, (*b*) a *cardiac* branch passing to the septum, between the auricles of the heart, and (*c*) *gastric* branches to the walls of the stomach.

The following points should be made out by practical dissection :—

1. Place the frog upon its side and remove the integument from the side of the head. On dissecting away the muscles just behind the tympanum, and carefully removing the anterior part of the supra-scapula, four nerve-threads will be seen, three taking their origin from the skull, the fourth from the spinal column. This last is the *hypo-glossal* (1st spinal nerve). Traced downwards it is found to be distributed to the muscles of the tongue. From the ventral aspect it is readily seen running beneath the thin transverse sheet of muscle (mylo-hyoid), between the two rami of the lower jaw. Near it, but somewhat deeper, and also distributed to the muscles of the tongue, is the *glosso-pharyngeal* (ninth cranial). This is continuous with one of the three cranial nerves seen just behind the tympanum. The other two are the *laryngeal* and *splanchnic* branches of the *vagus*.

2. Dissect away the integument and muscles from the rami of the lower jaw. The *mandibular* branches of the V. lie on their outer faces ; the *mandibular* branches of the VII. on the inner. Traced backwards, one of these, the mandibular, is found to run up beneath the squamosal bone to the skull. It is a branch of the trigeminal. The other may be traced to the angle of the lower jaw, and thence upwards over the columella and forwards into the skull. It is a branch of the facial.

3. Detaching or holding widely open the lower jaw, remove the mucous membrane of the palate. Two nerves (one on each side) are seen running from just in front of the guard of the para-sphenoid, where they emerge from the skull, to the vomers. They are the *palatine* divisions of the facial. By carefully dissecting the guard away on one side the *Gasserian ganglion* may be exposed.

4. Just on the lower edge of the orbit, beneath the eye

and above the mandible, is seen a nerve passing to the integument of the upper jaw. It is the *maxillary* branch of the trigeminal.

5. Carefully remove the eye. To do so the optic nerve supplying that organ must be severed. Passing along the upper border of the orbital cavity is a nerve running forwards to the nasal chamber. It is the *orbito-nasal* division of the trigeminal. It is not connected with the sense of smell.

In the fowl and rabbit there are twelve pairs of cranial nerves. Of these the first ten are essentially similar in distribution and origin to those of the frog. The eleventh is a spinal nerve, which (in the rabbit for instance) takes its origin a little posterior to the fifth cervical nerve. It then passes forwards in the fork between the two roots of the anterior spinal nerves, receiving fibres from the cord *en passant*, and leaves the skull, together with the ninth and tenth, to be distributed to some of the muscles in the neck. The twelfth, or hypo-glossal, answers to the first spinal of the frog, and has a similar distribution, but making its exit through the skull is reckoned as cranial.

Spinal Nerves.—In the frog the first spinal nerve forms the hypo-glossal. The second and third unite to form a *brachial plexus* for the innervation of the fore-limb. The fourth, fifth, and sixth spinal nerves pass to the body-walls. The seventh, eighth, and ninth form the *lumbo-sacral* or *sciatic plexus*, from which nerves are given off to the hinder region of the body, but especially to the hind-limb, dividing into a small *crural* to the front of the thigh, and a large *sciatic* to the back of the thigh. This latter divides into the *tibial*, which ends on the lower surface of the foot, and the *peroneal* which terminates on its dorsal surface. The tenth spinal nerve leaves the neural canal by the coxygeal foramen in the urostyle, and is distributed to the adjacent parts.

In the rabbit there is a brachial plexus formed by the union of the fifth to the eighth cervical and first thoracic nerves. It gives rise, as is shown in Fig. 72, to a *circumflex* (*cf.*), which dives into the muscles of the shoulder dorsal to the head of the humerus; a *median* (*m.*), which passes to the radial side of the fore-arm; an *ulnar* (*u.*), which passes to the ulnar side of the fore-arm; and a *musculo-spiral* (*m. s.*), which dives to the dorsal side of the humerus, and then passes to the radial side of the fore-arm. The other thoracic nerves run out between the ribs. In

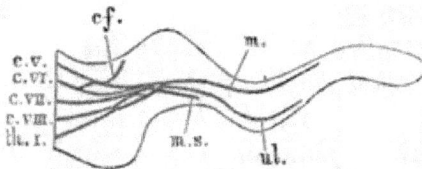

Fig. 72.—Brachial Plexus—Rabbit.

c. v.-c. viii. Fifth to eighth cervical nerves. *th. i.* First thoracic nerve. *cf.* Circumflex nerve. *m.* Median nerve. *m. s.* Musculo-spiral nerve. *ul.* Ulnar nerve.

the rabbit the lumbo-sacral plexus is formed by the union of the fifth to the seventh lumbar and first to the third sacral nerves. It gives rise to a *crural* nerve to the front of the thigh, an *obturator* nerve to the muscles near the acetabulum, and a great *sciatic* passing out between the ischium and the sacrum.

In the pigeon there is a brachial plexus which supplies the wing; a lumbar plexus, giving off the larger *femoral nerve* over the front edge of the ilium to the front of the thigh and the obturator nerve through the obturator fissure to the muscles near the acetabulum; and a sciatic plexus giving rise to the sciatic nerve which passes through the ilio-sacral foramen.

The Sympathetic System.—In the frog the sympathetic system has ten or more ganglia on either side, connected by longitudinal connectives These lie within the subvertebral lymph-space, and posteriorly are closely connected with the aorta. Each ganglion receives a filament from a spinal nerve. The most anterior ganglia are united by filaments with the ganglion of the ninth and tenth cranial nerves, from which a delicate thread passes into the cranial cavity and unites with the Gasserian ganglion. Branches of the sympathetic accompany the blood-vessels, a large *splanchnic* branch (on which is developed a *cœliac ganglion*) passing along the cœliaco-mesenteric artery.

In the rabbit the sympathetic may be seen in the thoracic

15

region on either side of the vertebral column, the ganglia lying on the heads of the ribs. Traced forwards into the neck, it is seen as a flat nerve, on which are developed a *posterior cervical ganglion* a little above the subclavian artery, and an *anterior cervical ganglion* just anterior to the larynx. Traced backwards, it gives off, about the level of the ninth thoracic ganglion, a large splanchnic nerve, which is much more conspicuous than the slender continuation of the chain. It enters a group of greyish pink transparent ganglia, constituting the *solar plexus*, of which the cœliac ganglion and the *mesenteric* ganglion, one on either side of the anterior mesenteric artery, are the most important.

In the pigeon there is a cervical ganglion near the brachial plexus. Anterior to this the sympathetic lies in the verte-brarterial canal. The two sympathetic chains unite posteriorly in a *ganglion impar* near the end of the caudal vertebræ.

The Functions of the Nervous System.—The function of the nervous system, as a whole, is to form a centralised means of intercommunication between the various parts of the organism, and thus to convert an aggregate of organs into a living unit. Closely connected with the nervous system are end-organs, (*a*) for receiving impressions from without, as in the organs of special sense ; and (*b*) for delivering impulses, as in the muscle-plates. Space need not here be occupied in repeating what has been said in Chapter IV. on reflex action ; nor is it within the scope of this volume to enter further into the nervous mechanism of responsive and original action. The following brief notes must here suffice :—

1. Afferent end-organs (*e.g.* special senses) receive impressions from without.

2. Efferent end-organs (*e.g.* muscle-plates) deliver stimuli to muscles or glands and other parts.

3. Afferent and efferent nerve-fibres transmit these impressions, the former towards, the latter away from, the chief nerve-centres.

4. The nerve-centres correlate these transmitted impressions.

5. The spinal cord contains, besides transmitting-fibres, many centres of reflex action for the performance of simple or complex organic actions.

6. The medulla oblongata contains a centre for such vital reflex actions as are connected with the regulation of the heart-beat, with breathing, and the distribution of the blood-supply.

7. The sympathetic is usually held to be a subsidiary system, largely occupied with regulating the calibre of certain blood-vessels.

8. The cerebellum is probably largely occupied with the co-ordination of muscular movements.

9. The optic lobes are probably centres for vision and the adjustment of the visual apparatus.

10. The thalami (of the rabbit) are regarded as centres by which sensory impressions are transmitted, while the corpora striata are centres by which motor impulses are transmitted. It is possible that purely responsive non-original acts require no higher centres than these.

11. The cerebral hemispheres are definitely involved in original and intelligent actions.

12. Accompanying and associated with, or perhaps identical with, certain nerve-actions (*neuroses*) going on within the body, there are, in all probability, certain feelings (*psychoses*) about which, in the frog, the pigeon, and the rabbit, we can know little.

PART II.

SOME INVERTEBRATE TYPES.

CHAPTER XII.

THE CRAYFISH.

THE Crayfish (*Astacus fluviatilis*) is a small animal, five or six inches, at most, in length, and somewhat resembling a diminutive brown or greenish-brown lobster. It inhabits fresh water in streams and rivulets, hiding beneath stones or in burrows of its own making. It walks on four pairs of jointed legs, of which the posterior pair are, during progression, directed backwards, the others having a forward direction; but when alarmed it darts backwards by a vigorous flap of its jointed abdomen, which ends in a flattened tail-fin. In front of the legs is a strong pair of jointed pincers; on either side of the mouth are jaws and "foot-jaws"; there are two pairs of feelers; and the eyes are placed on moveable eye-stalks.

The whole body, except where a joint renders mobility essential, is covered with a hard external shell; whence the crayfish and its relations are called *Crustacea*. This hard shell is periodically thrown off or shed (by *ecdysis*). The new skin is at first quite soft, and thus allows the animal to grow rapidly for a day or two, during which time the crayfish hides from its enemies. If one of the pincers or other limbs be seriously injured, or if the animal become fixed by one of the pincers, the limb can be cast off at the slenderest joint. A new limb slowly grows in place of that which is thus lost.

Beneath the abdomen, which is broader in the female than in the male, there may, in spring, be seen in the female dark berry-like eggs attached to the slender appendages or *swimmerets* of this region. These develop into young round-backed crayfishes which are somewhat unlike the adult. The adult male may be recognised by two pairs of imperfectly tubular appendages on the under side of the anterior end of the abdomen, pointing forwards, beneath the unjointed thoracic region.

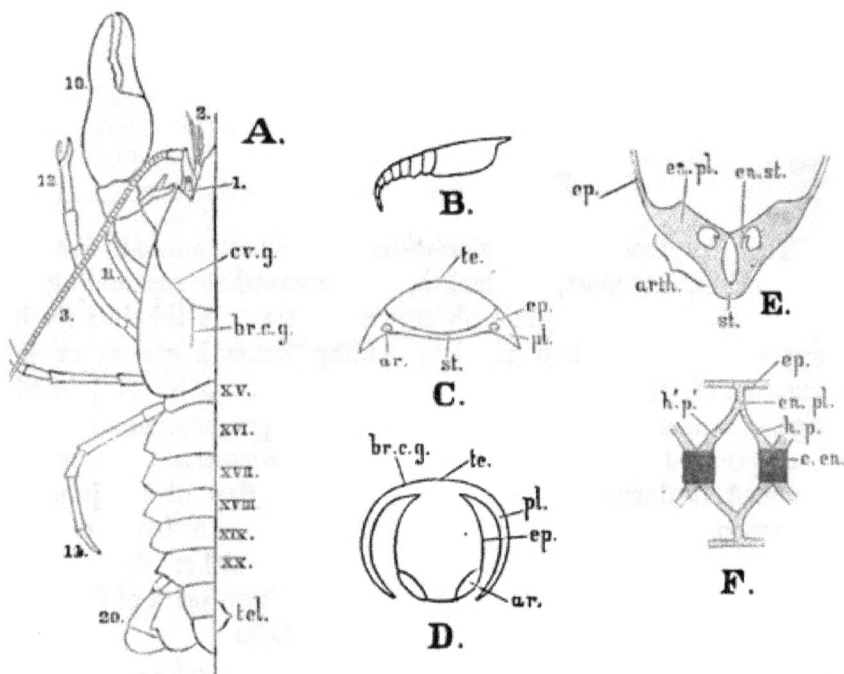

FIG. 73.—CRAYFISH EXOSKELETON.

A. Dorsal view of crayfish. B. Side view, with appendages removed.
C. Diagram of abdominal somite. D. Cross-section of cephalo-thorax.
E. Endophragmal system in cross-section. F. The same from above.

ar. Articular cavity. *arth.* Arthrophragm. *br. c. g.* Branchio-cardiac groove. *c. en. st.* Capital of endosternite. *cv. g.* Cervical groove. *en. pl.* Endopleurite. *en. st.* Endosternite. *ep.* Epimeron. *h'. p'.* Anterior horizontal process. *h. p.* Posterior horizontal process. *pl.* Pleuron. *st.* Sternum. *te.* Tergum. *tel.* Telson. xv.-xx. Abdominal somites. 1-3, 10-14, 20. Appendages visible in dorsal aspect.

External Characters.—The body of the crayfish is easily seen to have the following parts (Fig. 73, A.) :—

1. *The Cephalo-thorax,* occupying the anterior half of the body. The whole of this region is encased in a continuous shell, the *carapace,* ending anteriorly in a pointed *rostrum.*

2. *The Abdomen*, behind the cephalo-thorax; jointed, composed of six divisions or *somites.*

3. *The Telson*, forming the posterior end of the body in the mid-line; divided transversely into two parts, on the under side of the first of which is the vent.

4. *The Appendages*, of which there are 20 pairs; 6 being attached to the somites of the abdomen, and 14 to the cephalo-thorax. Of these 9 pairs are visible in Fig. 73, A., viz., eye-stalks, short two-whipped antennules, antennæ, pincers, 4 pairs of legs, and the large posterior swimmerets, which, with the telson, constitutes the terminal tail-flap.

Seen from the side in the normal position of rest, the abdominal region is slightly curved downwards, while the anterior end of the sternal surface is bent upwards, giving rise to what is termed the *cephalic flexure.* Hence the curve of the ventral or appendage-bearing surface is as represented in Fig. 73, B.

On the dorsal (or *tergal*) surface of the cephalo-thorax there is a curved **cervical groove**, from which there passes backwards on either side of the median line a short longitudinal *branchio-cardiac groove.* On the *tergal* surface of each abdominal segment there is a smooth anterior portion, seen best when the abdomen is in flexion or curved up beneath the body, which marks where the segment in front overlaps it.

Fig. 73, C., shows a posterior view of the fourth abdominal somite, from which the limbs and the soft parts have been removed. The tergal surface (*te.*), is, in a dorsal view, broadest in the mid-line, narrowing away to the pleura at each side. The *sternum* (*st.*) is very narrow, forming a mere transverse ventral bar. On each side of the sternum is seen the articular cavity (*ar.*) for the swimmeret. The projecting flaps at the sides are the pleura (*pl.*). Between the articulation of the limb and the pleuron is a narrow area called the *epimeron* (*ep.*).

The somites of the abdomen articulate together by means of pegs or pivots which fit into corresponding depressions, the peg being on the anterior edge of one segment and the socket on the posterior edge of the next. The pleura overlap, especially in flexion. The pleura of the second abdominal somite overlap those of the first and third; the third overlaps the fourth; the fourth, the fifth; and the fifth, the sixth. Thus lateral motion of the abdomen is prevented.

Fig. 73, D., shows a transverse section of the cephalo-thorax. The pleura are seen to be enormously enlarged, and are here called the *branchiostegites*. The space between the point of articulation of the limb and the pleuron is also much enlarged. Between this (epimeron) and the branchiostegite is enclosed the *branchial chamber*, in which lie the gills. In a dorsal aspect (Fig. 73, A.) the terga are seen to have coalesced into a continuous carapace.

Viewed from the ventral aspect, however, the sterna, though they tend to run together in the mid-line, give evidence that the thoracic region is compounded of coalesced somites homologous with those in the abdomen. Six sterna may be readily made out, of which the most posterior is a transverse plate, the next two triangular, and the other three longitudinally elongated. External to these sterna are the articular spaces for the four legs, the pincers, and, anterior to the pincers, the posterior maxillipeds. Between these articular spaces are transverse calcified inter-articular bars, or *arthrophragms*. The sterna in front of these six (*i.e.* between them and the mouth), are very delicate. In front of the mouth the sternal region forms a broad plate, with anterior spear-shaped process, the *epistoma*.

Projecting inwards from the ventral aspect of the thoracic region is a complicated system of "pillars and bulkheads," constituting the *endophragmal system*, which (1) gives attachment to certain muscles; (2) serves to protect the nerve-chain; and (3) assists in consolidating the cephalo-thorax. Fig. 73, E., shows diagrammatically the infoldings of the cuticle, which are seen in a transverse section of the region of the first pair of legs. Where the arthrophragm (*arth.*) joins the sternum (*st.*) there rises a little pillar, the *endosternite* (*en. st.*), with an expanded capital. From the outer end of the arthrophragm and the adjoining epimeron (*ep.*), there projects inward a plate, the *endopleurite* (*en. pl.*). Viewed from above the endopleurite is seen to split into two horizontal processes, as shown diagram

matically in F. Each of these processes meets the expanded
head or capital of an endosternite (*c. en. st.*). So that the head
of each endosternite is connected with the anterior horizontal
process of one endopleurite and the posterior horizontal process
of another endopleurite, as indicated in F., which, it must be
remembered, only represents the upper surface of the endophragmal system, and that quite diagrammatically. On comparing these diagrams with the real thing prepared by boiling
with dilute caustic potash, the essential structure will, it is
hoped, be understood. The minute details are beyond our
scope.

The Appendages.—There are twenty pairs of appendages
which are serially-homologous or homodynamous. They are—

1	pair of eyestalks or ophthalmites.[1]		
1	„	antennules.	
1	„	antennæ.	Cephalic.
1	„	mandibles.	
2	„	maxillæ.	
3	„	maxillipeds.	
1	„	chelate forceps or pincers.	
2	„	chelate legs.	Thoracic.
2	„	simple legs.	
6	„	swimmerets, of which the pos- terior are much enlarged.	Abdominal.

By comparison and careful study a typical or ideal appendage
may be conceived from which all the others may be derived,
according to Huxley's law, by suppression, by coalescence, or by
metamorphosis.

The student should carefully remove the appendages from
one side, and lay them in order on a sheet of white paper. In
removing the legs, forceps, and maxillipeds, care must be taken
not to tear away the gill from its attachment to the base of the
limb. The first maxilliped and the maxillæ are exceedingly thin

---

[1] The ophthalmite is, by some morphologists, not regarded as a homologue of
the other appendages.

and foliaceous. It is well to cut off the branchiostegite before removing the appendages.

The typical appendage consists of a basal part, or *protopodite*, from which arises an inner part or limb, the *endopodite*, and an outer part or palp, the *exopodite*, and a gill-bearing part or *epipodite*. In the legs, forceps, and second and third maxillipeds, the protopodite and endopodite together form a series of seven joints, two of which may, however be coalescent. The student

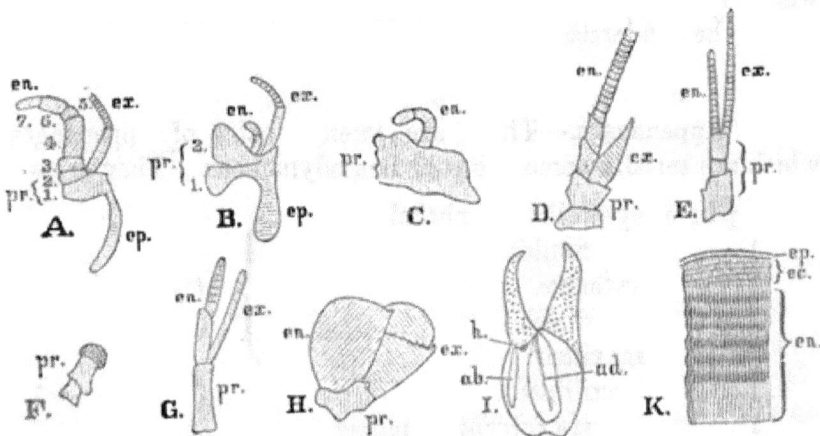

FIG. 74.—CRAYFISH—APPENDAGES.

A. Second maxilliped. B. First maxilliped. C. Mandible. D. Antenna.
E. Antennule. F. Eye-stalk (ophthalmite). G. Swimmeret. H. Twentieth
appendage. I. Chela, opened to show processes for attachment of muscles.
K. Microscopic section of exoskeleton.

en. Endopodite. *ep.* Epipodite. *ex.* Exopodite. *pr.* Protopodite.
1. Coxopodite. 2. Basipodite. 3. Ischiopodite. 4. Meropodite. 5. Carpopodite. 6. Propodite. 7. Dactylopodite.
In I.—*ab.* Process to which abductor muscle is attached. *ad.* Process to
which adductor muscle is attached. *h.* Hinge.
ec. Ectostracum. *end.* Endostracum. *ep.* Epiostracum.

should carefully study and draw the appendages. With the aid of the diagrams in Fig. 74, he will be able to trace the homologies of the parts. In the diagrams the seven joints are numbered. They are named—1. coxopodite; 2. basipodite; 3. ischiopodite; 4. meropodite; 5. carpopodite; 6. propodite; 7. dactylopodite.

Fig. 74, A. represents diagrammatically the second maxilliped. The third maxilliped is similar in principle, as is also the third lég. The forceps and the first two legs have the **dactylopodite** so disposed as to form with the propodite a chelate **termination** to the limb, as shown in I. In the fourth leg the epipodite is suppressed. B. is the first **maxilliped in** which, as in the maxillæ, the protopodite is very thin **and** flattened. In the second maxillæ the divisions 1 **and 2 are cleft;** the endopodite is small, and the epipodite **either coalesces with** the exopodite to form, or is itself modified **into, a** long **curved** plate, the *scaphognathite.* In the first **maxilla both** exopodite and epipodite are suppressed ; **the endopodite** is small. The mandible (C.) is stout and strong, **being** mainly protopodite, with a small-jointed palp, supposed **to be** the endopodite. In the antenna (D.) the endopodite is enormously produced, and the exopodite forms **a** flattened *squame.* In the antennule **(E.)** the **protopodite has three** joints; both endopodite **and exopodite are elongated.** In the ophthalmite (F.) we appear **to have all suppressed but the** protopodite.

In the abdominal region the typical appendage is as in G. The first two pairs **of** appendages in the male are specially modified, the endopodite of the first being so rolled upon itself as to enclose an incomplete canal. The first abdominal appendage in the female is very small, and sometimes wanting ; its exopodite **is** suppressed. In the posterior pair of abdominal appendages of both sexes we have much metamorphosis, the endopodite and exopodite (H.) being flattened and expanded, so **that the two** pairs, with the telson, form the broad tail-fin.

It will be noted that **many of the** appendages **are** jointed ; whence **a** large division of invertebrates with such joint⬛⬛limbs are known as the *Arthropoda* (**arthron,** a joint ; *pod⬛⬛⬛*). The joints are such as to allow the successive segments of the limb **free** play **in one** plane. **The third leg, for** example, is so hinged at its connection with the thorax as to allow the coxopodite to move backwards and forwards, but not up **and** down. **Referring to** the other divisions by numbers, **we may note that the** motion between 1 and 2 is up and

down, but not to and fro; between 2 and 3 there is little
motion; between 3 and 4 the motion is to and fro; between
4 and 5 up and down; between 5 and 6 again to and fro; and
between 6 and 7 once more up and down. The motions though
nearly, are, however, not quite at right angles; so that the
motion of the limb, as a whole, is tolerably free. The way in
which the several divisions are hinged will be readily seen on
examination of the limb. Fig. 74, I., shows the chela of the
forceps from which the soft parts have been removed, and which
has been opened out by the removal of its dorsal wall, except
where it is dotted in the diagram. The motion on the hinge-
line (*h.*), at two opposite points on the exoskeleton, is in the
plane of the paper. At two points of the exoskeleton, at right
angles to the hinge-points, there project inwards into the cavity
of the chela two long flat processes. To one of these (*ad.*)
is, during life, attached the adductor muscle; to the other and
smaller process (*ab.*) is attached the abductor muscle. By the
former the jaws are closed; by the latter they are opened. The
other joints of the limbs are similar in principle.

The exoskeleton consists of a tough animal substance, chitin,
the product of the epithelial cells. Except where a joint renders
flexibility essential it is impregnated with calcareous salts (carbonate, with about one-eighth phosphate, of lime). The structure
of the exoskeleton (from the ophthalmite) is shown in Fig. 74,
K. There is an external wrinkled layer (epiostracum), below
which is a laminated mass in which are a vast number of close-
set wavy canals running outwards through its thickness. The
laminæ are close-set externally (ectostracum, *ec.*), but wider
apart in the layers below (endostracum, *en.*). In the lowest
layers they again gradually become close-set and softer. A section of the dried exoskeleton is easily made by grinding down
on a file or on pumice-stone, and polishing on a fine-grained hone.

The Branchial Chamber.—If we cut with a pair of scissors from
behind forwards along the branchiocardiac groove, and then outwards along the cervical groove, a flap of exoskeleton, the
branchiostegite, will be removed, and the plume-like branchiæ

or gills will be exposed to view. They lie in the branchial cavity, which is bounded on the inner side by the delicate calcified outer wall of the thorax (epimeron), and opens in front by a canal in which a curved plate, the scaphognathite, in connection with the second maxilla, is constantly working in such a way as to produce a continuous current from behind forwards, over the gills and through the canal.

The branchial cavity is completely filled with the gills and with certain hair-like processes (setæ), developed on the coxopodites of the neighbouring limbs. Of branchiæ there are eighteen pairs :—

Six podobranchiæ ; these are the gill-bearing epipodites of the limbs from the second maxilliped to the third leg inclusive.

Eleven arthrobranchiæ ; attached to the inter-articular membranes between the basal joints of the appendages and the thoracic wall; one to that of each second maxilliped, and two to those of the third maxilliped to the third leg inclusive.

One pleurobranchia ; attached to the wall of the thorax above the fourth leg. There are two or three minute processes representing rudimentary pleurobranchiæ anterior to this.

The arthrobranchiæ consist of a stem beset with branchial filaments forming a plume. In the stem are two canals, an afferent and an efferent. The pleurobranchiæ are similar but larger. The podobranchiæ have a base beset with fine straight hair-like setæ. From this arises a stem which, above, carries a plume in front, and behind a V-shaped lamina, with the point of the V directed forwards. The lamina is plaited longitudinally, and is beset with short hooked setæ.

In examining the branchiæ practically, lay the crayfish on its side, and, after removing the branchiostegite, remove carefully, one by one, the six podobranchiæ. Then count the arthrobranchiæ, turning them carefully forwards as you do so. Note that they are attached to the soft inter-articular membranes. The last branchia is the pleurobranchia. Note that it is attached to the calcified epimeron above the inter-articular membrane.

The gills are highly vascular, the blood being separated from the surrounding water by a filmy chitinous membrane. Respiration is effected by the absorption of oxygen dissolved in the water, and the giving up of carbonic acid.

External Apertures.—1. On the concave upper face of the basal joint of the antennule is an elongated aperture, guarded by setæ, which opens into the auditory sac.

2. On the basal joint of the antenna is a conical process, near the apex of which is a small orifice, the opening of the green gland or renal organ.

3. Between the mandibles is a large longitudinal slit, the mouth.

4. On the basal joints of the second pair of legs are the generative apertures of the female.

5. On the basal joints of the fourth pair of legs are the generative apertures of the male.

6. Beneath the telson is the anal aperture.

General Internal Anatomy.—When the tergal portion of the cephalo-thorax and abdomen of the crayfish is removed by snipping it away with scissors, and the underlying skin is carefully stripped off, the following organs are visible without further dissection (Fig. 75). Anteriorly in the middle line is the rounded sac-like stomach (*st.*). A whitish bar crosses it transversely, the cardiac ossicle (*c. os.*), and to its anterior edge are inserted two anterior gastric muscles, which are attached to small (procephalic) processes at the base of the rostrum. Smaller muscles (cardio-pyloric) pass back from its hinder edge towards a second shorter transverse bar, and wing-like bars attached to it. This bar is the pyloric ossicle. From this posterior bar and its wings pass back two posterior gastric muscles (*pg.*), which are inserted in the carapace. Between them is a levator muscle (*l. m.*). On either side of the stomach are the cut ends of the powerful muscles (*m. m.*) by which the mandibles are closed. At the sides of the anterior end of the cephalo-thorax, situated low down, are two greenish oval masses, the green glands (*gr. gl.*), surmounted on the inner side by a delicate sac (*s.*).

Behind the levator muscle of the stomach are seen the anterior
lobes of the spermary or testis (*a. t.*), the median posterior lobe
being seen further back at *p. t.* Between these is the heart (*h.*),
with two dorsal openings or valves (*d. v.*). The heart is held
in position by fibrous bands, the alæ, fixed to its corners.

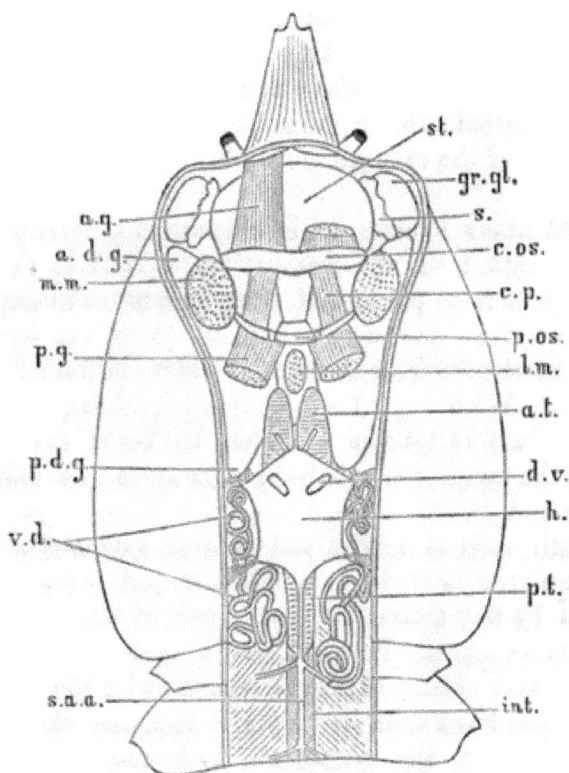

FIG. 75.—CRAYFISH—GENERAL ANATOMY.

a. d. g. Anterior part of digestive gland. *a. g.* Anterior gastric muscle.
a. t. Anterior lobe of testis. *c. os.* Cardiac ossicle. *c. p.* Cardiopyloric
muscle. *d. v.* Dorsal valve of heart. *gr. gl.* Green gland. *h.* Heart. *int.*
Intestine. *l. m.* Levator muscle. *m. m.* Mandibular muscle. *p. g.* Posterior
gastric muscle. *p. d. g.* Posterior part of digestive gland. *p. os.* Pyloric
ossicle. *p. t.* Posterior lobe of testis. *s.* Saccular part of green gland. *s. a. a.*
Superior abdominal artery. *st.* Stomach. *v. d.* Vas deferens.

Anteriorly it gives off a median ophthalmic artery and two
antennary arteries, the cut ends of which are figured in the

16

diagram. Posteriorly it gives off a superior abdominal artery (*s. a. a.*), which runs along the dorsal surface of the intestine (*int.*), which takes a straight course along the median line to the telson. The coiled tubes on either side of the heart are the vasa deferentia (*v. d.*), and the masses of brown cæca belong to the digestive gland or so-called liver (*a. d. g., p. d. g.*).

In the female there is in place of the testis an ovary of somewhat similar form, but shorter. In the breeding season it is large and mottled with berry-like eggs. The oviducts are straight thin-walled tubes running nearly vertically downwards.

The muscles of the crayfish are white, and the blood colourless.

The Alimentary System.—The crayfish is a voracious feeder. The food is seized by the forceps, transferred to the chelate legs, which tear it to pieces and then pass it on to the maxillipeds and maxillæ, the exopodites of which are in constant trembling motion. The mandibles form powerful crushing lateral jaws. With regard to this jaw-apparatus, as compared with that of the vertebrate, it should be noted (1) that it is external to the mouth ; (2) that the jaws move from side to side, and not up and down.

The mouth itself is median and ventral, bounded at the sides by the mandibles, in front by a shield-shaped plate or labrum, and behind by the united bases of a pair of fleshy lobes which form the metastoma. It leads into a short, wide œsophagus lined with a thin chitinous coat continuous with the exoskeleton. Salivary glands have been described as occurring in the walls of the œsophagus, on the metastoma and first maxilliped, the coxopodite of which is habitually tucked into the mouth.

The stomach is large and sac-like, and is divided into a larger anterior cardiac chamber and a smaller posterior pyloric chamber. The chitinous lining of the œsophagus is continued into the stomach, and in the dorsal region gives rise to a *gastric mill* composed of more or less calcified ossicles. Posterior to this it is folded inwards into the narrowed pyloric regions as valvular projections beset with hairlike setæ. This fold has been termed the *filter* or strainer.

To facilitate his study of the ossicles of the gastric mill, the student should remove the stomach with a portion of the œsophagus and intestine, and should boil it in a dilute solution of caustic potash to remove the softer parts. Making a ventral incision along its whole length, he should then pin it out under water, so as to expose the dorsal surface. Fig. 76 A. is a diagram of the ossicles thus seen; B. shows the ossicles separated in the mid-line, and seen from the ventral aspect; C. is a longitudinal section of the whole stomach.

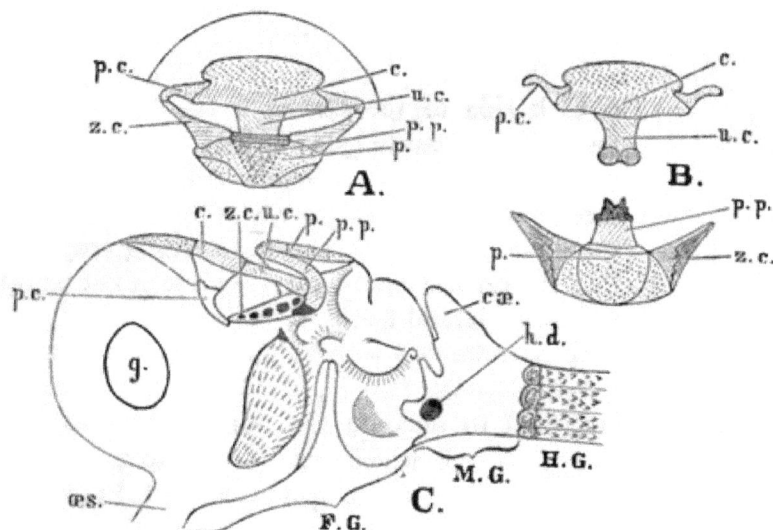

FIG. 76.—CRAYFISH—STOMACH.

A. Gastric ossicles from above. B. The same from below, separated transversely. C. The stomach in longitudinal section.

c. Cardiac ossicle. cœ. Cæcum. g. Mark of gastrolith. h. d. Duct of digestive gland. F. G. Fore-gut. H. G. Hind-gut. M. G. Mid-gut. œs. Œsophagus. p. Pyloric ossicle. p. c. Pterocardiac ossicle. p. p. Prepyloric ossicle. u. c. Urocardiac ossicle. z. c. Zygocardiac ossicle.

The essential parts of the gastric mill are :—

1. (a.) An anterior transverse plate, the *cardiac* ossicle (c.), the hinder part of which (shaded) is white and calcareous, the part in front (dotted) transparent.

   (b.) A smaller posterior transverse plate, the *pyloric* ossicle (p.), the greater part of which is transparent.

2. Between these two in the mid-line—
    (*c.*) An anterior median bar, the *urocardiac*, passing down-
        wards and backwards, the posterior end of which
        (see B.) has two oval brown convex surfaces.
    (*d.*) A posterior median bar, the *prepyloric*, passing
        downwards and backwards from the anterior edge
        of the pyloric to the posterior end of the uro-
        cardiac. Its ventral end carries a large brown
        bifurcated median tooth. This ossicle is shown
        in A. dotted, as seen through the transparent
        pyloric.
3. Between the transverse plates on either side there are—
    (*e.*) On each side of the cardiac a small curved an-
        terior lateral ossicle (*pterocardiac*).
    (*f.*) Between the ventral or outer end of each ptero-
        cardiac and the pyloric plate, a large posterior lateral
        bar—the *zygocardiac*. This has a strong projection
        into the cardiac cavity, carrying a series of power-
        ful brown lateral teeth.

The main muscles connected with this apparatus are (1) the
anterior and posterior gastric (see 75, *a. g.* and *p. g.*), which, by
their contraction, tend to separate the transverse plates; and
(2) the cardiopyloric (*c. p.*), which tend to approximate these
plates. When the plates are drawn apart (see 76, C.) the lower
end of the prepyloric, with its powerful median tooth, is drawn
downwards and forwards. At the same time the zygocardiacs,
with their strong lateral teeth, are swung inwards by the curved
pterocardiacs; so that the median and lateral teeth clash in the
middle line. This the student will readily see when he experiments with the stomach he has removed and cleaned with
potash.

The filter cannot here be minutely described. The student
will, however, readily see how the canal is closed in by valves
and by the approximation of its walls, being indeed reduced to
a narrow triradiate slit. At the opening of the pyloric portion
into the intestine there is a median dorsal, and on each side two
lateral, pyloric valves. Here the chitinous lining ceases, and a

short section of the canal has naked epithelial walls. This is the mid-gut (76, C., M. G.). Near its commencement there is on each side the wide opening of the duct (*h. d.*) of the digestive gland, and dorsally a short cæcum (*cæ.*).

Behind the mid-gut there are six squarish elevations with a chitinous investment. From them longitudinal ridges pass backwards, with a slight spiral twist, to the anus. The ridges are beset with small papillæ. All this part of the intestine is known as the hind-gut, and is lined throughout with a chitinous membrane continuous with the cuticle of the exterior.

The digestive gland is made up of an immense number of cæca. It occupies a large space on either side of the mid-gut, each lateral mass being slightly three-lobed. Each lobe has a main duct, the three uniting to form the common duct of that side. The organ, as a whole, is to be regarded as a much-divided diverticulum of the mid-gut.

Its product is a yellow fluid, which has a slightly acid reaction, converts proteids into peptones, acts on starch, converting it into sugar, and breaks up olive oil into an emulsion. The producing cells are of two kinds; (1) hepatic cells containing oil-globules, and giving rise to animal colouring matter; and (2) ferment cells giving rise to the digestive secretion. It has been proposed to call this digestive gland the hepato-pancreas. The student must not regard it as homologous with these glands in the vertebrata, but merely as partly performing the same functions.

On each side of the stomach (at *g.*, Fig. 76, C.), there is formed during the summer months a calcareous body called the gastrolith. During ecdysis these are shed into the interior of the stomach, and there broken up and dissolved. If they are not present, ecdysis would seem to be unsatisfactorily performed. It is said that during exuviation a layer of processes, or "casting hairs," is formed in such a way as to aid in loosening the old shell from the new.

The Heart and Circulation.—In a recently-killed crayfish the systole and diastole (contraction and relaxation) of the heart may be seen. This organ has the form shown in Fig. 75, *h.*, and Fig. 77, A. It lies in the pericardial sinus, being suspended therein by fibrous bands, the alæ. Six main valvular apertures admit the blood from the pericardial sinus, two dorsal (75, *d. v.*), two ventral, and two lateral (77, A., *l. v.*). The blood

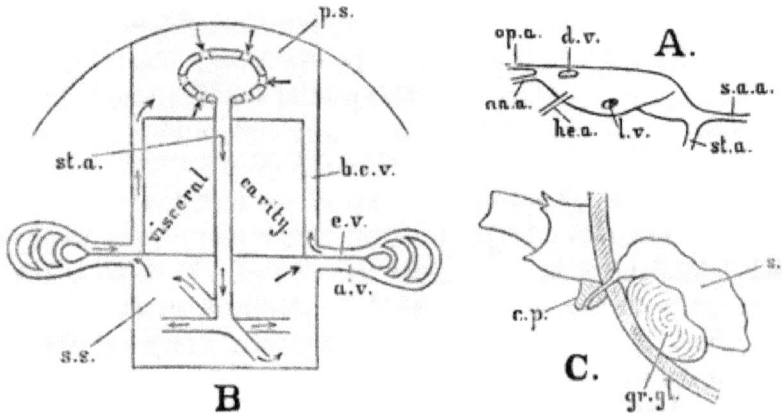

FIG. 77.—CRAYFISH—HEART: CIRCULATION: GREEN GLAND.
A. Heart from left side. B. Diagram of Circulation in the Thorax.
C. Green Gland.

a. v. Afferent vessel to gill. *an. a.* Antennary artery. *b. c. v.* Branchiocardiac vein. *c. p.* Conical process at base of antenna. *d. v.* Dorsal valve of heart. *e. v.* Efferent vessel from gill. *gr. gl.* Excretory part of green gland. *he. a.* Hepatic artery. *l. v.* Lateral valve of heart. *op. a.* Ophthalmic artery. *p. s.* Pericardial sinus. *s. a. a.* Superior abdominal artery. *s.* Saccular part of green gland. *s. s.* Sternal sinus. *st. a.* Sternal artery.

leaves the heart by seven arteries, the positions of which are seen in the side view of the heart (77, A.). The median anterior ophthalmic artery (*op. a.*) runs forward over the stomach, to supply the eyes and region adjacent. On either side of this are the antennary arteries (*an. a.*) which supply the stomach, mandibular muscles, green gland, and antennary appendages. Two hepatic arteries (*he. a.*) are mainly distributed to the liver. These five start from the anterior end of the heart. From the posterior end the superior abdominal (*s. a. a.*) runs back along the intestine, supplying it and the dorsal part of the abdomen. A sternal artery (*st. a.*) runs downwards and slightly forwards, gives off branches to the vas deferens or oviduct, pierces the ventral nerve-chain, and enters the sternal sinus within the endophragmal system. Thence one branch runs forward, supplying the ventral region and limbs of the cephalo-thorax so far as the mouth, another backward to supply the ventral region and limbs of the abdomen.

The arteries break up into capillary ramifications. From these, however, the blood is not collected into veins, as in the vertebrate, but passes into sinuses irregularly disposed among the internal organs. After passing through these sinuses it eventually reaches the sternal sinus (*s. s.*) beneath the cephalo-thorax. Thence it passes by afferent vessels (*a. v.* in Fig. 77, B., which shows diagrammatically the circulation in the thorax) to the gill-plumes; returning by efferent vessels (*e. v.*) to the base of the gill, whence it is carried to the pericardial sinus (*p. s.*) by branchio-cardiac vessels (*b. c. v.*). Thus the blood is aerated on its way to the heart. In the cod-fish it is aerated in the gills on its way from the heart.

The blood itself is a clear colourless fluid, which coagulates to form a clot. It contains colourless nucleated amœbiform corpuscles.

The Green Glands.—The excretory organs by which nitrogenous waste is eliminated from the blood of the crayfish are the green glands (**Fig.** 75, *gr. gl.*, and Fig. 77, C.). On the basal joint of the antenna there is a small conical prominence (77, C., *c. p.*), with an aperture on the inner side of its summit. This leads into the thin-walled sac (*s.*), which may be regarded as the much-expanded end of an immensely convoluted tube, which ends blindly in a triangular yellowish-brown lobule. This convoluted tube forms the body of the gland (*gr. gl.*) The secretion contains large quantities of uric acid, with very small traces of guanin—a substance analogous in some respects to uric acid, but less highly oxidised.

The Nervous System.—The nervous system of the crayfish presents us with a chain of ganglia running along the ventral aspect, connected by commissures.

After the crayfish has been opened up as shown in Fig. 75, and the relations of the alimentary system, the heart, and the organs of generation (p. 252) have been studied, the abdominal muscles should be carefully removed, the ganglia of this region being thus displayed. On removal of the viscera of the cephalo-thorax it is seen that the nerve-chain passes into the canal formed by the apodemata of the endophragmal system. The roof of this canal should be carefully cut away so as to expose the whole chain, the anterior ganglion of which lies between the eyes at the base of the rostrum. The chain is so easily exposed and so readily understood that a figure is here unnecessary; though the student should of course, in this as in all cases, draw one from his own dissection.

There are thirteen pairs of ganglia. In the abdomen there are six, slightly bi-lobed, connected by commissures which, on closer inspection, are seen to be double. In the thorax there are six. The most anterior is obviously multiple, the result of the fusion of two or more pairs. Lying behind and beneath the gullet it is called sub-œsophageal. This, the second, and the third thoracic ganglia, are connected by somewhat elongated double commissures. Between the third and fourth the commissures are short. Between the fourth and fifth they separate so as to allow of the passage of the sternal artery. The fifth and sixth lie close together, almost in contact. From the sub-œsophageal there pass forward two long commissures, which separate to pass one on each side of the gullet, uniting anteriorly in one large supra-œsophageal or cerebral ganglion.

From the ganglia and inter-ganglionic commissures nerves are given off. From the cerebral ganglion two antennary, two antennulary, and two optic nerves are readily traced. From the sub-œsophageal ganglion nerves may be traced to the muscles of the mandibles, maxillæ, and maxillipeds. Thus these two main ganglia innervate the first nine segments. The other eleven ganglia are segmentally arranged, and each innervates its own somite and appendages.

Just anterior to the gullet the commissures give origin to nerves, which meet in the mid-line and pass backwards as an unpaired nerve (anterior visceral), which breaks up into branches which innervate the stomach. From the last abdominal ganglion a nerve (posterior visceral) passes forwards to innervate the hind-gut.

It is probable that the original, *i.e.* not merely responsive, activities of the crayfish depend on the integrity of the supra-œsophageal ganglion, which also probably possesses an inhibitory and co-ordinating power over the lower centres. The sub-œsophageal ganglion, however, may have some co-ordinating power over lower centres. If both these higher centres be removed the crayfish, as an organism, is probably dead ; but the individual life and activity of the legs, chelæ, and swimmerets, continues vigorously but quite aimlessly.

Special Senses.— (1.) *Touch.*—Scattered over the surface of the body, or grouped in clusters at certain points, are a number of *setæ* (78, A. and B.) which are probably organs of touch. Each is a hollow diverticulum of the body-wall and exoskeleton. During life they are filled with delicate prolongations of the epithelial layer of the integument.

(2.) *Smell.*—From the seventh or eighth to the last joint but one of the exopodite of the antennule, there are, on the under side of each joint, two bundles of curiously-modified setæ (78, A. and C.). These are regarded as olfactory.

(3.) *Taste.*—No special organs of taste have been discovered. Such organs should be sought for on the metastoma, or perhaps on the maxillary appendages. It is scarcely probable that this sense is altogether wanting.

(4.) *Hearing.*—The auditory sacs are lodged in the basal joint of the antennule. The external aperture is in the concave upper face, its outer lip being guarded by closely-set setæ. Within the sac are delicate feathered auditory setæ (78, D.) implanted on either side of a ridge. Their tips are often imbedded in a gelatinous mass in which there are minute particles of sand (*s.*)

(5.) *Sight.*—The eye has a convex *cornea* (78, E., *co.*) continuous with the exoskeleton, and marked out into a number of facets, which are square, except near the edges of the eye where they become polygonal and eventually hexagonal. This is readily seen in a dried ophthalmite under a low power.   Beneath each facet is the end of a *crystalline cone* (*cr.*), which narrows in the deeper part of the eye, and then swells again to form a *striated*

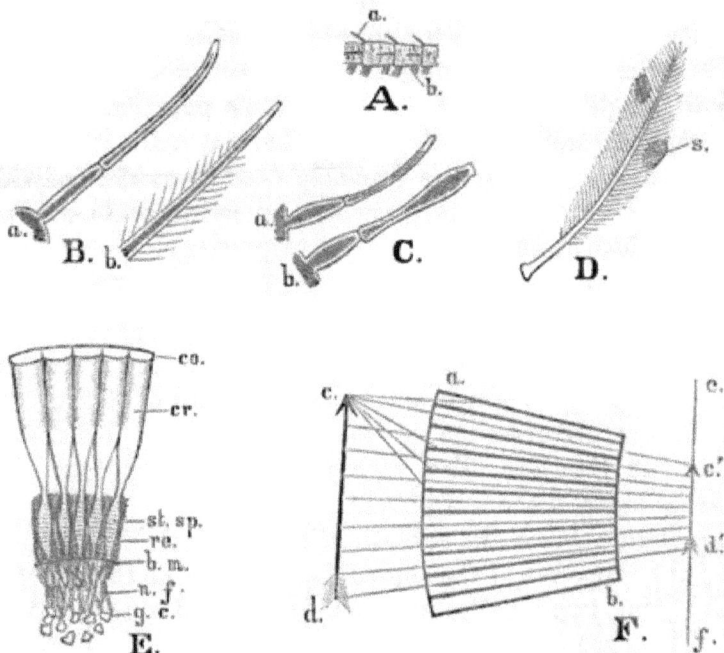

FIG. 78.—CRAYFISH—SENSE ORGANS.

A. Joints near the end of the exopodite of the antennule, showing tactile setæ at *a* and olfactory setæ at *b*.

B. Tactile setæ.—*a.* From the tip of the endopodite of the antennule.   *b.* From a tuft on the antennulary protopodite.

C. Olfactory setæ.—*a.* From the side.   *b.* From above.

D. Auditory seta, with grains of sand at *s*.

E. Portion of section of eye.—*b. m.* Basilar membrane.   *co.* Cornea.   *cr.* Crystalline cone.   *g. c.* Ganglion cells.   *n. f.* Nerve-fibres.   *re.* Pigment.   *st. sp.* Striated spindle.

F. Diagram of arthropod vision.—*a. b.* Transparent rods, with pigment between.   *c. d.* Object.   *c′. d′.* Its image on the screen *e. f.*

spindle or *rhabdom* (*st. sp.*) marked with delicate transverse lines. At the inner end of the striated spindles there seems to be a *basilar membrane* (*b. m.*), beneath which are the nerve fibrils and cells of the *optic bulb.* The *connective rods* between the crystalline cones and striated spindles may after treatment with osmic acid and teasing be resolved into four fibres. The inner ends of striated spindles are probably continuous with the nerve-fibrils of the optic bulb. Pigment is developed round the outer ends of the crystalline cones, the striated spindles, and the outer nerve-fibrils of the optic bulb.

Two theories of arthropod vision have been advocated: (1) that of compound vision; (2) that of mosaic vision. According to the former, a separate image of the object seen is formed at the end of each cone and spindle, the multitude of images being presumably co-ordinated by the arthropod cerebral ganglion, just as two images are in us by the human brain. This view finds few supporters to-day. The principle of the other view, that of mosaic vision, will be understood with the aid of the diagram 78, F. At *a. b.* are a number of transparent rods separated by pigmented material absorbent of light. At *c. d.* is an arrow placed in front of them. At *e. f.* is a screen placed behind them. Rays of light start in all directions from any point *c.* of the arrow; but of these only that which passes straight down one of the transparent rods reaches the screen. Those which pass obliquely into other rods are absorbed by the pigmented material. Similarly with rays starting from other points of the arrow. Only those which pass straight down one of the rods reach the screen. Wherefore there is thrown on the screen a reduced image *c'. d'.* of the arrow. The image is not continuous, but stippled.

So much for the general principle of so-called mosaic vision. According to this theory the several points of the object which excites vision stimulate corresponding points in the recipient organ. To the question, What are the true recipient end organs of the optic nerve? several answers have been suggested.

1. The crystalline cones are the end organs, and are continuous through the basilar membrane with the nerve-fibrils.

2. The cones and spindles are simply dioptric, and beneath the basilar membrane is a retina which is the true recipient.

3. The crystalline cones are dioptric, the connective rods and the striated spindles being the true end organs continuous with the nerve-fibrils.

The third answer is that most generally accepted. On this view the spindles, nerve-fibrils, and ganglion-cells answer to the vertebrate retina. Note that, if this be so, in the arthropod the recipient end-organs are turned towards the source of light, and not away from it as in the vertebrate.

Organs of Reproduction.—(1.) *In the Male.*—The spermary, or **testis,** has two anterior lateral lobes and one median posterior lobe. Its position is seen in Fig. 75, *a. t., p. t.* From the points where the anterior lobes join the median lobe there arise the much-coiled vasa deferentia, which, after a convoluted course, pass downwards **to open** on the basal joints of the last thoracic appendages.

In minute structure the testis shows an immense number of **vesicles attached to the ducts which lead** to the vasa deferentia. **These acini are** aggregated into lobules. The large nucleated cells which line the cavities of the acini multiply rapidly with karyokinesis (p. 89), and are eventually converted into spermatozoa of the peculiar and unusual form shown highly magnified in Fig. 79, A. In the vasa deferentia they become invested **with a viscid secretion.** The vermicelli-like matter thus produced is deposited by the male on the posterior thoracic and anterior abdominal somites of the female.

(2.) *In the Female.*—The **ovary is** in shape somewhat like a shortened thickened testis. The short wide oviducts pass downwards from its ventral side **to open on** the basal joint **of the second** leg (sixth thoracic appendage). Note, as something different from what we have seen in the frog, fowl, or rabbit, **that the lumen of the oviduct communicates** with the hollow **interior of the ovary.** The development of the ova may be thus summarised.

1. A group of cells forms a papilla projecting into the cavity, and covered by a structureless membrane, the *membrana propria* (*m. pr.*) Such a group forms an ovisac.

2. The central cell in the ovisac grows at the expense of the others, and becomes the ovum. Certain of the others form a layer round it, and constitute the epithelium of the ovisac (*ep.*)

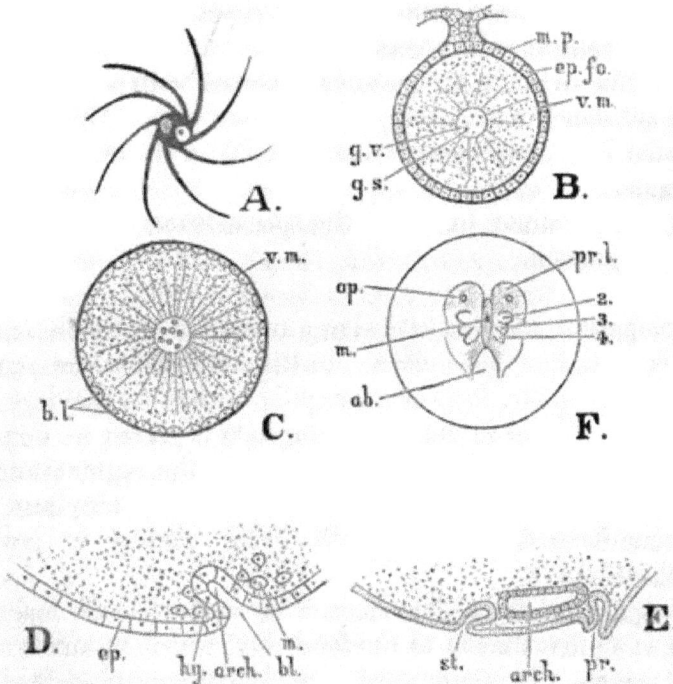

FIG. 79.—CRAYFISH—DEVELOPMENT.

A. Spermatozoon. B. **Ovum.**—*ep. fo.* Epithelium of follicle. *g. s.* Germinal spot. *g. v.* Germinal **vesicle.** *m. p.* Membrana propria. *v. m.* Vitelline membrane.

C. Ovum in which segmentation has produced a blastodermic layer, *b.l.* *v. m.* Vitelline membrane.

D. Invagination of blastodermic layer.—*arch.* Archenteron. *bl.* **Blastopore.** *ep.* Epiblast. *hy.* Hypoblast. *m.* Mesoblast.

E. Formation of fore and hind-gut.—*arch.* **Mesenteron.** *pr.* Proctodæum. *st.* Stomodæum.

F. View of developing embryo.—*ab.* Abdomen. *m.* Stomodæal invagination for the mouth. *op.* Optic pit. *pr. l.* Procephalic lobe. 2, 3, 4. Second, **third, and fourth** appendages.

3. The nucleus of the ovum becomes the germinal vesicle (*g. v.*).

4. Within it there appear numerous small corpuscles, the germinal spots (*g. s.*).

5. The protoplasm of the ovum is largely converted into deep brownish-yellow vitellus or yolk.

6. A vitelline membrane (*v. m.*) is formed around the ovum with its vitellus (Fig. 79, B.).

7. The ovisac bursts and the ovum is set free. The vitelline membrane continues to invest it.

8. In the oviduct it becomes invested with a viscid transparent substance which sets into a tougn case. This becomes suspended by a stalk of the same material to the swimmerets of the female.

The exact manner in which the spermatozoa unite with the ova is not known.

Development.—Owing to the large amount of food-yolk segmentation is partial or incomplete. In the fowl, where the segmentation is incomplete, it is also irregular, owing to the fact that the germinal disc of formative protoplasm is placed at one pole of the egg (telolecithal). In the crayfish the segmentation is regular, owing to the fact that the formative protoplasm is at the circumference, the food-yolk being central in position (centrolecithal).

By segmentation, a blastodermic layer (*b. l.*) is formed (Fig. 79, C.) as an investment to the food-yolk, which is arranged in conical masses. It is composed of nucleated protoplasmic plates, but the relation of the nuclei of these plates to the germinal vesicle is not well ascertained. Out of these protoplasmic plates the embryo is fashioned.

The first change is that the blastodermic layer becomes thickened over an area $\frac{1}{25}$ inch long in the neighbourhood of the pedicle, forming the germinal disc. Invagination takes place at the hinder end of this area (79, D.). Thus we have a differentiation of the blastodermic layer into epiblast (*ep.*) and hypoblast (*hy.*) with the formation of an archenteron (*arch.*) and a blastopore (*bl.*).

The mesoblast cells make their appearance (*m.*) in the neighbourhood of the blastopore between the hypoblast and the epiblast, its cells being probably derived from the former.

Thence it gradually spreads, first over the sternal and eventually over the tergal aspect of the embryo.

The blastopore completely closes. The archenteron gives rise to the mid-gut, from which the digestive gland arises as a diverticulum. The mouth and fore-gut arise as a stomodæal invagination (79, E., *st.*); the vent and hind-gut as a proctodæal invagination (*pr.*). These two invaginations do not at first communicate with the archenteron.

The anterior end of the embryo becomes marked out by the formation of procephalic lobes (Fig. 79, F., *pr. l.*), in the midst of each of which is an optic pit (*op.*). The posterior end is marked out by a rounded elevation, the rudiment of the abdomen. Between these two is a narrow groove, in the centre of which is the stomodæal invagination (*m.*). The sides of this groove grow inwards to form two solid cords of cells of epiblastic origin. At its front end a mass of cells grows inwards and connects the two cords in front of the mouth. The anterior mass gives rise to the cerebral ganglion ; the two cords which partially coalesce in the mid-line differentiate into the post-œsophageal chain of ganglia. Thus the central nervous system has an epiblastic origin.

The first appendages to appear are the rudiments of the antennæ (2), antennules (3), and mandibles (4). In certain other crustacea these are greatly developed as embryonic oar-like appendages, and the embryo at this stage becomes free. Such an embryo is called a *nauplius*. In the crayfish these appendages do not thus develop into embryonic organs, nor does the embryo become free till a much later stage. But the formation of a delicate cuticle after the appearance of these three appendages seems, in the crayfish or allied forms, to symbolise the nauplius condition. The other thoracic appendages appear as bud-like protuberances in regular order. All are similar at first, but subsequently differentiate.

When it is hatched, the young crayfish differs considerably from the adult. The cephalothorax is more convex and larger. The rostrum is short and bent down. The sterna of the thorax are relatively wider. The chelæ are more slender, and their

tips are incurved. The dactylopodites of the third and fourth
pairs of legs are hooked. The appendages of the first abdominal
somite are undeveloped. Those of the last are enclosed
within the telson. It is probably set free at the first ecdysis.
Thus, as we found to be the case in the vertebrate, after transformation is over there still is development.

The student should endeavour to obtain, in the spring
or early summer, a female crayfish with developing embryos.
He will find a study of the development of their external
features a matter of no great difficulty and of very great
interest.

CHAPTER XIII.

THE COCKROACH.

THE Common Cockroach (*Periplaneta* (Blatta) *orientalis*) is not an indigenous insect, but an immigrant from the East that has domesticated itself in European kitchens. Although commonly called a black-beetle, the cockroach does not belong to the true beetles (*Coleoptera*), but to a less modified group of insects, including the Locusts and Grasshoppers, known as *Orthoptera*. Unlike the higher and more modified insects (Butterflies, House-flies, Beetles, Bees, Ants), which undergo post-embryonic metamorphoses, passing through a soft-bodied, worm-like larva stage, and an inactive pupa or chrysalis stage, the cockroach, though it exhibits a certain amount of development after birth, undergoes no true post-embryonic metamorphosis. The minute cockroaches that emerge from the egg are little white six-legged creatures, with black eyes and without wings. They change their skin, or moult, seven times, rudiments of wings and wing-covers making their appearance in the later larval stages. After the seventh ecdysis (which is said to occur when the insect is about four years old) the cockroach is adult. Those insects which undergo complete metamorphosis are known as *Holometabola*, while those in which the metamorphosis is absent or, as in the cockroach, incomplete, are grouped together as *Ametabola*. In this respect, therefore, and in certain others, the cockroach is a little-specialised or primitive insect. Its family is also a very ancient one, remains of cockroaches being found in the Silurian strata.

External Characters.—The body is marked out into distinct regions. A well-developed head is followed by a narrow neck,

17

and this by a *thorax* divisible into a fore, mid, and hind segment (*pro-*, *meso-*, and *metathorax*).   Posterior to the thoracic region is the abdomen, in which ten segments may be made out, while two small triangular *podical plates*, one on each side of the vent, are held by some anatomists to represent an eleventh.

The surface of the body and limbs is invested by a *chitinous exoskeleton*, laminar in structure and secreted by a *chitinogenous layer* of cells (hypodermis).   Although the chitinous investment is continuous, it is not everywhere of the same thickness or consistency.   Where mobility is required, it forms a yielding cuticle, elsewhere it is thickened into harder and less yielding plates (*sclerites*) or tubes.   The whole cuticle, with its thickened elements, is shed at the periods of ecdysis, a median dorsal longitudinal fissure being formed in the thorax, through which the insect escapes from its old skin.

A front view of the head is shown in Fig. 80, A.   In the upper region is seen the *epicranial suture* (*ep.*), the lateral branches of which end in small rounded *white spaces* (*w. s.*).   Near these spring the long *antennæ* (*an.*) from an *antennary fossa* (*an. f.*). External to these are the large compound (*polymeniscous*) eyes. The mid-region of the face (*clypeus, cl.*) narrows below somewhat suddenly at the *ginglymus,* where the large *mandibles* (*l. mn., r. mn.*) are articulated.   The flap-like piece hanging from its lower edge is the *labrum* (*lb.*), behind which lies the mouth.   Below the mandibles, one of which is represented as widely open, are the *maxillæ* (*mx.*), to which are attached long *maxillary palps* (*mx. p.*). The shorter palps (*labial palps, l. p.*) belong to the *labium.* Looking down upon the head from above, there are seen on either side of the mid-line the *epicranial plates*, separated by the epicranial suture.   On either side of these are, anteriorly the eyes and posteriorly the *genæ*, which, passing down behind the eyes, form the sides of the exoskeleton of the head, reaching as far as the mandibles, and ending somewhat pointedly above the ginglymus.   A posterior view of the head shows a large shield-shaped *occipital foramen*, the lateral margin of which is strengthened by a thickened rim continuous with the tentorium or internal skeleton of the head.   The top and sides of this rim are bounded by the epicranial plates, which end below in facets for the articulation of the mandibles. At the inferior boundary of the foramen there is a large plate, the *sub-mentum*, which forms part of the mouth organs.

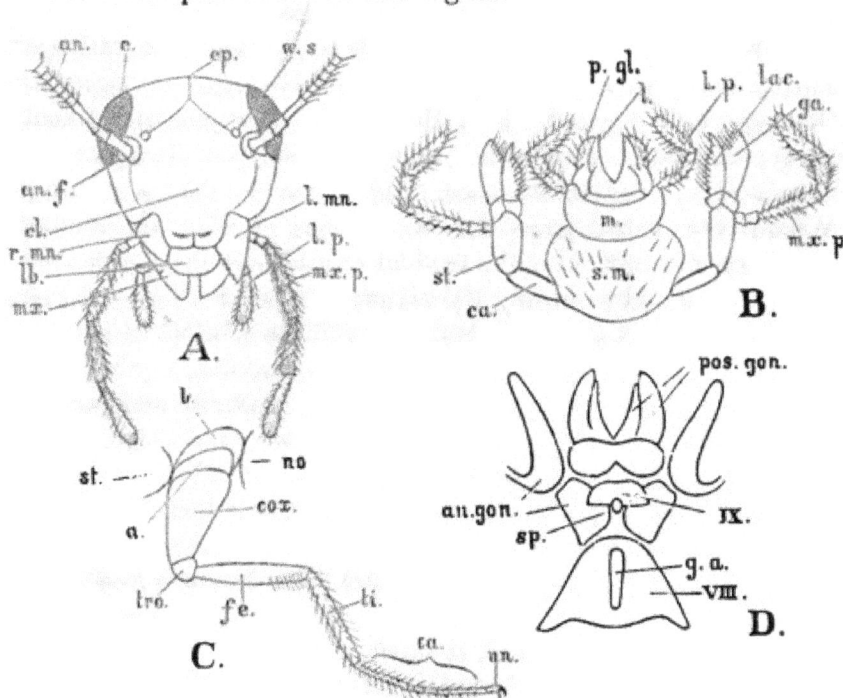

FIG. 80.—COCKROACH—APPENDAGES.

A. Head of cockroach; front view. B. Mouth organs removed. C. A leg. D. Gonapophyses of female.

a. and *b.* Lateral pieces. *an.* Antenna. *an. f.* Antennary fossa. *an. gon.* Simple gonapophysis. *ca.* Cardo. *cl.* Clypeus. *cox.* Coxa. *e.* Eye. *ep.* Epicranium. *fe.* Femur. *g. a.* Genital aperture. *ga.* Galea. *l.* Lacinia of labrum. *l. mn.* Left mandible. *l. p.* Labial palp. *lac.* Lacinia of maxilla. *lb.* Labrum. *m.* Mentum. *mx.* Maxilla. *mx. p.* Maxillary palp. *no.* Notum. *p. gl.* Paraglossa. *pos. gon.* Bifid gonapophysis. *r. mn.* Right mandible. *s. m.* Sub-mentum. *sp.* Opening of spermatheca. *st.* (in B.) Stipes. *st.* (in C.) Sternum. *ta.* Tarsus. *ti.* Tibia. *tro.* Trochanter. *un.* Ungues. *w. s.* White spot. viii. ix. Abdominal sterna.

The mouth organs posterior to the mandibles are shown in Fig. 80, B. The median part results from the fusion of a pair of labial appendages. It consists of a *sub-mentum* (*s. m.*) and *mentum* (*m.*), beyond which the labium is bifid, there being an inner part, the *lacinia* (*l.*), and an outer part, the *galea* or *paraglossa* (*p. gl.*), on each side, while *labial palps* (*l. p.*) spring from the mentum. External to the labium are the *maxillæ*. Each has a basal part, the *cardo* (*ca.*), which articulates with a pedicel arising from the occipital rim. Articulating with the cardo is the *stipes* (*st.*), beyond which the maxilla is composed of two parts: the inner, with sharp point and spinose blade, is the *lacinia* (*lac.*); the outer, softer and setose, is the *galea* (*ga.*). Where the galea joins the stipes springs the five-jointed *maxillary palp* (*mx. p.*) The student should find little difficulty in detaching and mounting the organs as figured. To get rid of the soft tissues it should be boiled in a dilute solution of potash.

It is not improbable that the mandibles, maxillæ, and labium,[1] are homologous with the mandibles, first maxillæ, and second maxillæ of the crayfish. It is not certain whether the antennæ of the cockroach and crayfish are homologous.

Within the head is a complex, somewhat cruciform, chitinous framework, the *tentorium,* for the support of the gullet and brain, and the attachment of muscles. Nerve cords pass through a central aperture.

In the neck there are eight small cervical plates or sclerites: two dorsal, somewhat triangular and apposed in the mid-line; four (two on each side) lateral, and somewhat oblique; two ventral, transverse and median, lying one behind the other.

In each somite of the thorax there is a distinct dorsal or tergal plate. The anterior of these, the *pronotum,* is the largest. The posterior plate (*metanotum*) is slightly smaller than the mid-plate (*mesonotum*). In the adult male the meso- and metanotum are thin and semi-transparent. In each somite of the thorax there is also a ventral plate, the sternum. The most anterior (*prosternum*) is coffin-shaped, with the long axis placed longitudinally. The mid-plate (*mesosternum*) shows indications in the male of

[1] The labrum is, in some insects (*e.g.* bees and their allies), regarded by certain recent observers as originally double, and as an extreme example of the coalescence of a pair of appendages. This is, however, doubtful.

being divided into **two lateral halves,** and is so divided in the female. The *metasternum* is the largest of the three sterna, and is in both sexes completely divided into two lateral divisions, between which is wedged in the base **of a** Y-shaped **chitinous** prop (*post-furca*) for the support of the nerve-chain **and for** attachment of muscles. A similar prop (*medi-furca*) is connected with the mesosternum. An anti-furca is, **if at all,** incompletely represented.

Each somite **of the thorax** carries a large pair of legs, of which the anterior **are** the smallest, the posterior the largest. Like those **of** the crayfish, they are jointed (arthropod type); the divisions being *coxa* (proximal), *trochanter*, *femur*, *tibia*, and *tarsus*, composed of six joints (or five joints, the last being sub-divided). The distal division of the tarsus bears claws or *ungues* (80, C.).

The limbs do not articulate directly **with the sterna or nota,** two lateral pieces intervening. These are shown in 80, C. **One** of these (*a*) is, in the third limb, triangular, and is attached to the coxa. **The other (*b*) is** larger, and **curves** round **the angle** of *a*. It articulates, **on the one hand, with the** sternum (*st.*), and on the **other with** the notum (*no.*). This latter piece (*b*) seems, in the **cockroach,** to belong to the somite ; the former (*a*) to the appendage. But there is some doubt about the homologies of these parts in the insecta.

Male cockroaches have, when adult, **two** pairs of wings. **The** anterior (wing-covers or *tegmina*) are brown and stiff, **and** attached to the second thoracic segment; the posterior pair, attached to the metathorax, are soft and membranous, and, when not used in flight, carried folded up beneath the wing-covers. Both tegmina and wings are strengthened by *nervures*, united by transverse ridges. The disposition of the nervures will be readily seen with **the aid of a** lens. It is a matter of importance **in the classification of these and allied insects. The** old-world cockroaches **seem to have had** transparent **mem-**branous wing-covers or anterior wings.

In the female the wings remain in a rudimentary condition. The anterior pair (wing-covers) are small, not reaching beyond the middle of the metanotum. The posterior pair are represented by a reticulated pattern on the lateral edges of the metanotum. This suggests or illustrates the view which is otherwise enforced by the study of development, that the wings of insects are homologous with the free edges of meso- and metathoracic terga. Such a wing arises as a sac-like fold of the chitinous cuticle, which passes into the condition of a double lamina, which, by the coalescence of its inner surfaces, becomes a continuous membranous expansion, strengthened by nervures which take their origin in connection with the tracheal tubes.

The abdomen of the female is broader than that of the male. There are ten terga, of which the eighth and ninth are hidden beneath the seventh. The tenth is large, of peculiar shape, and notched posteriorly. The first sternum is small and rudimentary; the second large, with an irregular anterior border; the third and fourth are the broadest. The seventh is very large, boat-shaped and cleft, a curious fold of integument being visible when the two halves are opened out laterally. The parts behind the seventh sternum are somewhat complicated. This is largely due to the infolding of the cuticle to form a genital pouch. The floor of this pouch is formed by the membrane between the seventh and eighth sterna; the back wall of the pouch is formed by the large eighth sternum, which is perforated by the large genital aperture. In the dorsal wall lies the ninth sternum, carrying the orifice of the spermatheca. Further back is the anal opening. The parts are best studied by removing all the ventral region posterior to, and including, the seventh sternum, boiling in dilute caustic potash, and examining under the simple microscope. When the relations of the parts have been made out, the seventh sternum may be carefully removed, and by cutting along the edges of the eighth sternum and bending it back, the remaining parts may be mounted, as shown in Fig. 80, D. In addition to the sterna VIII. and IX., two pairs of appendages, which belong to these somites, and which are termed *gonapophyses*, are seen. The anterior gonapophyses (*an. gon.*) of the eighth somite are somewhat thumb-shaped, and are normally applied together in the mid-line. Their basal moieties are broad, and serve to form part of the roof of the genital pouch. At the base of their distal moieties a chitinous slip passes to the eighth and ninth terga. The posterior gonapophyses (*pos. gon.*), which normally lie above the anterior pair, are divided into two portions, the inner slender and hard, the outer larger and soft. In the cockroach these appendages grasp the egg-capsule while it is being extruded from the genital pouch. The notched dorsal edge of the capsule bears the imprint of the inner portion of the posterior gonapophyses. In some insects this apparatus is curiously modified to form an *ovipositor,* by means of which holes are bored in earth, wood, or animal tissue, for the reception of the ova. In the bee the *sting* is a specialised form of the same apparatus.

Two cigar-shaped sixteen-jointed processes, the *cerci,* which probably have a sensory function, spring in both sexes from the edges of the tenth somite. On either side of, and slightly dorsal to, the vent, there are, in both sexes, triangular *podical plates,* regarded by some anatomists as the terga of an eleventh abdominal somite.

Ten terga and nine sterna can be made out in the male; but the first sternum is, as in the female, rudimentary. Attached to the ninth sternum are two slender *sub-anal styles.* There is also a complex system of plates and hooks, concealed within the ninth sternum, and constituting the male genital armature.

If we take the podical plates to represent an eleventh abdominal segment, the number of segments in the cockroach would seem to be seventeen, three cephalic, three thoracic, and eleven abdominal.

The *apertures* of the body are as follows :—

1. The *mouth,* lying between the labrum and the labium, and bounded laterally by the mandibles and maxillæ.

2. The *vent,* lying between the triangular podical plates.

3. Twenty *spiracles,* ten on each side of the body, forming the external orifices of the tracheal system. Eight on each side are abdominal, between the sterna and terga; two on each side are thoracic, between the pro- and meso-, and the meso- and meta-thorax. The thoracic stigmata are provided with external valves.

4. In the female, (*a*) the genital opening on the eighth sternum ; (*b*) the opening of the spermatheca on the ninth sternum.

5. In the male, (*a*) the opening of the ejaculatory duct behind the ninth sternum ; (*b*) the opening of a subsidiary gland on a double hook, which forms part of the external genital armature.

At these apertures the chitinous exoskeleton is inflected inwards.

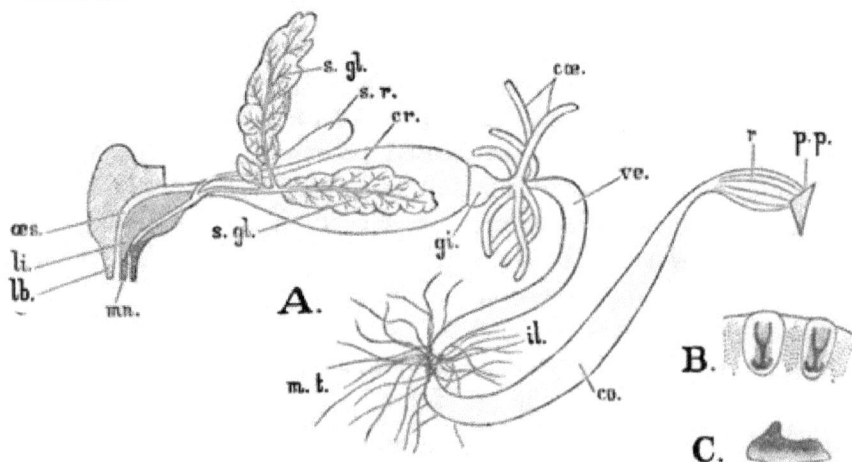

FIG. 81.—COCKROACH—ALIMENTARY CANAL.

A. Alimentary canal dissected out.   B. Part of gizzard under simple micro-scope.   C. One tooth of the same from the side.

b. c. Buccal cavity.   *cœ.* Pyloric cæca.   *co.* Colon.   *cr.* Crop.   *gi.* Gizzard. *il.* Ileum.   *lb.* Labrum.   *li.* Lingua.   *mn.* Mandibles.   *m. t.* Malpighian tubules.   *œs.* Œsophagus.   *p. p.* Podical plate.   *r.* Rectum.   *s. gl.* Salivary glands.   *s. r.* Salivary receptacle.   *ve.* Chylific ventricle.

The Alimentary Canal.[1]—The alimentary canal of the cockroach (Fig. 81) has three divisions: (1) a *stomodæal section* anteriorly, lined with a chitinous layer continuous with the

[1] The cockroach is conveniently dissected, under the simple microscope, or with the aid of a watch-maker's lens, in a shallow tin or glass vessel. In the bottom of such a vessel melt some paraffin, and mix in lamp-black. Allow the paraffin to cool and solidify as a firm cake at the bottom of the vessel. With a hot glass rod melt a convenient cavity and imbed therein the cockroach. Dissect under water or alcohol.

exoskeleton at the mouth; (2) the *mesenteron*, without chitinous lining; (3) a *proctodæal section*, lined with a chitinous infolding of the posterior integument. Each section is provided with appendages : (1) the *salivary glands and receptacles*, stomodæal; (2) the *cœcal tubes*, mesenteric; and (3) the *Malpighian tubules*, proctodæal.

The mouth leads into a buccal cavity, into which there projects a lingua (*li.*), or tongue-like process, supported by chitinous plates. Beneath the lingua opens the common duct of the salivary glands and their receptacles. Each salivary gland consists of two principal lobes and a small (anterior) accessory lobe on each side. There is also on each side an elongated salivary receptacle. The ducts of both glands and receptacles are lined with delicately-ribbed chitinous membrane, the finer branches of the salivary ducts end in secreting bulbs or acini. The student must examine a gland and its ducts under the microscope. The two glandular ducts unite in the middle line into a common duct; the ducts from the receptacles do the same; then both these unite into the salivary duct which opens beneath the lingua. The salivary secretion is neutral, and transforms starch into glucose.

The *gullet* is narrow, pierces the nerve-ring, passes through the occipital foramen, traverses the neck, and then, in the thorax, gradually widens into the large *crop* (*cr.*), which is at once followed by the muscular *gizzard* (*gi.*). The chitinous lining of the gizzard is differentially thickened, so as to form a circlet of six teeth, two of which are shown in 81, B., and one, in side view, at C. Between the main teeth are finely-toothed ridges. The toothed part of the gizzard is followed by six prominent setose cushions. The function of the toothed part of the gizzard is to act as a gastric mill, while the setose cushions act as strainers. Some observers, however, contend that, both in the cockroach and in the crayfish, the so-called gastric mill is to be regarded as part of the straining apparatus, and not as a divider of the food. In any case it would tend to squeeze the food pulp. The gizzard ends in a tubular portion which protrudes into the cavity of the succeeding portion of the canal.

This succeeding portion is the *chylific stomach* (*ve.*), constituting the mesenteron or mid-gut of the canal. At its anterior end there are eight *cæcal tubes* (*cæ.*), which have been regarded as a digestive gland, but are more probably diverticula of the chylific stomach, the lining cells of which have taken on the function of secreting a feebly acid fluid, which has the digestive properties of the pancreatic juice of vertebrates. The epithelial cells of the chylific stomach (in which there is no chitinous lining) are very probably absorbent in function.

Marking the junction between the mesenteric and proctodæal segments of the canal, but properly belonging to the latter, are a number (sixty or more) of long thread-like *Malpighian tubules* (*m. t.*), inconspicuously grouped into six clusters. They are lined with large epithelial cells, within which are found crystals containing uric acid, and said to consist of urate of sodium. Similar crystals are found in the lumen of the tube. The Malpighian tubules are therefore undoubtedly excretory in function. In another part of the body, however—namely, in the *fat body* (*corpus adiposum*), which appears as a white cellular mass surrounding the viscera when the cockroach is dissected—crystals containing uric acid have been found. It is conjectured that the urates here elaborated escape by the blood which bathes the perivisceral cavity, and are again taken up by the Malpighian tubules, to be thence finally discharged into the intestine.

Posterior to the Malpighian tubules is the short, narrow, *small intestine* or *ileum* (*il.*), which is separated by an annular fold from the more dilated *colon* (*co.*). This narrows and becomes somewhat constricted before it passes into the *rectum* (*r.*). The rectum is characterised by six longitudinal folds with largely-developed epithelium cells, beneath which are tracheal tubes. Between the longitudinal folds the epithelium is much thinned or absent. Posteriorly the alimentary canal opens by the vent, whence the inflected cuticle passes inwards to line the proctodæal portion of the canal (rectum, colon, and ileum). A transverse section of the stomodæal or proctodæal segment of the canal shows internally the chitinous lining, followed by an epithelial layer, external to which is a muscular layer with circular and longitudinal fibres. These fibres, unlike those of the alimentary canal in vertebrates, are striated. In the mesenteron the chitinous layer is absent, and the epithelial layer specially modified.

The Heart and Circulation.—The heart is a long delicate tube lying beneath the terga of the thorax and abdomen. It is divided into thirteen segments—three thoracic, and ten abdominal; but some of these are inconspicuous, and can hardly be made out in the common cockroach. At its posterior end each of the more conspicuous segments dilates into a median dorsal and two lateral lobes. The lateral lobes conceal paired lateral inlets. A pear-shaped mass of cells, hanging just in front of the constricted opening between any two segments, serves as a valve to close at once this opening and those of the lateral inlets. The heart lies in a pericardial chamber, the floor of which is constituted by a membrane in which are small oval openings. To this membrane fan-shaped muscular bands (*alary muscles*) arising from the anterior margin of each tergum, are attached.

The heart contracts in rhythmical waves from behind forwards; in a young cockroach, shortly after the ecdysis they have been reckoned at eighty per minute. The blood thus driven forwards passes by a slender *aorta* along the dorsal surface of the gullet, and terminates with a trumpet-shaped orifice just in front of the nerve-ring. The blood then passes throughout the body by a system of lacunar spaces. It therefore bathes the various organs, and is not injected into them; hence their diffuse form, seen in the salivary glands, cæcal tubes, and Malpighian tubules, in the reproductive organs, and in the fat-body which occupies the peri-visceral space or cœlom. The blood (which is colourless, with large nucleated amœboid corpuscles), after bathing these organs, reaches the pericardial cavity through the fenestrated membrane. It is doubtful whether there is any distinction into venous and arterial blood —a distinction perhaps rendered unnecessary by the tracheal system.

The Tracheal System.—Aerial respiration is, in the cock-roach, on quite a different plan from that which is found in the air-breathing vertebrate. There the blood is carried to a special sac, the lung, in which it is aerated ; and by the blood the oxygen thus absorbed is distributed through the system. In the insect a number of minute tubes (*tracheæ*) ramify throughout the body and carry air directly to the various organs.

From the spiracles there pass inwards tracheal trunks. Those in the abdominal region give off, (1) anterior and posterior branches, which unite together to form longitudinal lateral vessels ; (2) dorsal branches, which pass up into the pericardial cavity and branch right and left, thus giving rise, by union of the branches, to dorso-lateral arches; (3) ventral vessels running transversely, but coming into relation with ventro-lateral vessels, one on each side of the nerve-chain ; (4) median branches which ramify upon the alimentary and reproductive viscera. From the thoracic stigmata several ventral and dorsal transverse trunks take their origin, which are joined to each other by longitudinal connecting vessels, there being also a well-developed connection between these thoracic tracheæ and the anterior ventral transverse vessel in the abdominal region. From the anterior thoracic stigmata four large vessels, two dorso-lateral and two ventro-lateral, pass to the head.

Examined under the microscope the tracheæ are seen to be transversely striated, and in some cases, where a tube has been torn in mounting, a spiral thread of chitinous material is seen to be drawn out from the trachea. This chitinous lining strengthens the tubes, keeps them open, and gives them elasticity.

The thoracic stigmata are provided with external valves. Those in the abdominal region have internal valvular slits, which can be closed by the action of occlusor muscles on a chitinous bow developed along its anterior edge. Thus the tracheal system can be shut off from communication with the exterior, enabling the insect to live for some time in a noxious atmosphere.

Inspiration and expiration are effected by movements of

the abdomen, though the thoracic region in the cockroach also undergoes a slight change of form. In expiration **the terga** are flattened and depressed, the sterna being at the same time slightly raised. Opposite movements effect inspiration. **The** movements of expiration and inspiration may readily be watched in a tired bee. It must not be supposed **that the** tracheæ are alternately filled and **completely emptied during** this process. But little air **enters or leaves** the body ; in the ultimate ramifications of the vessels **carbonic acid** gas is exchanged for oxygen **by a** process of **diffusion.**

The Nervous System.—The central nervous system (Fig. 82, D.) is, in the cockroach as in the crayfish, ventral, and consists of a double chain, the coalesced ganglia of which are united **by** longitudinal connectives.

The coalescent *cerebral ganglia* (brain) are shown **in Fig. 82.** There are two large *hemispheres* (*h. s.*) from which **the** *optic nerves* (*op.*) pass **off** to the eyes. **Small nerves** proceed **from this** part **of the** brain to **the white spot** (*n. w. s.*). Below the hemispheres are the *antennary lobes* (*an. l.*), **from** which large *antennary nerves* (*an.*) are given off. **Below** these again a nerve is given off from the commencement of the commissure on each side, and these **unite in a** *frontal ganglion* (*f. g.*) which lies on the œsophagus ; thence the median *recurrent nerve* passes upwards and backwards under the hemispheres, and is seen in the dorsal view shown at C. (*r. n.*)

Large commissures **pass on either side of the gullet from the** cerebral ganglia to the *sub-œsophageal ganglia*, **which also lie in the** head. From their ventral aspect nerves **are given off to the** mandible, maxillæ, and labium. From these **ganglia** commissures (*con.*) pass backward through the neck to the first of the three coalescent pairs of the thoracic region, which innervate respectively the pro- meso- and metathorax and their appendages. **The** nerves of the wings have **a** Y-shaped junction with the nerve-chain, as shown in Fig. 82, D. In the abdomen there **are** six pairs of closely-united ganglia, **the** last being large. The abdominal commissures of **the male are looped.**

A *stomato-gastric* system innervates the crop and gizzard. The recurrent nerve comes into relation with *anterior* and *posterior stomato-gastric ganglia* (C, *a. stg., p. stg.*), the former of which is connected with the ventral surface of the hemispheres. Then running backwards along the dorsal surface of the crop, the re-

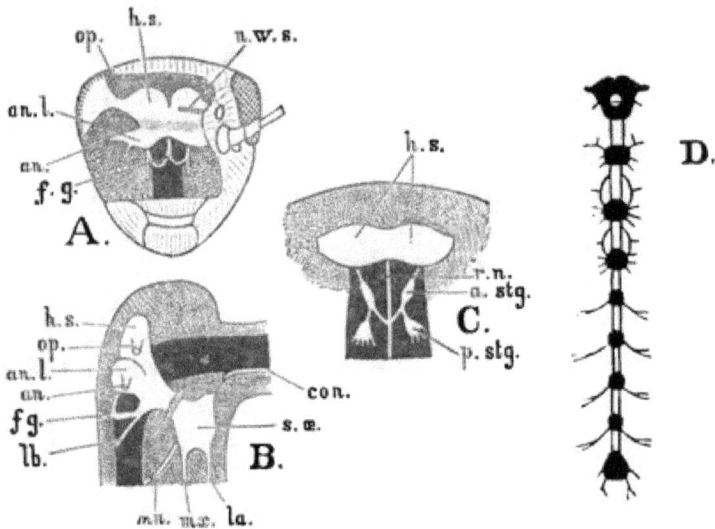

FIG. 82.—COCKROACH—BRAIN, ETC.

A. B. C. Head of cockroach opened so as to expose the brain—A. from the front; B. from the side; C. from above (after E. T. Newton). D. Nerve-chain.

an. Antennary nerve. *an. l.* Antennary lobe. *a. stg.* Anterior stomato-gastric ganglion. *con.* Commissure to thoracic ganglia. *f. g.* Frontal ganglion. *h. s.* Hemispheres of brain. *la.* Labial nerve. *lb.* Nerve to labrum. *mn.* Mandibular nerve. *mx.* Maxillary nerve. *n. w. s.* Nerve to white spot. *op.* Optic nerve. *p. stg.* Posterior stomato-gastric ganglion. *r. n.* Recurrent nerve. *s. œ.* Subœsophageal ganglion.

current nerve, about the middle of the crop, enters the apex of a triangular *ventricular ganglion.* From the lateral angles of this triangle nerves pass backwards and downwards on either side of the crop, and, breaking into branches, innervate this organ and the gizzard.

Each coalesced pair of ganglia has to some extent independent control over its somite and appendages. The general co-ordination of muscular movements is perhaps mainly the function of the so-called brain; but it would seem to be, in part at least, performed by the sub-œsophageal mass.

Special Senses in Insects.—The chitinous exoskeleton of insects rests upon a *hypodermic* or *chitinogenous layer.* Its free surface is, in the cockroach, divided into polygonal areas, and from the midst of some of these there rise large sense-hairs, developed from long flask-shaped hypodermic cells. These are *tactile* in character. The nerve filament that passes to such a seta expands at the base of the hair into a nucleated ganglion, a filament from which pierces the hypodermic cell, and is continued to the tip of the hair.

The paraglossæ and inner ends of the maxillæ may be in the cockroach, as they are in the wasp, supplied with end-organs of *taste.* No one who has seen a cockroach spitting, as he sucks from his antenna some distasteful substance, can doubt his being powerfully affected by sapid substances.

The sense of *smell* is located in the antennæ, in which there are developed a great number of minute sacculi, filled with serous fluid, the orifices of which are covered over with a delicate membrane. The nerve-fibres which proceed to these sacculi end in rod-like bodies. If the antennæ be extirpated or coated over with wax, the insect exhibits no repugnance to such substances as turpentine and carbolic acid, and does not eagerly rush to such food substances as bread sopped in beer.

No *auditory organs* have been described for the cockroach. But other insects possess such organs. In the grasshopper, for example, a thin chitinous membrane is stretched on a chitinous bow over a cavity in the tibia of the fore-leg. The cavity covered over by this membrane is in communication with the exterior, and within the cavity is a ganglionic mass at the end of a special nerve-fibre. In the antenna of the bee, at the bottom of certain open cavities, there are cone-like structures, which are regarded by some anatomists as auditory in function.

The many-faceted *eye* of the cockroach is constructed on the same principle as that of the crayfish. The organ is large, and occupies a considerable space on each side of the head. There are about 1800 *facets* on each eye. These facets are derived from the cuticle ; they are hexagonal (as a rule) in outline, and biconvex in cross-section. Beneath each facet is a short stout *crystalline cone* coated with dark pigment. The apex or deeper part of the cone is imbedded in a so-called nerve-rod or *rhabdom,* composed of four elements (*rhabdomeres*), which diverge above to receive the point of the cone. Around the rhabdom is a proto-plasmic sheath, inconspicuously divided into four segments or *retinulæ,* which have been found to contain "visual purple," such as is found in the vertebrate retina. Beneath this re-tinula layer is a *basilar membrane* perforated by a number of fine holes. The retinulæ and rhabdoms constitute the ultimate nerve-endings, and are connected through the fenestrations of the basilar membrane with nerve-fibrils. The basilar membrane is homologous with the basement membrane of the cuticle ; all between this and the facets is probably hypodermic in origin.

The optic nerve given off from the cerebral hemisphere of the brain gives rise, in the cockroach, before it reaches the basilar membrane, to two ganglionic enlargements. In the fly and bee there are three such ganglionic swellings. Each ganglion in the cockroach (and the inner two in the bee) is formed of a network of fine fibrils, surrounded by a cellular sheath of densely-packed nerve-cells, with large nuclei surrounded by a delicate proto-plasmic envelope. The fibres connecting the two ganglia decus-sate or cross ; and the fibres between the outer ganglion and the basilar membrane do so partially.

It has recently been urged that not only the layer of retinulæ and rhabdoms, but the inner and outer ganglion as well, should be regarded as constituting the insect retina ; in which case the retinulæ and rhabdoms answer to the layer of rods and cones in the vertebrate retina. The order of parts in the optical appar-atus of the cockroach is therefore : *Brain, optic nerve, retina, dioptric apparatus.* Of these the retina is constituted by *inner ganglion, decussating fibres, outer ganglion, partially decussating fibres, retinula layer.* While the dioptric apparatus may be represented by *corneal facets* and *crystalline cones.*

There has been much discussion as to the nature of insect vision. If we cut out the faceted cornea of a beetle's eye and clean its inner surface, mount for the microscope, and having focussed the facets raise the objective, we shall see, if the flat side of the mirror be used, a great number of images of the window or lamp, one for each facet. Some, relying on this fact, maintain that the compound eye is a multiple structure consisting, in the drone for example, of more than 13,000 eyes—the two compound eyes giving the brain or ganglia some 26,000 images to co-ordinate. It would seem, however, that in the water-beetle (*Hydrophilus*) the focus of a corneal lens is about three millimetres away and altogether behind the eye. It is on the whole more probable that each dioptric element transmits the image of a point, and that a number of the elements combine to form a stippled image, or a picture in mosaic.

In addition to the many-faceted eyes, some insects have simple eyes or *ocelli ;* the bee, for example, has three such, arranged in a triangle on the vertex of the head (queen) or on the forehead (drone). These would seem to be used for near vision, the many-faceted eyes being used for far vision. Each has a lens arising by differentiation and thickening of the cuticle ; the crystalline cones are not developed as such ; the hypodermic cells give rise to a retinal layer.

Reproductive Organs.—(1.) *In the Female.*—The paired ovaries consist of groups of eight beaded *ovarian tubules,* the apices of which are filamentous and united. In them the ova are developed (Fig. 83, A.). The upper part (*germogen*) of each tubule contains protoplasm, in which nuclei are imbedded ; further down separate nucleated masses are found ; and lower still these nucleated masses are arranged in single file, and are recognisable as ova. Around each ovum small nucleated cells arrange themselves as a single-layered follicle. A *vitelline membrane* is secreted by the inner surface of this egg-follicle, and a chitinous *chorion* by its outer surface. The lowest egg is much larger than the others   The two *oviducts* (*o. d.*) unite to

18

form a short *uterus*, which opens by the *genital aperture* (*e. o.*) on the eighth sternum into the genital pouch.

Above the eighth sternum, with its genital aperture, is the ninth sternum (see Fig. 80, D.), which carries at its anterior end the aperture of the *spermatheca*, which consists of a rounded sac and coiled cæcum. The spermatheca is a sperm-reservoir, and always contains spermatozoa in the fertile female.

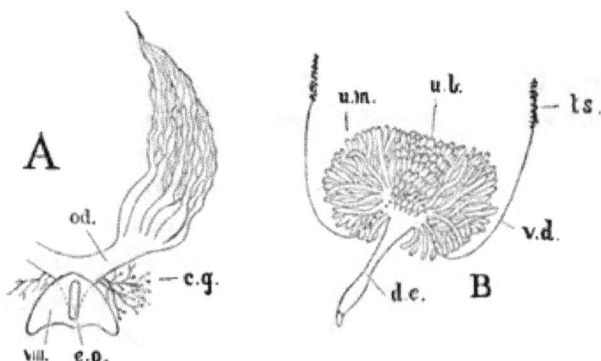

FIG. 83.—COCKROACH ; REPRODUCTIVE ORGANS.

A. Diagram of female organs of generation.—*c. g.* Colleterial glands. *e. o.* Genital aperture. *od.* Oviduct. viii. Sternum of eighth abdominal segment.

B. Mushroom gland of male.—*d. e.* Ejaculatory duct. *ts.* Testes. *u. b.* Utriculi breviores. *u. m.* Utriculi majores. *v. d.* Vas deferens.

Into the uterus, on its under side, there open the tubular accessory cement (*colleterial*) glands (*c. g.*), asymmetrically developed, the left being the larger. Within these long white cæcal tubes crystals of (?) oxalate of lime may be developed. Their function is to form an egg-capsule, within which sixteen ova, arranged in two rows, are normally laid. The capsule hardens and becomes brown by exposure to the air. It presents a longitudinal slit along its dorsal margin, the serrated edges of which bear tooth-like impressions, due to the grasp of the posterior gonapophyses.

The ova are about one-sixth of an inch in length, oval, with one face convex and the other slightly concave. The chorion which invests them is ornamented by being divided up into polygonal areas. In its convex moiety there are numerous *micropyles*, or minute holes for the admission of spermatozoa from the spermatheca.

(2.) *In the Male.*—On dissecting an adult male cockroach a curious "*mushroom-shaped gland*" (83, B.) is seen situated at the end of a short duct. This organ was for long regarded as the testis, since in the adult it contains developing spermatozoa. The true *testes* are liable to be missed, since they are only functionally active in the very young cockroach, becoming reduced, though they are not altogether obliterated, in the adult, in which they are vesicular bodies (83, B., *ts.*) lying beneath the fifth and sixth terga. *Vasa deferentia* (*v. d.*) pass from them into the coalescent cavities of the mushroom-shaped gland, or *vesiculæ seminales*, which bear a number of longer and shorter finger-like processes, the *utriculi majores* and *breviores* (*u. m., u. b.*). The epithelium of the testis in the young cockroach gives rise to *sperm cells*, which develop into hollow *spermatocysts*, around which spermotoblasts are placed radially. These are converted into spermatozoa. The later stages of the development of spermatozoa are, however, carried on in the vesiculæ seminales and their utricles, when the testes have become reduced. The spermatozoa pass from the vesiculæ seminales by an *ejaculatory duct* (*d. e.*), by which they are carried out of the body. A subsidiary gland of unknown function lies on the ventral surface of the ejaculatory duct.

Development.—The segmentation of the ovum in insects is on the centrolecithal system, as in the crayfish. The first segmentation nucleus undergoes division in such a way that a layer of cells forms a superficial *blastodermic layer*, while other *yolk-cells* remain scattered through the central mass.

In the blastodermic layer an elongated *ventral plate* is formed, in which the embryo begins to be developed. This broadens at the anterior end ; further back it is transversely divided into segments, of which the total number seems to be seventeen. Indications of the appendages appear early, limb-buds being formed, according to recent observers, on all the abdominal segments. These subsequently disappear.

In the mid-line of the ventral plate a *ventral groove* is formed, which, as it sinks below the general level of the epiblast, gradually detaches itself as a tube. The tube becomes a solid cellular mass, which splits into two longitudinal bands, the *mesoblastic bands.*

The *hypoblast* originates, in insects, according to some observers, by differentiation of the internal mass of yolk-cells; according to others, by differentiation of certain cells of the ventral groove.

The mesoblastic bands become divided into somites, each of which becomes hollow, the hollowed space separating an inner, dorsal, one-layered wall from an outer, ventral, several-layered wall adjoining the epiblast. By the coalescence of the cavities of contiguous somites a common cavity—the *cœlom*—is formed. The coalesced inner walls of the somites give rise to a *splanchnic layer ;* the coalesced outer walls to a somatic layer. Certain yolk-cells seem to pass into the cœlom, and it has been suggested that these give rise to the corpus adiposum, or fat body.

The nervous system is epiblastic in origin. The cerebral ganglia are developed separately as a pair of pre-oral thickenings of the epiblast. The post-oral chain is developed from longitudinal thickenings of the epiblast, which sink into the body and become separated from the superficial epiblastic layer.

All the anterior part of the alimentary canal (fore-gut), lined with a chitinous infolding from the edges of the mouth, is formed by a *stomodœal* invagination. All the posterior part of the alimentary canal (hind-gut), as far as, and including, the Malpighian tubules, with infolded cuticle from the vent, arises by a *proctodœal* invagination. The chylific stomach and its cæcal tubes (mid-gut), develops from the mesenteron. The muscular layer of the alimentary canal is contributed by the splanchnic layer of the mesoblast.

Perhaps the most striking peculiarity in the development of insects is the formation of an embryonic membrane similar to the amnion. A fold of membrane grows out from the blastodermic layer over the germinal plate, and, coalescing in the mid-line as does the vertebrate amnion fold, forms a double investment of the plate, an inner *true amnion* and an outer *serous membrane.* As the ventral plate increases in size and tends to enclose the embryo, the point of origin of these membranes is forced to retreat dorsally, until at last, when the plate envelopes the whole embryo, the amnion and the serous membrane become separated off, and form complete membranous investments of the embryo.

The heart is mesoblastic in origin. When the cavities of the lateral mesoblastic somites coalesce to form the cœlom, the inner wall becomes the splanchnic layer, the outer wall the somatic layer. At the edges of the mesoblastic bands the layer of cells which unites the somatic with the splanchnic wall forms, on each side, a grooved cardiac rudiment. As the hollowed mesoblastic bands grow upwards so as to approach the dorsal surface, these grooved rudiments tend to meet ; and the tubular heart arises by their ultimate coalescence.

The tracheal system arises from tubular infoldings of the epiblast. Each stigma has at first a distinct system, but these coalesce into a common system.

The reproductive organs are of mesoblastic origin.

CHAPTER XIV.

THE EARTHWORM.

THE body of the earthworm (*Lumbricus*) is long and ringed. The rings indicate a division into segments or somites, of which there may be as many as 350. There is no distinct head or neck, but the anterior end tapers off, while the posterior end terminates bluntly. There are no definite limbs or appendages, but nearly each segment is provided with chitinous bristles which aid it in progression, which is effected by alternate elongation and shortening of the body. The worm lives in moist earth, in which it forms burrows. It is omnivorous, swallows much earth, and normally consumes vegetal matter, such as leaves, fresh or half-decayed. It will, however, eat animal food, and at a pinch a dead comrade. Its power of regenerating excised parts is very remarkable. It is stated but hard to believe, that a worm not only survived the removal of the first five rings, including the brain, sub-œsophageal ganglion, mouth, and pharynx, but, within fifty-eight days, had completely regenerated these parts.

In the young worm, not yet sexually mature, the rings throughout the body are tolerably similar, those at the anterior end gradually lessening in diameter, and those at the posterior end becoming somewhat flattened. In the adult worm, however, certain of the rings between the twenty-fourth and thirty-sixth become lighter in colour, much swollen, and coalescent dorsally, and project downwards as ridges on either side of the mid-ventral line, giving rise to the *cingulum* or *clitellum*. The ventral aspect of segments eight to twelve also becomes swollen and tumid.

The mouth is in the anterior segment, situated beneath a

slightly projecting conical lobe, the *prostomium*. The vent is in the middle of the last body-ring. Other apertures to the body are numerous. Every segment of the body except the first has a dorsal pore opening into the anterior part of the ring in the mid-dorsal line, and two very minute pores, one on each side of the ventral line, which are the external orifices of the *nephridia* or *segmental* organs, whose function is excretory. In addition to these openings there are in each worm the orifices connected with the generative system; for the worm is hermaphrodite, each individual having both male and female genital organs. The male pores of the vasa deferentia open by tumid lips on the ventral aspect of the fifteenth ring. The female pores of the oviducts have a similar position in the fourteenth segment, and between the ninth and tenth and tenth and eleventh segments respectively, are the usually paired openings of the spermathecæ or receptacula seminis.

There are eight setæ or hair-like projecting spines to each segment, except a few of the most anterior and most posterior. They can be felt as a living worm is pulled backwards through the fingers, and can be readily seen with a lens, especially in a worm rigidly contracted by drowning in spirits. They are arranged in two double rows on each side. The outer double row is where the darker-coloured dorsal region shades off into the lighter-coloured ventral region. The inner double row is nearer the ventral line. Each seta is curved and pointed, and is lodged in a muscular sac, so that it can be protruded or retracted. On the tenth to the fifteenth rings, on the clitellum, and on a glandular prominence of the sixth ring in front of the clitellum, the spines are modified and often somewhat longer and thinner than elsewhere.

General Internal Anatomy.—The worm is readily killed by the vapour of chloroform; but it should not be too long exposed to this vapour, since it thus becomes contracted, and satisfactory dissection in this condition is impossible. It should, therefore, be soon removed and immersed in water for half-an-hour. The student will find it convenient to dissect the worm in a saucer

at the bottom of which is a layer of paraffin, poured in liquid
and allowed to set. The worm should be pinned out with the
first twenty-eight segments well stretched out, and the segments
should be numbered by marking the paraffin at the fifth, tenth,
fifteenth, and twentieth rings. The body should be opened by
a median incision from the twenty-fifth to the second segment.
Care must be taken not to cut so deeply as to injure the blood-
vessels. It will now be seen that the external constrictions
between the rings answer to internal septa which divide the
body into as many compartments as there are rings. These
must be carefully divided near the body-wall on each side, and
the walls pinned out laterally so as to display the internal
organs, as seen in Fig. 84, A., which represents the anatomy of
a sexually mature worm. After the worm is thus opened up
under water, the water should be poured away and replaced
with spirits of wine, which enables some of the parts to be more
readily seen, and hardens some of the organs so as to render
their dissection more readily accomplished.

The digestive canal is seen in the middle line with *buccal
cavity* (*b. c.*) anteriorly, opening into a large pharynx (*ph.*),
from which fine muscular fibres stretch to the body-walls. The
pharynx is succeeded by a long *œsophagus* (*œs.*), upon which are
three white protuberances in the tenth and twelfth segments.
These are the *calciferous glands* (*c. g.*). In the fifteenth ring the
œsophagus expands into a *crop* (*cr.*), which is followed by a
muscular *gizzard* (*gi.*); after which the *intestine* (*int.*) is carried
to the end of the body. If a living worm, selected for its trans-
lucency, be held up to the light, the intestine will be seen to
have a slightly coiled course, except when the worm is elongated
almost to the point of rupture. The intestine, in Fig. 84, is cut
at the twenty-second ring, and turned back to show the *nerve-
chain* (*n. ch.*), which is median and ventral in position. The walls
of the intestine are covered with soft yellowish-brown tissue.

Along the dorsal line of the alimentary canal runs a dorsal
blood-vessel, filled with red fluid. This gives off lateral vessels,
which in the sixth to the twelfth segments are much enlarged,
giving rise to the so-called *hearts* (*h.*).

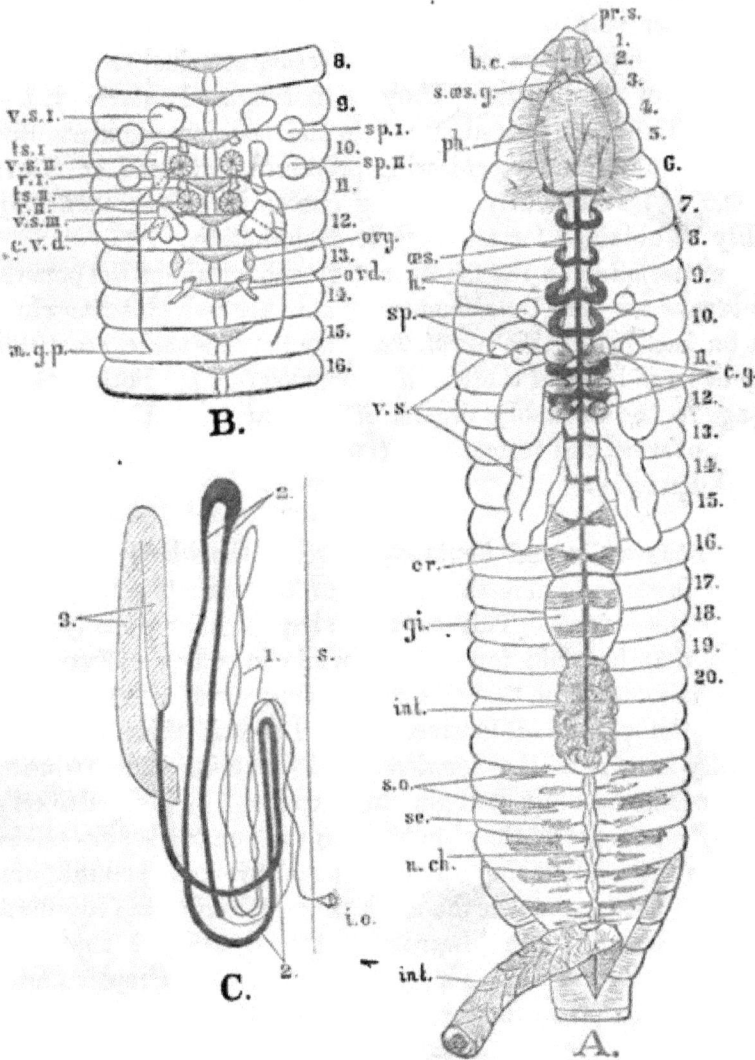

FIG. 84.—EARTHWORM: GENERAL ANATOMY.

A. Adult earthworm opened out from **dorsal side.** **B.** Sexually immature worm. **C.** Segmental organ (partly after **Gegenbaur**).

In **A.** or **B.**—*b. c.* Buccal cavity. *c. g.* **Calciferous** glands. *c. v. d.* Common vas deferens. *cr.* Crop. *gi.* Gizzard. *h.* Hearts. *int.* Intestine. *m. g. p.* Male genital pore. *n. ch.* Nerve-chain. *œs.* Œsophagus. *ovd.* Oviduct. *ovy.* Ovary. *ph.* Pharynx. *pr. s.* Prostomium. *r. i. ii.* Rosette-like openings of vasa deferentia. *se.* Seta sac. *s. o.* Segmental organ. *s. œs. g.* Supra-œsophageal ganglion. *sp. i. ii.* Spermathecæ. *ts. i. ii.* Testes. *v. s. i. ii. iii.* Vesiculæ seminales. **1-20.** Body-rings.

In **C.**—*i. o.* Internal opening. **1.** Thin-walled tube indicated by fine line. **2.** Thick-walled ciliated portion. **3.** Muscular portion.

On either side of each segment, behind the pharynx, is a coiled tube, *nephridium* or *segmental organ*, attached to the posterior face of the septum. They are not figured anteriorly, but their position is indicated at *s. o.* in the last few segments displayed. The inwardly projecting processes indicated at *se.* are the *sacs of the setæ*. By removing one of these sacs setæ may readily be obtained for microscopic examination.

From the ninth to the fifteenth segment certain of the generative organs are very conspicuous. There are two globular white sacs on each side, the *spermathecæ* (*sp.*); and there are three pairs of large lobes of the *vesiculæ seminales* (*v. s.*). Both of these belong to the accessory organs of generation. The essential organs, ovaries, and testes, with their ducts, can only be seen by careful dissection.

The Body-walls and Septa.—Investing the whole body, and turning inwards to line the sheaths of the setæ, is a thin transparent dense *cuticle*. This is easily stripped off, especially from a worm that has lain for a short while in spirits. Examined under the microscope it shows intersecting striæ, with refringent dots at the points of intersection. Beneath this is a layer generally known as the *hypodermis*. It contains columnar cells, among which are large oval glandular cells. The glandular cells are very numerous in the clitellum, on the ventral aspect of the eighth to the twelfth rings, at the lips of the male genital pore, and in the prominence on the sixth ring in front of the clitellum. The hypodermis turns inwards at the sheaths of the setæ. Beneath the hypodermis is a muscular layer, with circular fibres, between which pigment granules are scattered. Internal to this is a muscular layer with longitudinally-disposed fibres.

The septa are thin membranes of connective tissue, in which there are radiating and circular muscular fibres.

The Digestive System.—The mouth lies beneath the conical *prostomium*, the rest of the first ring forming a *peristomium*. It leads into a buccal cavity, divided by a slight constriction from the muscular pharynx, into which there projects a sucker-like fold. Salivary glands have been described as opening into the buccal cavity.

Behind the pharynx is the long narrowed *œsophagus*, on the side of which there are the three œsophageal or *calciferous glands*. These, or in any case the first pair, are diverticula of the œsophagus, the anterior pair being the largest. Their walls are lamellar in structure, containing deep follicles. If the hinder pairs be punctured and squeezed a number of white bodies exude, which contain in their centres a little fine granular matter, which, on treatment with acetic acid, disappears with effervescence, and which is carbonate of lime. The anterior pair of diverticula contain one or two, or a few larger calcareous concretions. In winter the glands seem to diminish in size. In young worms there are often but two pairs.

The œsophagus opens into an expanded *crop*, with muscular contractile walls ; and this is followed by the very thick-walled *gizzard ;* behind which the *intestine* passes, with but little change, to the end of the body. The intestine is laterally sacculated, and its lumen is much diminished by a thick longitudinal fold, the *typhlosole*, which projects inwards from its dorsal wall. Surrounding the intestine, especially at its anterior end, is a yellowish-brown, easily ruptured tissue, which is in close connection with the blood-vessels. Similar tissue may accompany the dorsal vessel above the œsophagus. It has been described as hepatic, and as producing a digestive secretion. It nowhere communicates, however, with the interior of the canal, and its close connection with the blood-vessels lends support to the view that it is vasifactive—concerned in the production of some constituent of the red fluid of the blood-vessels.

Little is known of the process of digestion. Worms seem to drag leaves, stalk forwards, into their burrows, and then moisten them with an alkaline secretion. This rapidly acts on the leaves, the cells with chlorophyll losing their green colour and becoming brown. It also acts on the starch granules within the cells, and on the protoplasmic contents of the cell-wall. The leaves are thus partially digested before they are taken into the canal. The function of the calciferous glands is not known for certain, but it has been suggested that the function is (1) excretory, to eliminate excess of carbonate of lime; (2) to neutralise within the canal the humous acids produced from the leaves. Absorption is mainly carried on, it may be presumed, in the intestine, the sacculations and the typhlosole giving additional absorbent surface.

A transverse section of the intestine shows: (1) an internal membrane of columnar epithelium; (2) a thin layer of vascular connective tissue; (3) a muscular layer, with circular fibres; (4) a longitudinal muscular layer; (5) the so-called hepatic, or by some observers, vasifactive tissue.

The Nephridia.—Excretion is effected by segmental organs or nephridia, of which there is a pair for each segment, except the anterior. Fig. 84, C. (after Gegenbaur and Howes) shows the arrangement of the coiled tube. The internal opening is by a ciliated funnel (*i. o.*) attached to the septum. Then follows a thin-walled tube, indicated by the thin line at 1. This is succeeded by a thick-walled glandular ciliated portion 2, which is followed by a muscular portion 3, which opens to the exterior just external to the inner row of setæ. The internal opening is in the segment anterior to that which contains the tube, which is connected to the posterior wall of the septum by a mesentery not shown in the figure. The student should examine *in situ* and mount several segmental organs, but must be content if he makes out the differentiation into three portions, and the ciliated funnel.

The Vascular System.—Two fluids in the earthworm claim the title of blood: (1) a colourless corpusculated fluid which occupies the cavity of the segments; (2) a red fluid, with minute corpuscles, which occupies definite vessels.

(1.) The *colourless fluid* resembles the blood of other invertebrates, and the lymph of vertebrates. It is found in abundance in the posterior segments of the body. Its corpuscles are amœbiform and nucleated. Not improbably its main function is nutritive.

(2.) The *red fluid* (pseudhæmal fluid) is contained in a definite system of vessels.

The typical arrangement in segments, posterior to the twelfth, seems to be as follows :—

1. The *dorsal vessel* running above the alimentary canal gives off, on either side—

2. *Commissural vessels* of two kinds—

    (a.) *Intestinal*, two or sometimes (?) three in each segment, which break up into capillaries on the walls of the alimentary canal or in the "hepatic" tissue ;

    (b.) *Lateral*, one in each segment, to the septa and muscles of the body-wall, and thus to the subneural.

3. *Sub-intestinal vessels ;* a pair of closely-connected longitudinal vessels lie one above the other, between the alimentary canal and the nerve-chain. The superior collects red fluid from the intestinal walls, the inferior gives off vessels to the segmental organs.

4. A *sub-neural vessel*, lying beneath the nerve-chain, collects blood from the body-walls and septa, and from the segmental organs.

5. Delicate *latero-neural vessels* run one on each side of the nerve-chain, and give off branches along the nerves.

From the sixth to the eleventh segments single commissural vessels pass from the dorsal to the single sub-intestinal vessel. These commissural vessels are enlarged to form so-called hearts.

In the twelfth segment the commissural vessels do not reach the sub-intestinal, but bend forwards and become lateral vessels running forward along the œsophagus.

In the first six segments, the dorsal, the sub-intestinal, and these lateral vessels branch out to form a net-work over the pharynx.

In the "hearts" the contraction is from above downwards. But it would seem that, elsewhere, there is no definite circulation of red fluid in one direction. Not improbably it is forced hither and thither during the alternate contraction and extension of the body.

The red fluid contains minute corpuscles, which would seem to be free nuclei. The fluid has not improbably a respiratory function.

Respiration.—There are no specially differentiated respiratory organs, respiration being apparently effected by the surface of the body.

The Nervous System.—The "brain" or supra-œsophageal mass is divided into two pear-shaped masses, the broad ends of which unite in the middle line. A pair of nerves run forwards from it along the buccal mass to the prostomium. Commissures pass round the gullet to the sub-œsophageal ganglia, which lies in the third segment. Posterior to this there runs a nerve-chain of two conjoined strands, with ganglionic swellings, one for each segment. The swellings are not well marked, and nerve-cells are not restricted to them, but occur throughout the cord. Its sheath is muscular. A pair of nerves is given out on each side from each ganglion, and a single interganglionic nerve on each side in relation to the septa. Around the nerve-chain the septa are incomplete, allowing of communication between the segments.

The Senses.—There are no eyes, but the worm is sensitive to light which falls upon its anterior segments. When suddenly illuminated it dashes like a rabbit into its burrow. It is also somewhat sensitive to radiant heat. There is no organ of hearing at present known, but the worm is very sensitive to vibrations. It is exceedingly sensitive in the matter of touch, being affected by a very light puff of breath. It does not seem highly sensitive to unnatural odours, but finds out the presence of favourite foods if buried. No organs of taste have been discovered, but the worm can apparently distinguish red from green cabbage, and exhibits a decided preference for certain foods, such as carrot, cherry, celery, onion, and horse-radish.

The Organs of Generation.—The earthworm is hermaphrodite, each individual having both male organs and female organs.

The Female Organs.—Minute *ovaries* (Fig. 84, B., *ovy.*), about $\frac{1}{16}$ of an inch in length are attached to the posterior face of the

septum, between the twelfth and thirteenth segments. The ova
increase in size from the attached to the free portion of the
organ. The *oviducts* (*ovd.*) open by ciliated funnels into the
thirteenth segment, pass through the septum between the
thirteenth and fourteenth segments, and open on to the exterior
on the latter segment near the exterior row of setæ. Connected
with the funnel of each oviduct is a small sac-like *receptaculum
ovorum.*

Between the ninth and tenth, and tenth and eleventh segments
are the *spermathecæ* or copulatory pouches, white spherical bodies
which are accessory female organs.

The Male Organs.—The state of the male organs differs very
much in mature and immature worms. The student should not
fail to dissect both.

Fig. 84, B., shows the organs in a young worm in which the
cingulum was not swollen. There are two pairs of *spermaries*
or *testes* (*ts.* i. and ii.), minute and difficult to see, attached
to the posterior face of the septa between the ninth and
tenth, and tenth and eleventh segments. In the tenth and
eleventh segments are seen two pairs of rosette-like funnels
(*r.* i. and ii.), which form the internal openings of the *vasa
deferentia.* The two vasa deferentia of each side unite into a
common vas deferens (*c. v. d.*), which opens by the male genital
pore (*m. g. p.*) in the fifteenth segment. In addition to the
essential organs and their ducts there are important accessory
organs. These are the seminal reservoirs or *vesiculæ seminales.*
In the young worm there are three pairs. The anterior (*v. s.* i.)
grow forwards into the ninth segment. The mid-pair (*v. s.* ii.)
grow into the tenth and eleventh segments from the septum
that divides them. The posterior (*v. s.* iii.) grow backward into
the twelfth segment.

So much for the young worm. In the worm that is sexually
mature the vesiculæ seminales increase enormously, and form
the large-lobed organs shown in 84, A., *v. s.* The anterior pair
and the mid pair unite in such a way as to form a central body
which envelopes the anterior testes and rosettes, and which has
two anterior and two posterior lobes. The posterior pair also

unite across the mid-line so as to envelope the posterior testes
and rosettes. Their lobes become much enlarged and grow out
backwards. Since they contain vast numbers of spermatozoa
in various stages of development, it is not surprising that these
large seminal reservoirs used to be regarded as the testes, and
that the true testes were overlooked.

The true testes are largely made up of spherical *sperm cells.*
Of these cells the most superficial and furthest developed
separate from the testes and pass into the seminal vesicles or
reservoirs which envelop them. These reservoirs are not
merely hollow sacs, but are divided by septa and vascular
trabeculæ into a vast number of incomplete compartments.
Here the products of the testes are received and undergo further
development.

The mode of development of the spermatozoa is as follows :—
The sperm cells, ere they leave the testis, undergo segmentation, passing through stages in which there are 2, 4, 8, 16
nuclei, and so on. The nucleated masses thus formed tend to
stand out like buds from the cell that generates them, in the
midst of which there remains a passive unsegmented mass.
At this stage the whole is called a *spermato-sphere,* the bud-like
projections *spermato-blasts,* and the central bud-bearing mass the
blastophor. By this time the spermato-sphere has been detached
and is lodged in the seminal reservoir. As development here
proceeds each spermato-blast gives rise to a whip-like filament,
which continues to elongate, while the bud from which it proceeds becomes rod-like. Eventually each rod with its whip
separates off as a *spermatozoon,* the rod forming the *head,* and the
whip the *tail.* The blastophor probably atrophies. Although
there are differences in details, the principle involved in this
mode of development would seem to be typical. If, as is probable, the sperm cell is to be regarded as equivalent to the germ
cell or ovarian ovum, then it must be noted that all the spermatozoa derived from one such sperm cell are homologous with the
single ovum.

The fully-developed spermatozoa are conveyed outwards by

the vasa deferentia at the period of congress. For though the worm is hermaphrodite, the ova of one individual are fertilised by the spermatozoa of a second. At such time the spermatozoa are received by each worm from the other into the spermathecæ, and are there stored until the ova shall be laid.

It has been suggested that the oviducts, vasa deferentia, and spermathecæ are homologous with segmental organs; which have in the generative segments been specially metamorphosed in subservience to the process of reproduction. The evidence for this view, however, involves careful comparison with other allied forms of life, and is therefore outside the limits of this work.

The eggs are laid in special chitinous cases, probably secreted by the cingulum. Certain so-called *capsulogenous organs*, modified from the setigerous capsules of the inner series in the generative segments have, however, been assigned this function. Into each egg-case seminal fluid is poured from the spermathecæ.

Development.—Segmentation is complete and nearly regular. It results in a flattened blastosphere forming a disc, composed of an upper layer of smaller cells and a lower layer of larger clearer cells. The disc becomes folded over by a process of invagination, the larger and clearer cells becoming internal to the others. Thus an elongated, two layered, more or less ciliated gastrula is produced. The blastopore at first forms an elongated slit at the folded edges, along the ventral surface; but it gradually narrows to a pore which becomes the mouth. The central or enteric cavity is lined by hypoblast cells, but at the region of the mouth the epiblast becomes inflected so as to line the œsophageal region. Thus there is no stomodæal invagination. The anus and hind-gut are, however, formed by a proctodæal invagination.

The exact mode of origin of the mesoblast is a matter of uncertainty. When formed it segments into quadrate masses

19

somewhat resembling the mesoblastic somites of the vertebrate, which become hollowed, and by apposition give rise to the body segments.

The nervous system arises by two distinct thickenings of the epiblast : (1) in the supra-œsophageal region, (2) in the region of the ventral cord. The junction between the two takes place comparatively late.

CHAPTER XV.

THE SNAIL.

THE Common Snail (*Helix aspersa*) is an air-breathing terrestrial mollusk. In its possession of an external shell into which it can withdraw itself, in the complete absence of lateral appendages to serve as organs of locomotion, and in the curious want of bilateral symmetry, this organism differs from those which we have hitherto considered.

In the fully-extended snail (Fig. 85, A.) the head is seen to be fairly differentiated, though it shades off into that part, the *foot*, on which the organism creeps, which is in turn surmounted by the shell, containing the visceral sac. The head bears two pairs of tentacles, a shorter anterior pair (*a. t.*) and a longer posterior pair (*p. t.*), at the expanded ends of which are the pigmented eyes. The mouth (*m.*) is bounded by a circular lip and lateral lips, and bears a brown horny jaw above. The thickened ridge which is seen just within the lip of the shell is the *collar* (*co.*). The shell itself is spiral, and consists of several *whorls* gradually diminishing in size to the apex. The line of junction between adjacent whorls is a *suture*. The hollow spindle around which the whorls are wound is the *columella*. The lip of the shell is called the *peritreme*. Externally the shell is covered by an organic *cuticle*. Beneath this the calcareous shell is composed of two layers. The first layer has a thin superficial portion showing confused striation, and a thicker portion (*honeycomb layer*) formed of vertical prisms. The second or deeper layer (*nacre* or *mother of pearl*) is colourless, and consists of several strata of prisms arranged horizontally, the axes of the prisms in successive layers being set at right angles.

Both cuticle and calcareous layers are secreted by the posterior edge of the collar.

There are six external apertures. Beneath (1), the mouth, is (2), the opening of the *pedal gland* (*p. g.*). Below the posterior

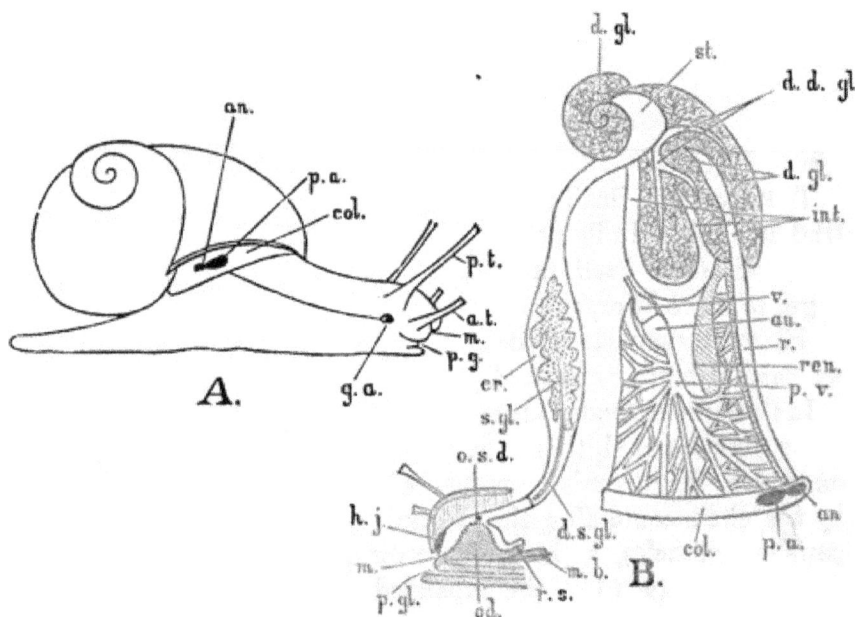

FIG. 85.—SNAIL: DIGESTIVE ORGANS, HEART, AND PULMONARY VESSELS.

A. External characters of snail. B. Snail dissected to show digestive system, pulmonary chamber, heart, and renal organ.

a. t. Anterior tentacle. *an.* Anus. *au.* Auricle. *col.* Collar. *cr.* Crop. *d. d. gl.* Duct of the digestive gland. *d. gl.* Digestive gland. *d. s. gl.* Duct of salivary gland. *g. a.* Genital aperture. *h. j.* Horny jaw. *int.* Intestine. *m.* Mouth. *m. b.* Muscular band. *o. s. d.* Opening of salivary duct. *od.* Odontophore. *p. a.* Pulmonary aperture. *p. gl.* Pedal gland. *p. t.* Posterior tentacle. *p. v.* Pulmonary vessel. *r.* Rectum. *r. s.* Radula sac. *ren.* Renal organ. *s. gl.* Salivary gland. *st.* Stomach. *v.* Ventricle.

tentacle on the right side is (3) the *genital aperture* (*g. a.*). In the collar there is (4) a large *pulmonary aperture* (*p. a.*) leading into the lung ; and close to this, on its right side, a larger aperture (5), the *vent*, and closely associated with this (6), a smaller orifice, by which the *duct of the renal organ* terminates.

The snail can, on occasion, completely retract itself into its shell. At such times the collar is exposed so that respiration can be carried on through the pulmonary aperture, which rhythmically opens and shuts. When the snail retreats into its shell for a protracted period, *e.g.* during the winter months, there is formed across the opening of the shell an *epiphragm* or *hybernaculum*, consisting of hardened mucous impregnated with calcareous matter.

If a snail be completely immersed its tissues absorb a large quantity of water. It may then be rapidly killed by placing in moderately hot water. It should thus die well extended, with its tissues flaccid. After making out the external apertures (except the minute orifice of the renal duct), and inserting a probe into the pedal gland—which extends into the foot, and may be opened out now or at a later period—the student should remove the shell bit by bit, being careful not to injure the delicate tissue of the digestive gland, and should then pin out the snail under water. A probe inserted into the pulmonary aperture is seen to pass into the *pulmonary chamber*, which is roofed over by the *mantle*. This mantle-roof, in which a number of blood-vessels are seen to ramify, unites along the collar with the dorsal body-walls, except at one point which remains open as the pulmonary aperture. Along the right-hand side of the pulmonary chamber runs the rectum.

Dissection is best begun by cutting along the base of the collar, so as to sever the connection between the mantle and collar and the dorsal body-wall. An incision should then be carried up just to the right of the rectum, so as to leave the rectum attached to the mantle. The roof of the pulmonary cavity may now be examined from its ventral aspect. In it is seen a white tongue-shaped mass, the *renal organ* (Fig. 85, B., *ren.*), to the left of which (in this position) is the pericardial cavity, on opening which the *auricle* (*au.*) and *ventricle* (*v.*) of the *heart* will be seen. The floor of the pulmonary chamber is seen to be constituted by a white muscular septum, which separates it from the body-cavity beneath. This should now be opened, and an incision should be carried forward with scissors to the anterior end of the head. A large *crop*, and anteriorly a *buccal mass*, will now be seen. Just posterior to the buccal mass, and encircling the œsophagus, is the *nerve-collar*. Besides these organs there are a number of white tubes (Fig. 87), which belong to the organs of generation. We will now take the systems of organs in detail.

The Alimentary System.—The student who has dissected so far will, with care, not find much difficulty in displaying the digestive organs as shown in Fig. 85, B. The genital organs are removed, the mantle dissected from the foot, certain white glistening bands (columella muscles) have been cut away, and the whole of the foot posterior to the head removed. The head and buccal mass are shown in section, the nerve-collar not being represented. The stomach, intestine, and digestive gland readily assume this position when the integument is carefully stripped off.

The mouth (*m.*), on the dorsal side of which is the horny jaw (*h. j.*), leads into a buccal cavity. Into this there projects from below a rounded *odontophore* (*od.*), which carries on its upper surface a tooth-bearing lingual ribbon or *radula*. The radula is developed in a diverticulum of the buccal chamber, the radula sac. In this there is a layer of large cells, from each of which several hard chitinous teeth are developed. Development is continuous, and the whole radula grows forward over the odontophore, just as our nails grow forward over their beds from a groove in the skin of the finger answering to the radula sac. The radula itself should be stripped off from the odontophore of another snail, and examined under the microscope. There is a central row of sharp teeth, on either side of which are a number of lateral rows. The way in which this lingual ribbon is used may be seen if a pond-snail (*Limnæus*) be watched as it licks the conferva from the sides of an aquarium tank.

The buccal cavity leads into an œsophagus, which is soon followed by an expanded *crop* (*cr.*), at the sides of which are large *salivary glands* (*s. gl.*). From these the ducts (*d. s. gl.*) lead downwards and open into the buccal cavity above the odontophore (*o. s. d.*). The product of the salivary gland converts starch

into sugar. Beyond the crop the diameter of the canal narrows,
but soon expands again to form the *stomach* (*st.*). Here the
digestive tube makes a sharp turn and proceeds as the intestine
(*int.*), which makes an S-shaped loop through the tissue of the
digestive gland (*d. gl.*), and then passes into the *rectum* (*r.*). The
large digestive gland is divided into two lobes, one of which
lies in the apex of the shell. This lobe pours its secretion into
the stomach by a special duct. The ducts of the three divisions
of the left lobe (*d. d. gl.*) unite to form a common duct. Three
kinds of cells are found in the digestive gland : (1) *ferment cells*,
which blacken rapidly with one per cent. osmic acid; (2) *hepatic
cells*, which contain spherical globules soluble in ether, and
excrete small vesicles which have yellowish contents, and are
evacuated with the fæces ; and (3) *calciferous cells*, containing
granules of carbonate or phosphate of lime, which are said to be
utilised in the formation of the hybernaculum.

Some of the cylindrical epithelium cells lining the buccal cavity
are ciliated. A section of the œsophagus gives the following
layers : (1) an internal lining cuticle ; (2) an epithelial layer ;
(3) a lacunar mucosa layer ; (4) a muscular layer, with (*a*) circular and (*b*) longitudinal fibres; (6) an external peritoneal
layer. The epithelium which lines certain depressions in the
stomach has been regarded as glandular, that covering certain
conical processes of the mucosa layer as absorptive ; and from
the distribution of the glandular and absorptive organs it would
seem that, with the exception of the buccal portion, which is
used for collecting food, and the terminal portion for expelling
fæces, no separate functions are assignable to the other divisions
—crop, stomach, intestine—of the alimentary canal. The ferment secreted in the digestive gland would seem to be capable of
converting starch into sugar, and of digesting proteids.

The Heart and Circulation.—The heart (Fig. 85, B.) has a
single *auricle* (*au.*) and single *ventricle* (*v.*), between which there
is a valvular orifice. It lies in a *pericardial chamber*, which is in
communication with the renal organ. From the ventricle there
passes forward a single *aorta.* This runs back, passes dorsal to

the intestine, and almost at once **splits into two trunks**—a *posterior aorta*, which follows the **course** of the intestine on its
convex side, and an *anterior aorta* which, after **giving off** branches
to the genital organs, passes into the body-cavity, where it may
readily be traced, giving off branches to the salivary glands and
mid-region of the foot, and proceeding to the head, piercing the
nerve-collar (Fig. 86, A., *ao.*), and being distributed to the buccal
mass and neighbouring parts.   The ultimate branches of **the**
arteries would seem to end in funnel-shaped dilated openings into
the lacunæ which lie within and around the tissues, there being
no true capillaries.   The blood finds its way into large blood-
sinuses, there being a **large** *lateral sinus* on **each side of the foot.**
There is **also a** large *visceral sinus* running along the **concave**
side of **the body whorl.**   This latter **is in direct** communication
with the *circular pulmonary sinus* which runs round the floor of
the **pulmonary** chamber; **but the** communication between the
lateral sinus and this pulmonary **circle must** be indirect.   From
the pulmonary circle *afferent* **pulmonary** *vessels* lead the blood to
the roof of the lung-chamber, whence efferent pulmonary vessels
form factors of the large pulmonary **vein** (*p. v.*), which conveys
the blood to the auricle.   **It** will be seen that the renal organ
(*ren.*) is in the course of the blood which is finding its way **to the**
pulmonary vein, on the **right side** of the roof of the lung.

The blood is colourless, **and** contains nucleated corpuscles.

Respiration.—The **afferent and** efferent pulmonary vessels in
the roof **of the** pulmonary chamber form a network, in which
the blood is separated from the air by the thin partition of their
walls.   Here, therefore, carbonic acid gas is given off by the
blood and **oxygen** absorbed.   Air enters and leaves the lung
by the pulmonary **aperture,** being drawn in by the contraction
of **the** muscular fibres **of the** septum which **forms** the floor of
the chamber, and being expelled by the relaxation of **these**
fibres and consequent rise of the septum by the pressure of the
viscera which lie beneath it.

The Renal Organ.—The renal organ consists of two parts, the

distinction between which may readily be seen in a snail that
has been drowned in cold or lukewarm water.   A thin-walled
saccular part (shaded in Fig. 85, B., *ren.*) will then be distended
with water, and its duct running alongside the rectum will be
visible for at least the upper part of its course.   The other part of
the organ is yellowish white and lamellar in structure.   Within
the lamellæ are great numbers of rounded refringent granules.

The Nerve-collar.—Lying over the œsophagus, as shown in
Fig. 86, A., are united *cerebral ganglia* (*c. g.*).   Lying below the
œsophagus are two ganglionic masses tolerably close together.   Of

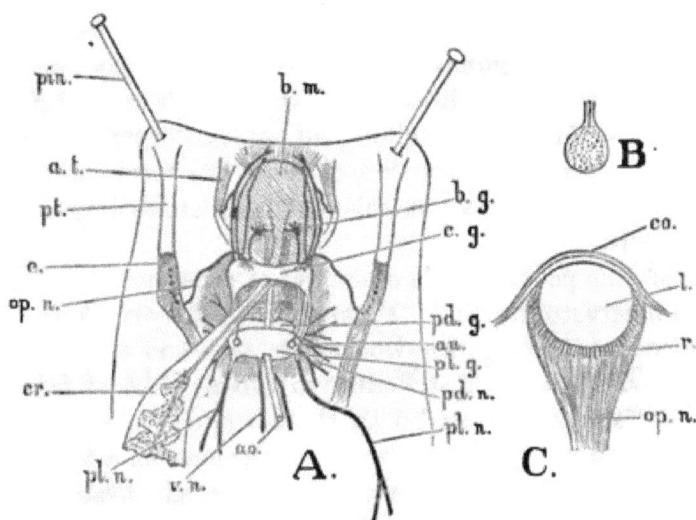

FIG. 86.—SNAIL: NERVOUS SYSTEM, EAR, AND EYE.

A. Dissection to show nerve-collar, its sheath being removed.   B. Otocyst,
or auditory organ.   C. Eye.

In A.—*ao.* Aorta.   *au.* Auditory organ.   *a. t.* Anterior tentacle (retracted)
b. g. Buccal ganglion.   *b. m.* Buccal mass.   *c. g.* Cerebral ganglion.   *cr.*
Crop.   *e.* Eye.   *op. n.* Optic nerve.   *p. t.* Posterior tentacle (retracted).
pd. g. Pedal ganglion.   *pl. g.* Parieto-splanchnic ganglion.   *pd. n.* Pedal nerve.
pl. n. Pallial nerve.   *v. n.* Visceral nerve.

In C.—*co.* Cornea.   *l.* Lens.   *op. n.* Optic nerve.   *r.* Retina.

these the anterior is formed by the fused *pedal ganglia* (*pd. g.*),
while the posterior is formed by the fused *parieto-splanchnic*

ganglia (*pl. g.*). Commissures pass on either side of the œso-phagus to both these ganglia. The three united ganglionic masses are called the *nerve-collar*, which is invested by a loose sheath. In the dissection from which the diagram was taken, the anterior (*a. t.*) and posterior tentacles (*p. t.*) were both in the retracted position. The *optic nerve* (*op. n.*) and the nerve to the anterior tentacle have therefore a somewhat backward direc-tion. The optic is distinct, and arises from a knot on the cerebral ganglion. The nerve to the anterior tentacle is a branch of the *labial nerve* running forward to the lip. A second nerve has a somewhat similar course, and a small nerve passes over the buccal mass to the integument in the neighbourhood of the horny jaw. In addition to these an *auditory nerve* runs between the two commissures to the auditory organ (*au.*), which is supposed to be seen in the diagram through the substance of the parieto-splanchnic ganglion ; and a *buccal nerve* runs forward to a *buccal ganglion* which lies close to the termination of the salivary duct. The two buccal ganglia are united by a delicate commissure.

From the pedal ganglia a number of pedal nerves (*pd. n.*) run out to innervate the foot. From the parieto-splanchnic ganglia *pallial nerves* (*pl. n.*) proceed on either side, and sooner or later branch before being distributed to the neighbourhood of the collar. Closely associated with the aorta (*ao.*) is a *visceral nerve* (*v. n.*), which innervates the organs of generation, a twig passing to the heart. Another nerve, taking its origin near the aorta, passes to the floor of the pulmonary chamber.

Special Senses.—The *eyes* are borne by the posterior tentacles. One of these organs is seen in diagrammatic section in Fig. 86, C. Beneath the external corneal layer (*co.*), is the crystalline lens (*l.*). Beneath this is the retina (*r.*), the rods of which are sur-rounded by pigmented material. The retina is continuous with the optic nerve (*op. n.*) Notwithstanding the possession of so well-developed an eye, the snail does not seem to be aware of the presence of a small object, such as the head of a pin, until it is within from $\frac{1}{4}$th to $\frac{1}{8}$th of an inch from the end of the posterior tentacle. If in a vigorous and healthy snail the posterior tentacle be removed, the eye is regenerated. In some cases it has been found that several perfect eyes have regenerated in the same tentacle.

The *auditory organ* is a minute sac (Fig. 86, A. and B.), with which the auditory nerve is connected. The sac contains a number of otoliths aggregated into a mass.

With regard to the *sense of smell* there has been much discussion. The anterior tentacle is often spoken of as olfactory. Near its end is a large ganglion from which fibres pass to the epithelium, in which there are said to be developed sensory knobs. Snails, however, in which these tentacles have been removed by excision, are apparently still possessed of a sense of smell. The lobate processes round the mouth have been regarded as the seat of olfactory sensation. Recent investigation, however, seems to point to the glandular rather than the sensory nature of these organs. Finally it has recently been contended that the pedal gland is the olfactory organ. Some of the cells are glandular and mucus-secreting. Others are deeper-seated and oval, and give off a delicate rod which enlarges at its free end into a ciliated knob. These are regarded as sensory. Thus it is maintained that we have in this organ the three necessary factors of an olfactory organ, the presence of a layer of sensory cells, the entrance of air, and the addition of a secretion from a glandular organ.

With regard to the seat of a sense of *taste* little or nothing definite is known.

The whole body is sensitive to touch, but in the tentacles this sensitiveness is much more delicate than over the general surface. The nerve that proceeds to the posterior tentacle bifurcates within that organ. One branch goes to the eye. The other passes to a ganglion, from which fine threads pass to a highly sensitive prominence at the end of the tentacle. It is very probable that we have here a delicate tactile organ.

The Organs of Generation.—The snail is hermaphrodite, male and female organs being found in the same individual.

The essential organ is lodged in the right lobe of the digestive gland in the apex of the shell (Fig. 87, *h. gl.*), from which
it has not been extricated in the dissection from which the
diagram is taken. It is known as the *hermaphrodite gland,*

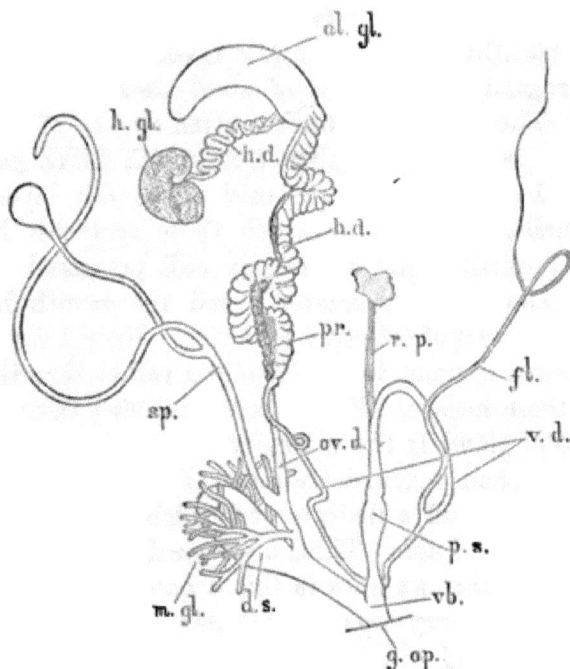

FIG. 87.—SNAIL: ORGANS OF GENERATION.

al. gl. Albumen gland. *d. s.* Dart Sac. *fl.* Flagellum. *g. ap.* Genital Aperture.
h. d. Hermaphrodite duct. *h. gl.* Hermaphrodite gland imbedded in terminal
lobe of digestive gland. *m. gl.* Muciparous gland. *ov.d.* Oviduct. *pr.*
Prostate. *p. s.* Penis sac. *r. p.* Retractor muscle of penis sac. *sp.* Spermatheca. *vb.* Vestibule. *v.d.* Vas deferens.

or *ovotestis.* Within the minute acini, of which it is an aggre-
gation, both ova and spermatozoa are developed. And if an
ovotestis be teased with needles, and examined under the microscope, developing or ripe ova, and sperm cells, spermatoblasts
attached to a blastophore, blastophores with bundles of almost
fully developed spermatozoa, or separated male elements may be
seen.

Much **larger** than the essential organ or ovotestis **is the** white tongue-shaped *albumen gland (al. gl.)*, between the base **of** which and the hermaphrodite gland runs a delicate convoluted tube, constituting the upper part of the *hermaphrodite duct (h. d.)*. Into this duct the albumen **gland pours its secretion. Below** the albumen gland the duct is **much folded** and puckered, **and** closely connected with its lower **part is the** white glandular *prostate.* **Towards its lower end it becomes** partially divided into two tubes **by an infolded** partition. Lower still it gives rise to a shorter *oviduct (ov. d.)*, and a longer *vas deferens (v. d.)*. The oviduct is joined by another tube, the *spermatheca (sp.)*, which higher up **bifurcates** into two tubes, one ending blindly without dilatation, and in its natural position closely attached to **the genital** duct, the other ending in a rounded dilatation, which **in** its natural position lies in the loop of **the** intestine **at** the bifurcation of the aorta. Below the point **of** junction of the oviduct and spermatheca the common **tube is** joined **by the** ducts of two *muciparous glands (m. gl.)*, just below **which there** is a **large white** oval **sac**, the *dart sac*, which, if **it be carefully** opened, **will** sometimes be found to contain a **sharp calcareous** dart, the so-called *spicula amoris*, on which are **four slightly-** twisted ridge-like blades. The tube common **to the oviduct,** spermatheca, **muciparous glands, and dart sac, now passes to the** *vestibule (vb.)*, **which opens on to the exterior by the** *genital aperture (g. ap.)*.

The vas deferens, after diverging from the common genital duct, takes a curved course. From it a long whip-like diverticulum, the *flagellum (fl.)*, is given off. Near the vestibule the **tube is** swollen, and if it be carefully opened it will be found **to contain a** conical intromittent organ (*p. s.*). From this point downwards the tube may be everted. A retractor muscle (*r. p.*) passes **back** from the tube above this point, and is inserted into the muscular floor of the pulmonary cavity.

The spermatozoa, after descending the vas deferens, would seem to be enclosed in a sperm capsule, or spermatophore, the flagellum not improbably secreting the material out **of** which this **is** moulded. It is a delicate spindle, two or three inches in length ; the clear transparent ends taper off to fine points. In the middle is a white opaque portion, in which are the spermatozoa. At the time of congress the two snails exchange spermatophores, which pass into the spermatheca, where they seem to break up, giving exit to the spermatozoa, by which the ova are fertilised, as they pass down the oviduct. The muciparous glands, which give rise to a milky secretion, perhaps supply them with cases.

The darts are extruded at such times. They may sometimes be found sticking in the skin of the snail ; while in dissecting the snail their remains may sometimes be found in or near the spermatheca or the genital duct.

The ova are laid in groups, generally in June or July, and in about 20 days hatch into young snails.

Development of Pond Snail.—The development of the pond snail (*Limnæus*) is more typical than that of *Helix*, and its ova and embryos are readily obtained, and studied without difficulty. If pond snails be kept in a glass aquarium tank, there will, in summer, be found adhering to the glass, or to water-weed, small oval patches, or elongated strips of jelly-like substance, divided into a number of compartments, within each of which is an ovum or an embryo.

Segmentation is holoblastic and fairly regular. First two and then four segmentation spheres are formed. Then from the upper surface of these four spheres there separate off smaller spheres, which, by a continuation of this process and by sub-division amongst themselves, become comparatively numerous. In Fig. 88, A., there are four large spheres, *macromeres* (*mac.*), and several smaller spheres, *micromeres* (*mic.*). Subsequently both macromeres and micromeres increase in number, a segmentation cavity is formed, and invagination takes place in such a way that the macromeres are pushed in (Fig. 88, B.) and become hypoblast (*hy.*), the micromeres giving rise to epiblast (*ep.*).

The blastopore (*bl.*) so formed would seem to elongate before it finally closes, the position of the future mouth and anus being at the anterior and posterior end of the slit. Both mouth and anus, however, are generally held to be formed by subsequent stomodæal and proctodæal invaginations.

At an early stage of development there is formed round the embryo an equatorial circlet of cilia, by means of which the embryo is caused to rotate within the jelly-like walls of the com-

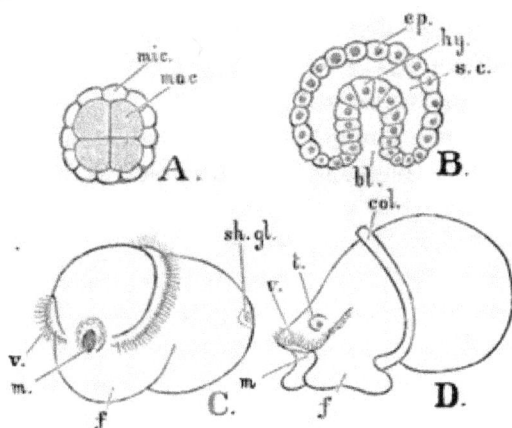

FIG. 88.—POND SNAIL: DEVELOPMENT.

A. Ovum after segmentation into macromeres (*mac.*) and micromeres (*mic.*).

B. Gastrula Stage.—*bl.* Blastopore. *ep.* Epiblast. *hy.* Hypoblast. *s. c.* Segmentation cavity.

C. Trochosphere stage.—*f.* Foot. *m.* Mouth. *sh. gl.* Embryonic shell gland. *v.* Velum.

D. Veliger stage.—*col.* Collar. *f.* Foot. *m.* Mouth. *t.* Tentacle. *v.* Velum.

partment in which it lies. Subsequently growth consists in the gradual enlargement of one hemisphere, and the reduction of the other. In the larger hemisphere both mouth and anus will be formed. In Fig. 88, C., the embryo is in what is known as the *trochosphere* stage. The incomplete circlet of cilia is seen at *v.*; the stomodæum at *m.*; and in the larger hemisphere the foot (*f.*) is becoming marked off from the visceral hump. At the posterior end of this is shown (at *sh. gl.*) the position of the so-called *shell gland*. It is found in many molluscan embryos. In *Limnæus* it is merely an embryonic organ, the true shell being formed, as in the snail, by the cells in the neighbourhood of the collar.

In Fig. 88, D., the embryo is shown at a later stage, known as the *veliger* stage. The incomplete circlet of cilia has become the velum (*v.*) in front of the mouth (*m.*). The foot has become bi-lobed, and the collar is developed. The visceral hump is covered with a delicate shell. At *t.* the tentacle is being developed.

The common snail does not seem to pass through a trochosphere and veliger phase, these stages in molluscan existence being shortened or suppressed.

Mesoblast takes its origin at the lips of the blastopore.

The nerve-collar arises as several epiblastic thickenings.

CHAPTER XVI.

THE FRESH-WATER MUSSEL.

THE Common Fresh-water Mussel (*Anodonta cygnea*) is enclosed in a bivalve shell which constitutes its exoskeleton. This shell (Fig. 89) is greenish brown in colour and rough ; of an elongated oval form blunter at one end, more tapering at the other ; and

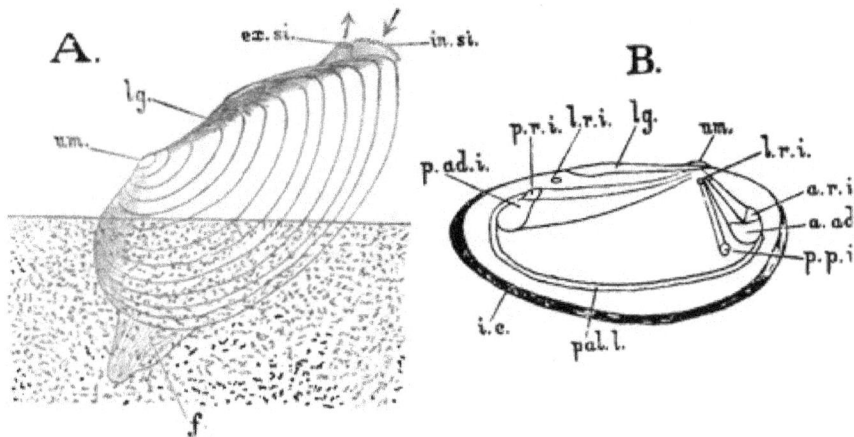

FIG. 89.—MUSSEL : SHELL AND EXTERNAL CHARACTERS.

a. ad. i. Anterior adductor impression. *a. r. i.* Anterior retractor impression. *ex. si.* Exhalent siphon. *f.* Foot. *i. e.* Inturned edge. *in. si.* Inhalent siphon. *l. r. i.* Lesser retractor impressions. *lg.* Ligament. *p. ad. i.* Posterior adductor impression. *p. p. i.* Protractor pedis impression. *p. r. i.* Posterior retractor impression. *pal. l.* Pallial line. *um.* Umbo.

marked with a series of concentric lines, each of which was at one time the margin of the shell.   The starting-point or origin of this series of curved lines, the spaces between which represent

20

successive additions to the shell during growth, forms, on the dorsal border of the valve nearer the more rounded anterior end, a small blunt eminence, the *umbo* (*um.*). Posterior to this, the exoskeleton is uncalcified for some distance along the dorsal border. This uncalcified part is called the *ligament* (*lg.*) It serves by its elasticity to keep the shell slightly agape. Fig. 89, A., shows the mussel in its normal position partially buried in the mud or sand. A fleshy part of the animal, the *foot* (*f.*), protrudes at the anterior end. Posteriorly a short double tube projects from the shell. The dorsal tubular part is smaller and smoother, the ventral part is larger and provided with a number of papillary processes. If the water be slightly turbid, a current will be seen continually setting in at the lower part of the tube (*inhalent siphon, in. si.*) and out through the upper part (*exhalent siphon, ex. si.*).

The mussel may be killed by placing it in cold water and gradually raising the temperature of the water until it is as hot as the hand can bear without discomfort. The shell will be slightly gaping and the foot protruded. On removing the mussel from the water, pulling the valves somewhat further apart, and looking in between them from the ventral aspect, their inner surface will be seen to be lined with a delicate yellowish *mantle* or *pallium*. Between the mantle and the fleshy foot are seen on each side two flattened lamellar organs, the *gills.* These lie in a large chamber (*infra-branchial chamber*), normally closed below by the meeting across the mid-ventral line of the mantle folds, but open posteriorly by the inhalent siphon, the papillary walls of which are now seen to be a specially differentiated part of the mantle. A probe passed into the exhalent siphon enters a smaller chamber (*supra-branchial chamber*) partially separated from the infra-branchial chamber by the growing together, or concrescence, of the posterior portion of the inner gill-lamellæ behind the foot. The probe (Fig. 90, A., *pr.*) will, however, emerge from the upper into the lower chamber by a wide slit which separates the inner gill from the foot. Anterior to the foot, between it and a strong band of muscle (*anterior adductor*) which passes from one valve of the shell to the other, is seen the *mouth*. The *vent* opens into the supra-branchial chamber close to its exhalent aperture.

With the handle of a scalpel the mantle may now be separated from its attachment to the left valve of the shell by a thickened ridge, the *pallial muscle*. The anterior **adductor muscle** must be cut close to the shell, and a similar muscle (*posterior adductor*) must be cut at the hinder end. In cutting these muscles, other smaller muscles will also be severed. By cutting the ligament the left valve of the shell may now be removed without difficulty.

Its inner surface is white and glistening, with a play of iridescent colours. The scars or impressions of the adductor muscles (Fig. 89, B., *a. ad. i.*, *p. ad. i.*) will be readily seen, and smaller impressions of the *anterior retractor* (*a. r. i.*, confluent with *a. ad. i.*), *posterior retractor* (*p. r. i.*), and *protractor pedis* (*p. p. i.*). From each of these impressions there runs towards the umbo a fainter tapering impression which marks the gradual shifting of the muscles during the growth of the shell. The impression of attachment of the pallial muscle of the mantle is seen as the pallial line (*pal. l.*).

If a thin section of a fragment of the shell be ground down and examined under the microscope, the three layers of which it is composed will be readily made out. Externally is the horny *periostracum*; beneath this is a *prismatic layer* of pallisade-like prisms; internal to this is a laminated *nacreous layer*. The nacreous layer is secreted by the whole external epithelium of the mantle. The outer layers are formed only at the free edge of the mantle.

The mussel now lies exposed to view in its right valve. The left lobe of the mantle should be cut away close to its origin, and the underlying parts exposed as in Fig. 90, A. The cut ends of the muscles (*a. ad.*, *p. p.*, *a. r. m.*, *p. r. m.*, *p. ad.*), the mouth (*m.*), foot (*f.*), and gills (*i. g.*, *o. g.*), the right mantle lobe (*pal.*), and the inhalent and exhalent siphonal apertures (*in. si.*, *ex. si.*) will be readily made out. A probe (*pr.*) has been passed through the supra-branchial chamber into the infra-branchial chamber, between the left inner gill and the foot. Two labial palps, or labial tentacles (*l. p.*) are seen as triangular backward-directed processes, on either side, just beneath the mouth. They are highly vascular and richly supplied with nerves. Dorsal to the gills is seen a large transparent sac filled with fluid. the

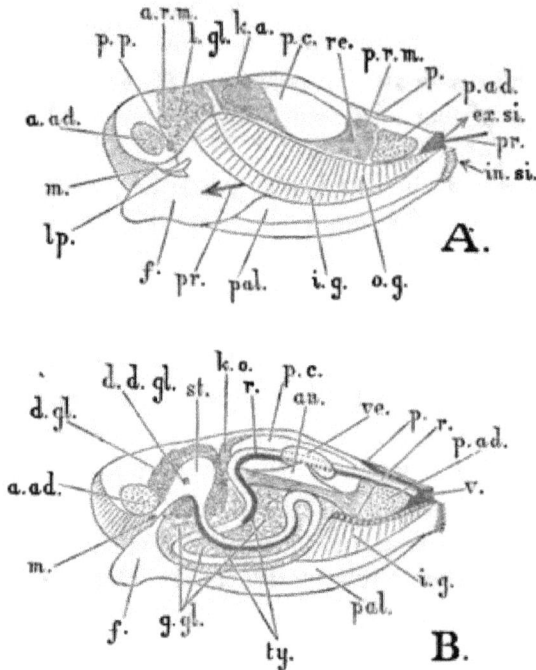

FIG. 90.—MUSSEL—GENERAL ANATOMY: DIGESTIVE ORGANS.

a. ad. Anterior adductor muscle. *a. r. m.* Anterior retractor muscle. *au.* Auricle. *d. gl.* Digestive gland. *d. d. gl.* Duct of digestive gland. *ex. si.* Exhalent siphon. *f.* Foot. *g. gl.* Generative gland. *i. g.* Inner gill. *in. si.* Inhalent siphon. *k. o.* Organ of Keber. *l. p.* Labial palp. *m.* Mouth. *o. g.* Outer gill. *p.* Pore. *p. ad.* Posterior adductor muscle. *p. c.* Pericardial chamber. *p. p.* Protractor pedis muscle. *p. r. m.* Posterior retractor muscle. *pal.* Mantle. *pr.* Probe passed from exhalent siphon between the inner gill and the foot. *r.* Rectum. *re.* Renal organ (organ of Bojanus). *st.* Stomach. *ty.* Typhlosole. *v.* Vent. *ve.* Ventricle.

pericardium (*p. c.*), within which the heart may be dimly seen. Beneath the pericardium is seen through the body-wall the greenish brown tissue of the *organ of Bojanus* or *renal organ* (*re.*).

At the anterior end of the pericardium is seen the brownish tissue of the *organ of Keber* (*k. o.*).

Alimentary System (Fig. 90, B.).—For the dissection of the alimentary canal it is well to work on a specimen that has lain for two or three days in spirit.   With the aid of the figure little difficulty should be found in exposing the canal, and slitting it open by passing a guarded bristle (tipped with sealing-wax), and following it with scissors.

The mouth, which lies between the anterior adductor and the foot, leads by a short œsophagus into the so-called stomach (*st.*) ; an irregular sac imbedded in the brownish tissue of the digestive gland (*d. gl.*), the secretion of which enters by ducts (*d. d. gl.*) near its anterior end.   From the stomach the *intestine* passes downwards and takes the coiled course indicated in the figure.   Into its cavity (at *ty.*) there projects an infolding of the wall or *typhlosole*.   The intestine passes upwards and enters the pericardium at its anterior end, running backwards thence as the *rectum* (*r.*), which passes through the ventricle (*ve.*) of the heart. The rectum, in which the typhlosole is well marked, leaves the pericardium near the dorsal side of its posterior end, and passes over the posterior adductor muscle and along the dorsal side of the supra-branchial chamber, where it ends in an *anus* (*v.*) placed on a prominent papilla.

On scraping the lining membrane of the intestine, ciliated and columnar cells will be found.   Some of the columnar cells would seem to be secreting cells.

On teasing out a small fragment of the digestive gland, it is seen to be composed of branched cæcal tubes.   Their lining epithelial cells, examined under a high power, show, in addition to the protoplasm and nucleus, a vesicle which encloses spherical brownish-green granules.

The food of the mussel consists of minute organisms swept into the infra-branchial chamber by the inhalent current caused by the lashing of the innumerable cilia on the gills.

The Gill-Plates.—The plate-like gills, which give to the class of *Mollusca* to which Anodon belongs the name of *Lamelli-branchiata*, lie, as we have seen, on either side of the foot anteriorly, and hang down into the infra-branchial cavity posteriorly. Each gill-plate consists of two lamellæ, an inner and an outer, or preferably a *descending* and an *ascending lamella*, and each lamella may be regarded as arising from the union of a great number of *gill-filaments*, descending and ascending, which have become fused together into a perforated plate-like structure. Fig. 91 will exhibit the relations of the lamellæ. Fig. 91, A., is a section passing through the middle of the ventricle (*ve.*) and the posterior part of the foot (*f.*). The gill-filaments take their origin from a *gill axis* (*g. a.*), forming in this part of the body a ridge on the body-wall. From this axis there pass downwards two *descending filaments*, an inner, forming part of the inner lamella of the outer gill, and an outer, forming part of the outer lamella of the inner gill. After proceeding downwards the filament turns sharp round and forms an *ascending filament*, running parallel with the descending filament. The ascending filament of the outer gill forms part of its outer lamella; that of the inner gills forms part of its inner lamella. The upper ends of the ascending filaments of the outer gills become united by concrescence along their whole length with the mantle. The upper ends of the ascending filaments of the inner gills are in this region free, thus giving rise to the slit-like communication between the supra-branchial chamber (*s. b. c.*) and the infra-branchial chamber (*i. b. c.*).

Fig. 91, B., is a diagrammatic section taken through a region anterior to that shown in A., nearer the anterior end of the gill-plates. It shows that here the ascending lamina of the inner gill has united by concrescence with the foot. Fig. 91, C., is a similar section taken through the posterior adductor muscle. It shows that here, in the region posterior to the foot, the ascending lamellæ of the inner gills have united with each other by concrescence. Fig. 91, D., is taken through the siphonal cavity, behind the posterior adductor. It shows that the gill-axis, which in A. and B. was closely adherent to the body-wall, and in C. was suspended by a suspensory ligament, has here become free.

Thus in the region through which D. passes, there is a common supra-branchial chamber for the inner and outer gills of each side; in the region of C. this is sub-divided into three

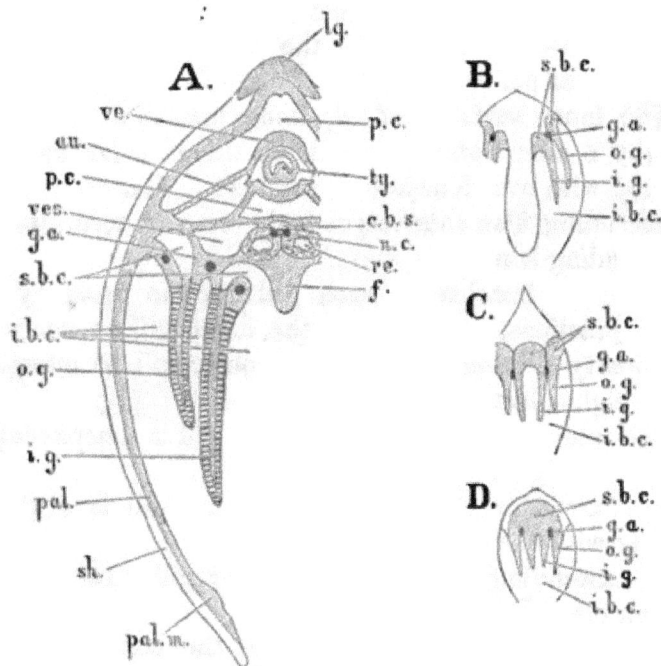

Fig. 91.—Mussel—Transverse Sections: Gills.

A. Through the heart and posterior end of the foot.
B. Through the middle of the foot near the anterior end of the gills.
C. Through the posterior adductor muscle.
D. Just behind the posterior adductor muscle.

au. Auricle. *c. b. s.* Central blood sinus. *f.* Foot. *g. a.* Gill axis. *i. b. c.* Infra-branchial chamber. *i. g.* Inner gill. *lg.* Ligament. *n. c.* Nerve commissure. *o. g.* Outer gill. *p. c.* Pericardial chamber. *pal.* Mantle. *pal. m.* Pallial muscle. *re.* Renal organ. *s. b. c.* Supra-branchial chamber. *sh.* Shell. *ty.* Typhlosole of rectum. *ve.* Ventricle. *ves.* Vestibule of renal organ.

In B. C. D. the cavity of the supra-branchial chamber is shaded.

parts, one median, common to the two inner gills, and two lateral, one for each outer gill: in the region of A. the median chamber is (*a*) becoming sub-divided by the foot, and (*b*) is in communication with the infra-branchial chamber. Finally, in the region of B. the supra-branchial chamber is sub-divided into four parts, one for each gill-plate.

Regarding, then, the gills as composed of a great number of filaments with descending and ascending moieties—and we are justified by comparative morphology and development in so regarding them—we have to note the large amount of fusion or concrescence of parts.

1. The inner surfaces of adjoining filaments have become united by a membranous expansion, the lamellar membrane, perforated with oval fenestræ or windows. Thus the ascending and descending filaments respectively come to form ascending and descending lamellæ.

2. The two lamellæ of each gill become fused by inter-lamellar junctions, occurring in the inner gill at intervals of about twenty filaments, and in the outer gill at intervals of seven or eight filaments.

3. The ascending lamella of the outer gill is concrescent with the mantle along its whole length.

4. The ascending lamella of the inner gill is concrescent anteriorly with the foot.

5. The ascending lamellæ of the inner gills are concrescent with each other posteriorly.

6. The union of the gill-axis with the body-wall may be partly due to concrescence.

Two little square patches should be cut out from the descending (outer) lamella of the inner gill of a recently-killed mussel, and be mounted in water (without cover glass), the one with its outer and the other with its inner face uppermost. Examined under a moderate power of the microscope, that with its outer side uppermost shows a number of parallel bars—the gill-filaments—strengthened with yellowish chitinous rods. Where there is a fenestra in the underlying lamellar membrane, the play of numberless cilia, working in one direction on one filament, and the opposite direction on the other, may readily be watched. The patch with its inner face uppermost, shows (especially with reflected light) the oval fenestræ in the lamellar membrane. If some of the gill substance be teased and examined under a high power, there will be seen: (1) fragments of the chitinous rods, (2) ciliated cells, (3) irregular or branched nucleated cells of the lacunar tissue, which supports the ciliated cells and forms the ground substance of the gill, and (4) colourless amœbiform blood corpuscles.

If the gill from which the square patches have been cut be examined with a hand lens, the inter-lamellar junctions will be seen. They are best seen, however, in the outer gill of a mussel in which this gill has been converted into a brood-pouch for embryos.

The ceaseless lashing of the cilia of the gills produces the current of water which sets in through the inhalent siphon and escapes through the exhalent siphon. This current passes from the infra-branchial into the supra-branchial chamber, partly through the fenestræ of the gill-lamellæ, partly through the large slit-like aperture between the foot and the ascending lamella of the inner gill. It subserves two functions: (1) alimentation by sweeping *infusoria, diatoms,* and other minute organisms within the reach of the cilia of the mouth cavity; (2) respiration as the water is driven past the gill-filaments and through the fenestræ of the lamellar membrane.

The Heart and Circulation.—The heart (Figs. 91 and 92) consists of a median chamber, the *ventricle* (*ve.*), and two lateral chambers, the *auricles* (*au.*). These parts lie within the *pericardial chamber* (*p. c.*). In a recently killed mussel the slow wave-like contractions of the heart (five or six a minute) may be observed. Seen from above, by opening up the pericardial chamber from the dorsal aspect, the ventricle is seen as a thick-walled oval sac, with a tendency to two lateral lobes posteriorly. Through its midst passes the rectum. The auricles are thin-walled conical sacs, the bases of the flattened cones being applied to the sides of the pericardial chamber, and their apices communicating with the median ventricle. The aperture of communication between the auricle and ventricle is a longitudinally elongated valvular slit. If the rectum, a little anterior to the ventricle, be severed, and the

cut end examined with a lens, the *anterior aorta* will be seen
running along its dorsal surface. And if the rectum, a little
posterior to the ventricle, be similarly cut and examined, the
posterior aorta will be seen running beneath the rectum (along
its ventral surface). By the anterior aorta the blood is distributed to the foot, labial palps, intestine, and the anterior
part of the mantle. By the posterior aorta the adductor
muscles of this region and the posterior part of the mantle are
supplied. These vessels break up into irregular chambers or
lacunar spaces, whence the blood collects into a large vessel, the
so-called *vena cava* or *central blood sinus* (Fig. 91, A., c. b. s.),
which may readily be seen beneath the floor of the pericardial
chamber. Thence it passes by a number of vessels to an
irregular plexus in the renal organ ; and thence on each side to
a longitudinal vessel which lies in the gill-axis. From this
longitudinal vessel the blood passes into vertical vessels which
run downwards into the descending lamellæ at the points of
union of the inter-lamellar junctions with the lamellar membrane. There are similar vertical vessels in the ascending
lamellæ ; but there does not seem to be a distinct capillary
system with definite walls between these two, but rather a
loosening or incoherence of the walls of the vessels, such as to
admit the passage of blood from them into the substance of the
lacunar tissue. The vertical vessels in the ascending lamellæ
carry the blood upwards to horizontal vessels which run along
their margins, by which the purified and aerated blood is conveyed to the auricles. A plexus of vessels or spaces arises near
the junction of these afferent vessels with the auricles, and
ramifies in the so-called *organ of Keber* (Figs. 90, A., and 92, k. o.),
which embraces the anterior end of the pericardial cavity. The
function of this organ is not well understood.

Thus the general order and direction of circulation in the
mussel is : ventricle, aortæ, smaller vessels and lacunar spaces
in the foot and mantle, central blood sinus, renal organ, gills,
auricles, ventricle.

The blood is colourless, with numerous white amœbiform
nucleated corpuscles.

The protrusion of the foot is largely effected by the forcing of blood into the lacunar spaces that abound therein. According to some observers there are—(1) three ciliated openings in the ventral surface of the foot by means of which water can enter and be admitted into the blood system, and (2) pores of communication between the blood system and the pericardial chamber. This provides for an ingenious mode of flushing the renal organ and sweeping out its solid contents. The writer is, however, in common with other observers, unable to satisfy himself of the existence of these communicating pores, internal or external.

The Renal Organs.—The renal organs or so-called organs of Bojanus of the mussel are paired structures lying beneath the pericardial cavity. Their relations are shown in the dissection from the side, Fig. 91, and in the cross-section, Fig. 92, A. Each consists of two parts : (1) a *glandular part* (*re.*), which communicates anteriorly with the pericardial cavity, and posteriorly with (2) a *vestibule*, with delicate thin walls, communicating anteriorly with the exterior by the *external renal aperture* (*e. re. a.*). To show this aperture the gills must be cut away as shown in Fig. 92. Just anterior to it is the external aperture of the generative gland.

Open the pericardial cavity from the dorsal aspect, divide the rectum at the posterior end of the cavity, cut through the auricles near their junction with the walls of the chamber, and turn the rectum and heart forwards. The floor of the pericardium will then be displayed. Beneath it in the mid-line is seen the central blood sinus, and on either side the vestibule of the renal organ. At its anterior end two small apertures will be seen. They are the *reno-pericardial apertures.* Pass a guarded bristle into one as far as it will go. Remove the floor of the pericardial chamber and roof of the vestibule (coalescent). The glandular part of the renal organ will be seen projecting like a ridge into the vestibule. At the posterior end of the vestibule, which does not extend so far back as the glandular part, the guarded bristle, which was passed from the pericardial chamber

into the renal gland, may be seen to emerge by an irregular
opening.   The vestibules communicate by an inter-renal aperture.   Anteriorly there is a small aperture in the floor of each
vestibule.   Through it may be passed a guarded bristle, which

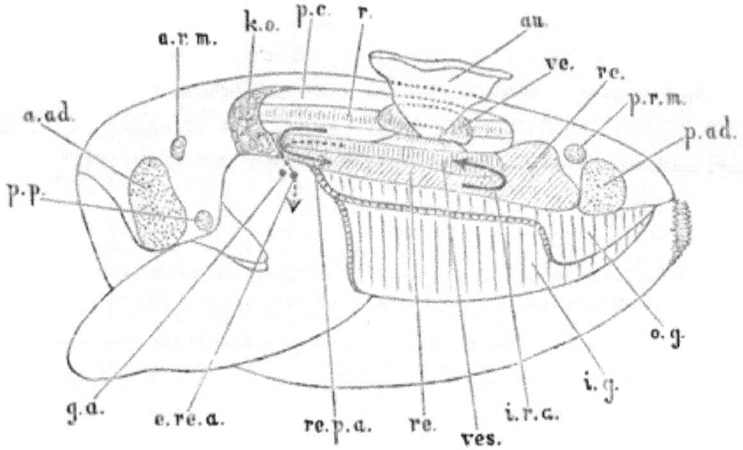

FIG. 92.—MUSSEL: HEART AND RENAL ORGAN.

a. ad. Anterior adductor muscle. *a. r. m.* Anterior retractor muscle.
au. Auricle. *e. re. a.* External renal aperture. *g. a.* Generative aperture.
i. g. Inner gill. *i. r. a.* Aperture between glandular portion and vestibule.
k. o. Organ of Keber. *o. g.* Outer gill. *p. ad.* Posterior adductor muscle.
p. c. Pericardial chamber. *p. p.* Protractor pedis muscle. *p. r. m.* Posterior
retractor muscle. *r.* Rectum. *re.* Glandular part of renal organ. *re. p. a.*
Reno-pericardial aperture. *ve.* Ventricle. *ves.* Vestibule of renal organ.

will emerge by the external renal aperture close to the anterior
end of the inner gill.   (The dotted arrow in Fig. 92 passes
through this aperture.)   The cavity of the glandular part may
now be opened up, its brownish solid contents washed out, and
the irregular walls noted.

In a second mussel the renal organ may be dissected from
the side, and the parts laid open and carefully drawn with the
guidance of the diagram, Fig. 92.

The secretion of the renal organ has recently been shown to
contain uric acid and urea.   These substances were also found
in the blood of the central blood sinus, but none was found in
blood taken from the gills.

The Nervous System (Fig. 93).—The nervous system of the mussel consists of three pairs of ganglia with their commissures. On the dorsal side of the mouth there are two somewhat triangular *cerebral ganglia* (*c. g.*) united by a short commissure which passes over the mouth. From each of these ganglia, besides local branches to the labial palps and the neighbouring muscles, there passes downwards and slightly backwards a *cerebro - pedal* commissure (*c. p. c.*), ending in the *pedal ganglion* (*p. g.*). The two pedal ganglia are orange bodies applied to each other in the mid-line. Each gives off a branch to the minute auditory organ, and branches to the neighbouring parts.

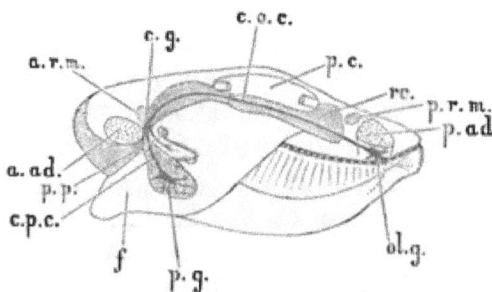

Fig. 93.—MUSSEL: NERVOUS SYSTEM.

a. ad. Anterior adductor muscle. *a. r. m.* Anterior retractor muscle. *c. g.* Cerebral ganglion. *c. o. c.* Cerebro-olfactory commissure. *c. p. c.* Cerebro-pedal commissure. *f.* Foot. *ol. g.* Olfactory ganglion. *p. ad.* Posterior adductor muscle. *p. c.* Pericardial chamber. *p. g.* Pedal ganglion. *p. p.* Protractor pedis muscle. *p. r. m.* Posterior retractor muscle. *re.* Renal organ.

From each cerebral ganglion there also passes off, upwards and backwards, a *cerebro-olfactory* commissure (*c. o. c.*) (cerebro-parieto-splanchnic of some authors), which is readily traced in the region of the renal organ. It proceeds to an *olfactory ganglion* (*ol. g.*) (parieto-splanchnic of some authors), which lies near the surface beneath the posterior adductor muscle.

Recent observations seem to show that there are two classes of nerve-fibres connected with the great adductor muscles—the one class motor, giving rise to contraction, the other class inhibitory, producing relaxation, when mainly by the elasticity of the ligament the shell gapes. The motor nerves for each muscle spring from the ganglion next it; the inhibitory fibres, on the other hand, are stated to take their origin entirely from the cerebral ganglia.

Special Senses.—We know little about the special senses in anodon. The general surface is sensitive to *touch*, but no special end organs have been described. The labial palps have a rich nerve-supply ; we may conjecture that they minister to *taste*— we cannot at present do more. The papillary processes of the inhalent siphon are sensory, but to what sense they minister we are uncertain. The thin layer of elongated epithelial cells in connection with the olfactory ganglion is held to be an organ of *smell* (*osphradium*)—and the olfactory ganglion is regarded as a centre for the co-ordination of sensations received in connection with the inhalent current. The auditory organ is a small sac lying posterior to the pedal ganglion of each side. It is difficult to dissect out in the mussel. But in the small fresh-water mollusc *Cyclas corneœ* it may without difficulty be seen if the transparent foot be mounted for the microscope and examined under a low power. It is a small sac, containing a mobile particle or *otolith*. The sense of *sight* is apparently absent in the mussel.

The Reproductive Organs.—The mussel is diœcious. The *generative opening* (Fig. 92, *g. o.*) is just anterior to the external opening of the renal organ. It opens into a duct which gives off a great number of cæcal branches. The reproductive organs are thus similar in the two sexes, and occupy in winter and spring much space in the upper part of the foot. Unless a little of the organ be examined microscopically it is not easy to distinguish the testis from the ovary.

If a little of the *testis* be teased in normal saline vast numbers of *spermatozoa* will (in winter) be seen dancing in the field of the microscope, especially if the slide be warmed. They have short rounded heads, and long vibratile tails.

If a little of the *ovary* be teased, *ova* in various stages of development will be seen. Fig. 94, 1-3, shows such ova. Externally is a vitelline membrane, produced in immature ova into a spout-like tube, through the aperture of which (*micropyle*) the protoplasm of the ovum was originally in connection with the ovary. In developing ova the protoplasm is granular, and stains readily. But an ovum that has been dehisced, or set free, is wonderfully clear and glassy (4). In such ova—which may be obtained in quantity in December by making a clean incision in the upper part of the foot of a recently killed mussel, and mounting the

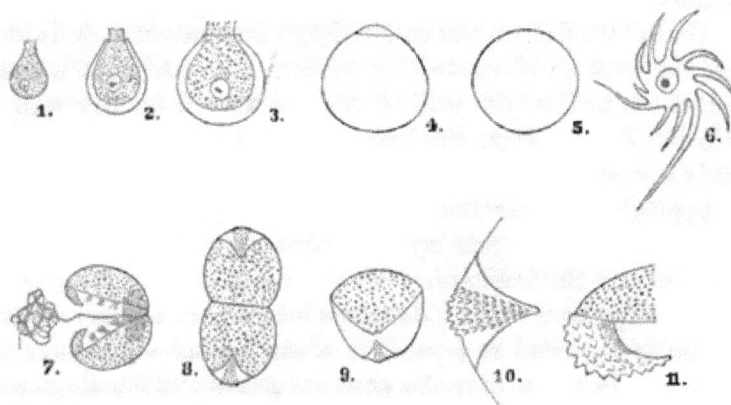

FIG. 94.—MUSSEL : OVA, SPIRAL STARS, GLOCHIDIA.

1, 2, 3. Young ova torn from the ovary. 4. Ripe ovum dehisced. 5. Ovum, showing spiral star perhaps accompanying the formation of the polar cell. 6. Spiral star under high power (Gundlach's $\frac{1}{18}$ inch immersion). 7. Glochidium with byssus and tactile organs. 8, 9. Glochidia shells. 10, 11. Beak of Glochidium under high power.

fluid which exudes—beautiful spiral stars may be seen (5 and 6). One such star—rarely a second—may be seen in each ovum. They perhaps accompany the formation of the polar body.

Development.—The ova undergo the **early stages of their** development in the external gills of the **mother. Here, in all** probability, they are fertilised **by spermatozoa shed into the** surrounding water by a male **anodon, and carried inwards by** the inhalent current.

Segmentation is unequal. The granular protoplasm of the impregnated **ovum** buds off in succession a number of **small clear** cells. These small cells are free from yolk, the mother cell retaining all the food-yolk. Between this large yolk-segment and the small clear cells **a** cavity—the segmentation cavity—is formed. From the small clear segments epiblast will be formed.

The single yolk-sphere will give rise to hypoblast and mesoblast. The large yolk-segment next undergoes cleavage, until about ten blastomeres are formed from it, two of which are larger than the others. These two slip into the cavity of the blastosphere, while the others flatten out and undergo involution so as to form a gastrula, with an elongated blastopore. A considerable space separates the archenteric wall of the gastrula from the wall of the blastosphere. From the two larger cells which first slipped inwards a number of small cells are budded off, which form a layer applied to the archenteric wall, and subsequently develop into a network which fills up the space between this wall and the wall of the blastosphere.

The embryo now rapidly develops into a peculiar larval form —originally regarded as a parasite of the mussel—known as the *glochidium.* In the winter the external gill of the female mussel will be found to be distended with immense numbers of these glochidia, together with yellow granular food material. Fig. 94, 7-11, shows their form and structure. Each has a pair of valves perforated with minute apertures, united along a straight hinge line, and connected by a single adductor muscle. When examined alive they may be seen to open and shut as they lie within the vitelline membrane. The edge of the shell opposite the hinge is produced into an incurved beak, set with sharp spines. Long coiled threads—the *byssus*—may be seen in some cases hanging down from the glochidium. Within the valves is seen a mass of cells forming the embryonic mantle-lobes, and careful examination will disclose three pairs of peculiar *tactile organs*, consisting of columnar cells, bearing numerous fine bristles.

The glochidia remain for a long while in the gill of the parent. But if some fish (*e.g.* perch or sticklebacks) be placed in the tank in which such brood-mothers are living, the embryos are ejected in great numbers. In some cases (in all ?) they are ejected through the pore marked *p.* in Fig. 90, A. and B., which is placed at the dorsal end of a canal, the ventral end of which opens into the exhalent siphon of the supra-branchial chamber. The glochidia are then set free from the egg-membrane, and swim by flapping the valves. **Ere long** they become attached by the byssus-thread to a fish, and hang there snapping **their** valves until they bury them in the epidermis of the fish. **There** they become encysted by a morbid growth of **the** epiderm **cells** of their host; and there **they undergo a post-embryonic meta-** morphosis. The **byssus** and its secreting **organ atrophy,** and the tactile organs disappear. The **single adductor also** atrophies; but before it **has** disappeared **rudiments of the** anterior and posterior adductors **of the adult become** visible. The foot arises from the **point at which the** byssus disappeared. The gills arise as papillæ covered **with** richly ciliated epidermis. The mantle-lobes undergo great change, or are perhaps formed *de novo.* The permanent shell is formed as two plates on the surface of the still parasitic larva. Finally, the little anodon, now resembling in all essential features the adult, is set free and dropped to the bottom.

The **glochidium embryo is peculiar to** fresh-water mussels **(anodons and unios).** The **great majority of** lamellibranch **molluscs are hatched as free-swimming** embryos with a ciliated **velum. It is clear, however, that** such free-swimming embryos **of fresh-water mussels, did** they exist, would be liable to be swept out to sea **by the flow of** the river water in which they lived. **On the other hand, if** they were developed under the protection **of the parent until** they acquired the adult condition and sedentary **habit, there** would be little dispersal of the genus. The attachment of the glochidium to the epidermis of the fish, while it (1) prevents the embryos being swept away by the seaward current of the river, at the same time (2) provides an efficient means of dispersal.

21

CHAPTER XVII

THE LIVER-FLUKE AND TAPEWORM.

THE Liver-fluke (*Fasciola* (*Distoma*) *hepatica*), a parasitic organism which is the cause of the liver-rot in sheep—a disease which, during the winter of 1879-80, caused in England the death of some three million sheep—is a flattened oval animal about an inch in length, and provided on its ventral surface with two suckers, from which it received its original name of Distoma.

General Summary of Life-history.—The liver-fluke, as such, is found in the liver of sheep. Here it reaches sexual maturity, each individual producing many thousands of eggs, which pass with the bile into the alimentary canal of the *host*, and are distributed over the fields with the excreta. Here, in damp places, pools, and ditches, free and active embryos are hatched out of the eggs. Each embryo (Fig. 97, C., much enlarged) is covered with cilia, except at the anterior end, which is provided with a head-papilla (*h. p.*). When the embryo comes in contact with any object it, as a rule, pauses for a moment, and then darts off again. But if that object be the minute water-snail, *Limnæus truncatulus* (Fig. 97, B., nat. size), instead of darting off the embryo bores its way into the tissues until it reaches the pulmonary chamber, or more rarely the body-cavity. Here its activity ceases. It passes into a quiescent state, and is now known as a *sporocyst* (Fig. 97, E.). The active embryo has degenerated into a mere brood-sac, in which the next generation is to be produced. For within the sporocyst special cells undergo division, and become converted into embryos of a new type, which are known as *rediæ* (F.), and which, so soon as they are sufficiently developed, break through the wall of the sporocyst. They then increase rapidly in size, and browse on the digestive gland of the water-snail (known as the *intermediate host*), to which congenial spot they have in the meantime migrated. The series of developmental changes is even yet not complete. For within the rediæ (besides at times daughter rediæ) embryos of yet another type are produced by a process of cell-division. These are known as *cercariæ* (Fig. 97, G.). Each has a long tail, by means of which it can swim freely in water. It leaves the intermediate host, and after leading a short active life, becomes encysted on blades of grass. The cyst is formed by a special larval organ, and is glistening snowy white. Within the cyst lies the transparent embryonic distoma, which has lost its tail in the process of encystment.

The last chapter in this life-history is that in which the sheep crops the blade of grass on which the parasite lies encysted; whereupon the cyst is dissolved in the stomach of the host, the little liver-fluke becomes active, passes through the bile-duct into the liver of the sheep, and there, growing rapidly, reaches sexual maturity, and lays its thousands of eggs, from each of which a fresh cycle may take its origin. The sequence of phenomena which constitute such a cycle is known as *alternation of generations*, or *heterogamy*, of which there are many modes among the lower members of both the animal and vegetable kingdoms. It is characterised by discontinuity of development. Instead of the embryo growing up continuously into the adult, with only the atrophy of provisional organs, it produces germs from which the adult is developed. Not merely provisional organs but provisional organisms undergo atrophy. In the case of the liver-fluke there are two such provisional organisms, the embryo-sporocyst and the redia.

We may summarise the life-cycle thus :—

(1) *Ovum* laid in liver of sheep, passes with bile into intestine, and thence out with the excreta.

(2) *Free Ciliated Embryo* in water or on damp earth ; passes into pulmonary cavity of Limnæus truncatulus, and develops into

(3) *Sporocyst*, in which secondary embryos are developed, known as

(4) *Rediæ*, which pass into the digestive glands of Limnæus, and within which, besides daughter rediæ, there are developed tertiary embryos, or

(5) *Cercariæ*, which pass out of the intermediate host and become

(6) *Encysted* on blades of grass, which are eaten by sheep. The cyst dissolves, and the young flukes pass into the liver of their host, each developing there into an adult hermaphrodite

(7) *Fasciola hepatica.*

The Distoma Stage.—The liver-fluke, which is the terminal product of this long series of changes, is somewhat oval in form,

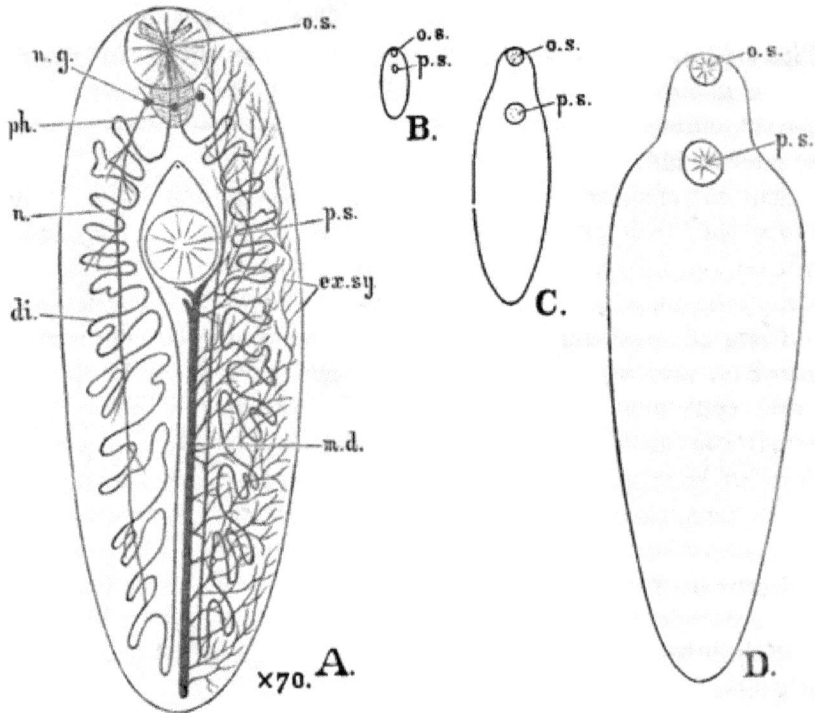

FIG. 95.—LIVER-FLUKE: DIGESTIVE SYSTEM.

(After A. P. Thomas and La Valette.)

A. Young distoma (× 70).—*di.* Branched digestive canal. *ex. sy.* Excretory system. *m.d.* Its mid duct. *n.* Nerve. *n. g.* Nerve ganglion. *o. s.* Oral sucker. *p. s.* Posterior sucker. *ph.* Pharynx.

B. C. D.—Stages in the growth of distoma (× 7½) to show relative positions of suckers.

and is provided with two suckers. In Fig. 95, B., C., and D., three stages of growth (after A. P. Thomas × 7½) are shown. The relative shifting of the suckers should be noted. The anterior or *oral sucker* (*o. s.*) is at the anterior end of the body, and in this the mouth occupies a central position. The other (*p. s.*) is ventral, and relatively nearer the posterior end in young flukes than in the adult organism. The cuticle has minute spines.

The mouth leads into a muscular *pharynx* (Fig. 95, A., *ph.*), which is followed by a short œsophagus. The alimentary canal then bifurcates, one limb passing down either side of the body and ending blindly. Each limb has a number of lateral branches. In Fig. 95, A. (partly after A. P. Thomas × 70), these are represented at an early stage, viz., that shown in B. In later stages the branches are more complicated, though they always end somewhat bluntly, and the digestive organs are masked by the organs of reproduction. There is no anus. Living, as does the fluke, upon the prepared juices of its host, there is presumably little indigestible matter in its stolen food-stuff.

Excretory System.—At the posterior end of the body there is an excretory pore which leads into a median tube (Fig. 95, *m. d.*). From this tube are given off a number of ducts which branch repeatedly, and form a net-work (*ex. sy.*). The fine terminal vessels end in minute ciliated funnels, opening into the lacunar spaces which represent the cœlom.

Nervous System.—This consists of a nerve-ring round the pharynx, with a single ventral and paired lateral ganglia (*n. g.*). From the lateral ganglia arise lateral nerves (*n.*), running down the sides of the body.

The Generative System.—The liver-fluke is hermaphrodite, male and female organs being developed in each individual. It would seem, however, that there is cross-fertilisation, so that the ova of one individual are fertilised by the spermatozoa of another.

The *testes* (spermaries) are large branched and multi-lobed organs occupying much of the central part of the body (Fig. 96, *ts.*). Two *vasa deferentia* (*v. d.*) pass forward and unite into a reservoir and looped duct (*ductus ejaculatorius*), which terminates in

a protrusible *cirrus* (*ci.*) lodged in a cirrus sac.   This is situated
ventrally between the two suckers.

The *ovary* (*ovy.*) is branched and tubular.   From it an *ovarian
duct* meets the ducts of the large yolk-glands or *vitellaria*, which

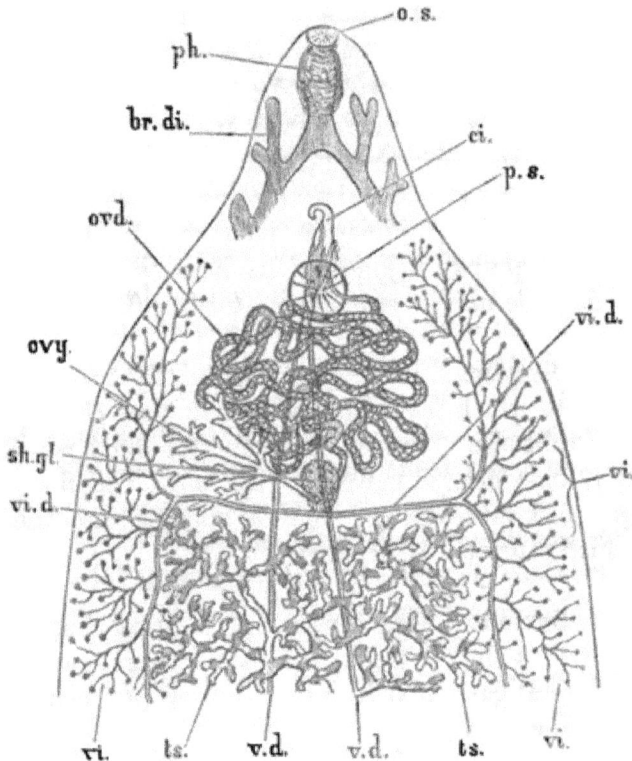

Fig. 96.—Liver-Fluke : Organs of Generation.

br. di., Branches of digestive system. *ci.* Cirrus. *o. s.* Oral sucker. *ovd.*
Oviduct crowded with ova. *ovy.* Ovary. *p. s.* Posterior sucker. *ph.*
Pharynx. *sh. gl.* Shell gland. *ts.* Testis. *vi.* Vitellarium. *vi. d.* Duct of
vitellarium.

lie one on each side of the body, and are composed of a multitude of small branches terminating in rounded masses.   Beyond
the points of union of ovarian and vitelline ducts, an *oviduct* or
uterus conveys the ova to an aperture near the base of the cirrus.
A *shell gland* (*sh. gl.*) is situated at the junction of the ovarian

and vitelline ducts. The oviduct is usually crowded with eggs, which are here undergoing the early stages of their development. A canal (Laurer's) from the vitelline duct is said to open dorsally.

The Ovum.—The egg is small (·13 by ·08 mm.), oval, smooth, and transparent, with a yellowish-brown chitinous shell. At the anterior end is an *operculum* (Fig. 97, A., *op.*). The embryo (*em.*) occupies but a small space, the rest being occupied with yolk-cells. Segmentation goes on within the uterus ; but after the egg is laid, further development can only go on at a reduced temperature of 23°-26° C. All the eggs laid under the same conditions do not appear to produce embryos at the same time, some embryos emerging weeks, or even months earlier than others. What conditions determine this difference is not known, but such a state of things is clearly of advantage to the species. The embryo within the egg increases by absorption of yolk, develops a head-papilla and a double eye-spot, and becomes ciliated though the cilia are at present motionless. Eventually the operculum springs open and the embryo emerges.

The Free Embryo.—The embryo is conical in form (Fig. 97, C.) with a retractile head-papilla (*h. p.*) at the anterior end. The whole surface (head-papilla excepted) is covered with long cilia, borne by flattened ectoderm cells arranged in five or six transverse rings around the body. The four cells which compose the first ring are thicker than the others. Beneath the ectodermic ciliated layer is an endodermic body-wall, within which are a number of delicate vesicular germinal cells (*g. c.*). A rudimentary digestive tract is visible behind the head-papilla. Over this in the figure (after A. P. Thomas) is seen the eye-spot (*e. s.*), shaped like two crescents united back to back, and developed in the body-wall. In the body-wall there are also two ciliated funnel-shaped spaces of the excretory system (*ex.*), in each of which there is a single large flame-shaped cilium.

The embryo is exceedingly active, darting backwards and forwards, and rotating on its long axis. The duration of its free and active life is, however, not more than some eight hours. If it does not meet with a water-snail within that time it dies.

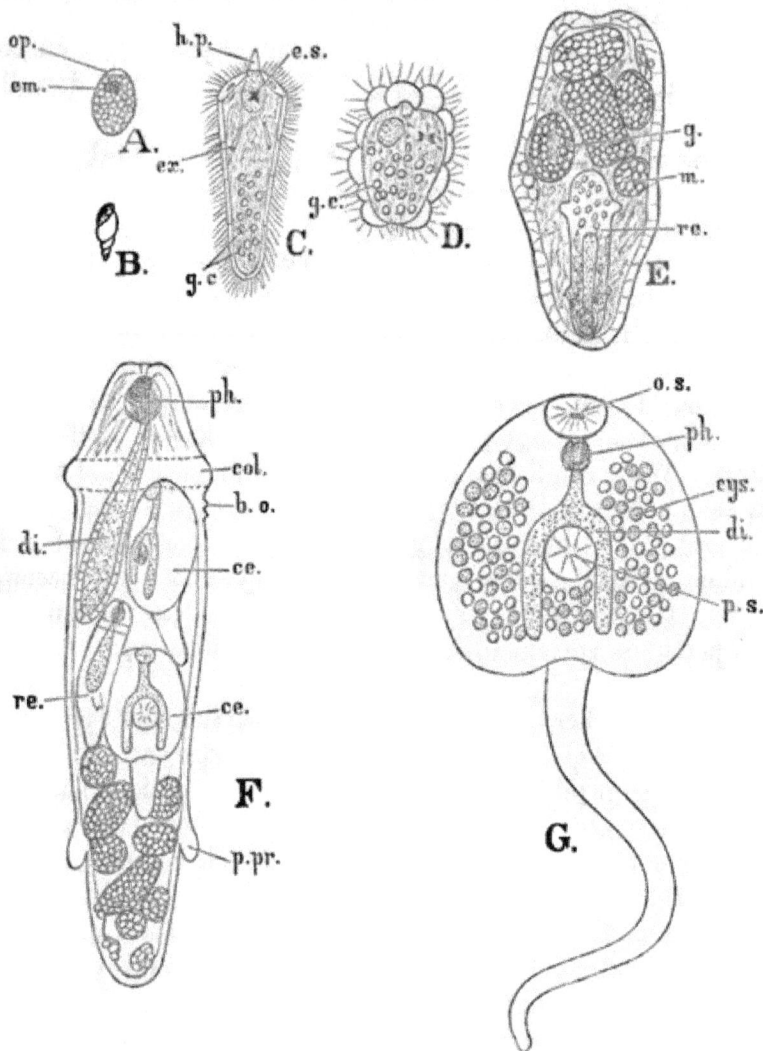

FIG. 97.—LIVER-FLUKE : EMBRYONIC STAGES. (After A. P. Thomas.)

A. Ovum.—*em.* Embryo. *op.* Operculum.

B. Limnæus truncatulus (natural size).

C. Free Embryo.—*e. s.* Eye-spot. *ex.* Excretory vessel. *g. c.* Germinal cells. *h. p.* Head-papilla.

D. Embryo preparing to become a sporocyst. *g. c.* Germinal cells.

E. Sporocyst—*g.* Gastrula. *m.* Morula. *re.* Redia.

F. Redia.—*b. o.* Birth opening. *ce.* Cercaria. *col.* Collar. *di.* Digestive sac. *ph.* Pharynx. *p. pr.* Posterior processes. *re.* Daughter redia.

G. Cercaria.—*cys.* Cystogenous organ. *di.* Digestive sac. *o. s.* Oral sucker. *p s.* Posterior sucker. *ph.* Pharynx.

There are thus many chances to one against further development; hence the vast number of ova. If it meets with a Limnæus truncatulus it spins round upon its axis, head-papilla protruded, alternately shortening and suddenly extending its body, and thus bores its way through the soft tissues into the pulmonary chamber. It now becomes quiescent. The external ciliated cells become swollen by absorption of water, their cilia sticking out stiffly from them. The body becomes elliptical, the eye-spot breaks up (Fig. 97, D.), and the organism passes into the condition of a sporocyst.

The Sporocyst.—This now increases in size until it becomes ·6 mm. long, while the germinal cells within it increase in number, partly by division of the previously contained germinal cells, partly by the multiplication and division of the cells of the body-wall, some of which become detached. The cells within the sporocyst thus pass into the *morula stage* (E., m.). One side then becomes flattened, and the cells here appear to be invaginated, so that a *gastrula* (g.) is constituted; but the walls of the gastrula are in contact, so that there is no primitive digestive cavity or archenteron. Each spore is now surrounded by a delicate membrane, and becoming more quadrate in shape, passes into the redia condition (E., re., and F.). At one end a number of cells are arranged to form a spherical pharynx (F., ph.), which leads into a blind digestive sac (di.). A little behind the pharynx the surface of the body is raised into a ridge, forming a ring (col.) surrounding the anterior end; whilst near the opposite end, two short processes grow out (p. pr.). This secondary embryo has now passed into the redia stage.

The Redia.—Within such a sporocyst as is shown in Fig. 97, E., there are generally one or two rediæ ready to leave, and others in a more or less advanced stage of development. When it is about ·26 mm. long the redia bursts through the sporocyst, the wall rapidly closing up again after the exit. The active little redia finds its way into the digestive gland of the Limnæus, and then feeds upon that organ, the annular ring or collar (Fig. 97, F., *col.*) enabling it to fix itself while it browses. There is an excretory system (not figured) consisting of two groups of ciliated infundibula, one just behind the collar, the other just behind the posterior processes. The mouth is surrounded by lips, and there is a muscular pharynx (*ph.*) opening into the digestive sac (*di.*), consisting of a blind tube lined with clear nucleated cells. There is also a definite birth-opening (*b. o.*).

At the posterior end of the redia, in the lining of the body-cavity, there are large clear cells which undergo segmentation, and form mulberry masses (*morulæ*) enclosed in a delicate membrane. The morula is detached into the body-cavity, and is invaginated to form a gastrula, the walls of which are, however, in contact, so that the primitive cavity (*archenteron*) is obliterated. Some of these would seem to develop into rediæ (*re.*) like the parent redia. Others become at first oval, and then more and more elongated, whilst one end becomes narrower, and is partly constricted from the remainder to become the tail of the cercaria (*ce.*), while the rest of the tertiary embryo becomes its body.

The Cercaria.—A sucker appears at the anterior end of the cercaria (Fig. 97, G., *o. s.*), in the midst of which is the mouth-opening. A second sucker (*p. s.*) appears about the middle of the ventral surface of the flattened body. The digestive tract has a less marked pharynx (*ph.*) than the redia, and is forked; but the limbs have no lateral branches. When it is about ·3 mm. long the cercaria makes its exit from the redia by the birth-opening. In mature cercariæ there are very minute spines on the anterior end of the body. Within the body, at the sides, are a number of opaque white granules, forming the cystogenous organs (*cys.*).

The cercaria passes out of the intermediate host, and after a short active life comes to rest on a blade of grass or other object, assumes a spherical form, and is encased in a white cyst composed of opaque white granules from the cystogenous organs. The tail is pinched off during the process of encystment.

The Cyst is thus round and snowy white; but within it lies the minute transparent embryo—a tailless cercaria—which, if it pass into the alimentary canal of a sheep, will be carried thence to the liver, and there gradually assume the form of an adult liver-fluke.

The fluke may be developed in other forms than the sheep, such as the goat, and occasionally in the larger cattle, but since they are less close feeders they less frequently take in the encysted cercaria.[1]

By way of appendix to the foregoing life-history of the liver-fluke, the life-history of another parasite, the tapeworm, may here be given.

One form of tapeworm (*Tænia solium*) infests man as its host. The adult worm may be many yards in length. Anteriorly there is a minute *scolex* or head (resembling the upper part of Fig. 98, C.). It has a double circlet of spines set close together, and behind this four suckers. Then follow a great number of joints or metameres. These are constantly forming behind the scolex, and as new ones are budded off those previously formed are pushed further away from the head. Thus the metameres are progressively older as we proceed from the anterior to the posterior end. The youngest are scarcely differentiated, and form the so-called neck. Behind this they have the form shown in Fig. 98, D.; further down they are as shown in E.; near the posterior end they are as represented at F. The scolex and metameres is called a *strobila*. From time to time some of the posterior metameres are detached as *proglottides*, and pass out with the excreta.

Running down the whole strobila are canals of the excretory or water-vascular system. In the scolex there is a circular vessel. Four longitudinal trunks (Fig. 98, G., *l. v.*) run down, two on each side, the whole series of metameres. At the hinder end of each metamere there is a transverse vessel (*tr. v.*). At the extremity of the strobila is a *porus excretorius*. From the trunks fine branched vessels, with ciliated junctions, open out.

---

[1] Mr. Bolton, of Birmingham, has supplied me with liver-flukes and with cercariæ of allied forms. Prepared and mounted flukes may be obtained of most dealers in microscopic objects. Messrs. Watson, of High Holborn, supplied me with a very well-mounted specimen.

Nerve ganglia have been described in connection with the water-vascular ring of the scolex.

The scolex itself contains no reproductive organs. In the metameres these organs are progressively developed, each metamere being hermaphrodite, with cross-fertilisation. They are

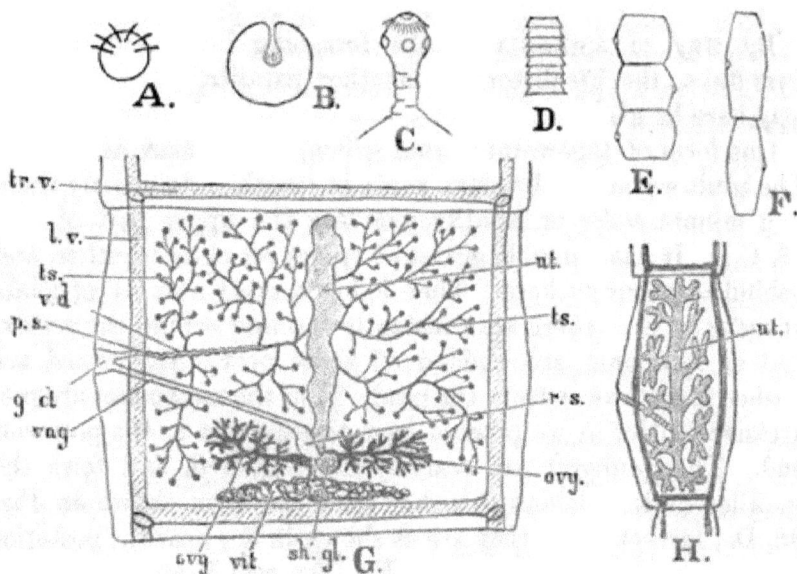

Fio. 98.—TAPEWORM.

A. Embryo with six stylets.   D. Anterior metameres.
B. Invaginated blastosphere.   E. Median metameres.
C. Scolex of bladder-worm.   F. Posterior metameres.

G. A median metamere.—*g. cl.* Generative chamber. *l. v.* Longitudinal water-vascular vessel. *ovy.* Ovary. *p. s.* Penis sac. *r. s.* Receptaculum seminis. *sh. gl.* Shell gland. *tr. v.* Transverse water-vascular vessel. *ts.* Testis. *ut.* Uterus. *vag.* Vagina. *v. d.* Vas deferens. *vit.* Vitellarium.

H. A terminal segment in which much of the genital apparatus has atrophied, and the uterus (*ut.*) is distended with ova.

seen in their most perfect condition in metameres of the stage shown at E. The *spermaries* or *testes* (Fig. 98, G., *ts.*) consist of a great number of sperm-balls. These pour their contents into efferent vessels, by which they are conveyed to the *vas deferens* (*v. d.*), which opens by a protrusible termination (*p. s.*) into a genital cloaca (*g. cl.*), opening into the exterior by a genital pore, which in successive metameres is alternately on the one side and the other (see E.). Into the genital cloaca there also opens a vagina (*vag.*), which expands to form a *receptaculum seminis* (*r. s.*). Closely connected with this are the apertures of four structures : (1) the *ovary* (*ovy.*); (2) the large *uterus* (*ut.*); (3) the *yolk gland* (*vitellarium*) (*vi.*) ; and (4) the small *shell gland* (*sh. gl.*).

The ova passing from the ovary are impregnated by spermatozoa in the receptaculum. They also receive yolk (or albumen ?) from the yolk gland, and are encased in a shell from the shell gland. Then they pass into the uterus. As the uterus becomes distended with eggs it assumes the form shown in Fig. 98, H. (*ut.*), which represents a metamere in the stage figured at F. The other reproductive organs are absorbed or completely masked. When these ripe proglottides break away and are expelled with the excreta, they die and decay ; but the eggs which they contain in vast numbers do not readily lose their vitality, and only await favourable conditions for development.

Before showing what these conditions are, it will be well to suggest the question whether the strobila is one organism or many. Is the scolex an asexual zooid which gives rise by budding to a great number of sexual zooids ? or have we one segmented organism, of which the anterior portion is specially differentiated while all the others are serially homologous, and each provided with male and female reproductive organs ? The former view is that generally held. In which case we have here an instance of alternation of generations in which the proglottides are to be regarded as provisional embryonic organisms. On the other view we have provisional embryonic body-segments, a state of things curiously intermediate between embryonic organs and embryonic organisms.

Within the uterus the ova undergo segmentation with the formation of a morula provided with six stylets (Fig. 98, A.), and enclosed in a shell. If now a portion of a proglottis, or some of the embryos, be swallowed by a pig, the egg-shell is dissolved by the gastric juice of this *intermediate host*, and further development is rendered possible. The embryo, by means of the stylets with which it is provided, penetrates the tissues of the intermediate host, in the muscles of which it becomes encysted, having assumed the form of a hollow bladder-like blastosphere. The blastosphere seems then to become invaginated, and within the cæcum so formed a pre-scolex is developed (Fig. 98, B.), which by evagination becomes a scolex (C.), attached to which is the blastosphere. This is known as the *cysticercus* stage, and the organism is called a bladder-worm. The pork which contains such bladder-worms is called measly. Such pork, if eaten by man in an uncooked or incompletely cooked state, gives rise to tapeworm. The scolex attaches itself to the intestine by its suckers and circlet of hooks. The bladder seems to be lost, and in its place there bud out the series of metameres which convert the organism into a fully-developed tapeworm strobila.

CHAPTER XVIII.

THE HYDRA.

THE fresh-water polyp (*Hydra viridis*) may be found in summer attached to duckweed in ponds. If some of the duckweed be placed in a glass vessel near a window, the little green polyps will congregate to that side of the vessel which is turned towards the light. They are minute gelatinous bodies, half an inch or less in length. Each is attached to the glass or to the duckweed by a flat disc, known as the foot. When the hydra is fully extended the foot is more expanded than the body, but when the organism is in a state of contraction the foot is narrower than the rounded and swollen body. At the end of the body opposite to the foot is the *hypostome* (Fig. 99, *h. s.*), bearing at its summit the mouth. Encircling the hypostome are the *tentacles* (*t.*), varying in number according to the age of the individual, but seldom exceeding eight. By alternately applying the foot and the tentacles, the hydra can creep with a looping motion. The body surface generally, and the tentacles especially, are irritable. If one of the tentacles be touched, it is rapidly withdrawn, the irritation spreads to the others, and they too are drawn towards the body, which itself contracts, so that the whole organism, to the naked eye, looks like a rounded speck of green jelly. In Fig. 99, C., the hydra is represented in its fully elongated condition; in A. it is moderately contracted; in B. in the condition of utmost contraction. After irritation the contraction relaxes, and the hydra resumes its fully expanded state. Should any minute animalcule or small water-flea come in contact with one of the tentacles, it seems to stick, and after a short struggle, its activity ceases. The tentacle is meanwhile retracted (Fig. 99, C.) ; other tentacles bend round to lend assistance ; and the animalcule is tucked into the mouth, and thus passes into the body-cavity, which occupies the whole

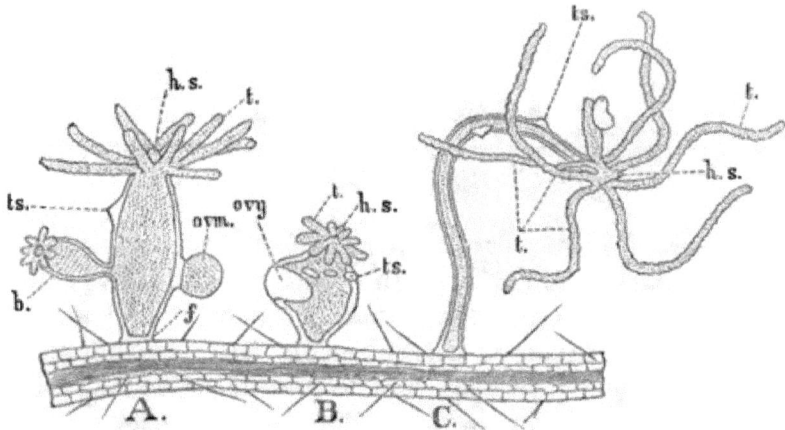

Fig. 99.—HYDRA VIRIDIS.

A. Hydra half retracted, with a bud and an ovum attached to the shrunken ovary.
B. A small hydra firmly retracted.   C. A hydra fully extended.
b. Bud. f. Foot. h. s. Hypostome. ovm. Ovum. ovy. Ovary. t. Tentacles. ts. Testis.

of the interior of the polyp. Here ere long it breaks up ; the nutritive matter is absorbed by the lining cells of the cavity, and the indigestible residue is got rid of through the mouth, for there is no separate vent.

Histology.—In Fig. 100, A., a diagrammatic longitudinal section through the body-wall of hydra is given. It is divided into two main layers, *ectoderm* (*ec.*) external, and *endoderm* (*en.*) internal. In the ectoderm there are large conical neuro-muscular cells (*n. m.*). If a hydra be teased (fresh, or after treatment with 20 per cent. acetic acid, or 1 per cent. osmic, or dilute chromic and osmic) some of these cells at C. will be readily seen. At their pointed ends they give off one or two delicate *muscle processes* (*m. p.*), which come into close relation with the thin supporting lamella (*s. l.*), a portion of which adheres to one of the fibres in C. 2. In the interstices between the neuro-muscular cells, especially in the deeper part of the ectoderm, there is an *interstitial tissue* (*int.*), composed of small

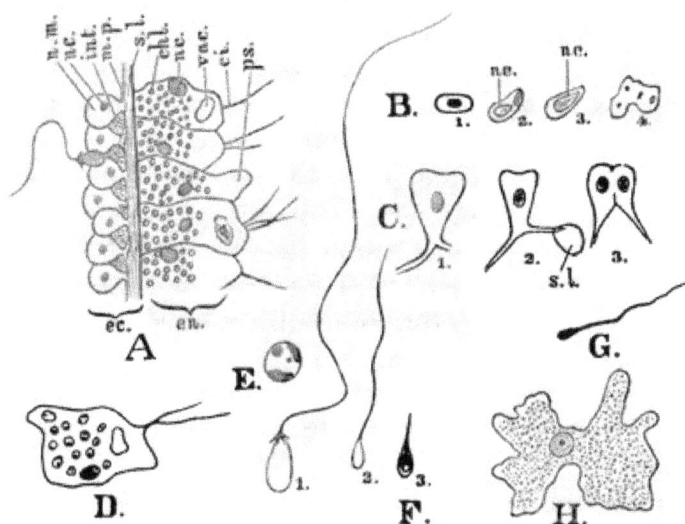

FIG. 100.—HYDRA HISTOLOGY.

A. Diagrammatic longitudinal section through a part of the body-wall.— *ec.* Ectoderm. *en.* Endoderm. *chl.* Chlorophyll corpuscles. *ci.* Cilium. *int.* Interstitial cells. *m. p.* Muscle processes. *nc.* Nucleus. *n. m.* Neuro-muscular cell. *ps.* Pseudopodial process. *s. l.* Supporting lamella. *vac.* Vacuole. (The regularity of the ectoderm is diagrammatic).

B. Interstitial cells.—*ne.* Nematocysts.   E. Chlorophyll corpuscle.
C. Neuro-muscular cells.—*s. l.* Sup-   F. Thread-cells (nematocysts).
    porting lamella.   G. Spermatozoon.
D. Endoderm cell.   H. Ovum.

cells, some of which are shown at B. In the patch of cells at 4 the boundaries (after osmic) could not be made out. The interstitial cells are absent in the foot.

Within these interstitial cells are developed the curious organs known as thread-cells or *nematocysts*. When the tentacle of a living hydra is examined under a high power, long fine

22

threads may be seen protruding here and there. On addition of an irritant, such as dilute acetic acid, great numbers shoot forth. They are of use in capturing (and perhaps paralysing) prey. Some of these cells are shown in Fig. 100, F. The largest (1) have spines at the base of the very long whip. Others have no spines, and may be incompletely developed thread-cells. In the resting stage (before eversion) the thread-cell is rounder, and resembles a flask in which the tapering neck has been pushed inwards so as to lie within the flask. Within the neck the barbs appear as arrow-heads. The neck is continued into a long filament regularly coiled like a ship's cable. It would seem that during eversion the neck and the whole length of filament is turned inside out. This process would, moreover, seem to be a vital one, and within the command of the *Hydra*. Infusorians (*Euplotes*) have been observed wandering over the body and tentacles of *Hydra fusca* without causing any discharge of nematocysts. The free end of a cell bearing a nematocyst is generally prolonged into an elementary sensiferous process or *cnidocil*. Connected with these cells are small stellate *nerve-cells*.

Internal to the muscle processes and the delicate membrane (supporting lamella) with which they enter into close relation, is the endoderm layer, consisting of large irregular cells (Fig. 100, A. *en.*, and D.). Their inner free surfaces carry cilia, while their deeper parts are crowded, in *Hydra viridis*, with the green *chlorophyll corpuscles* (*chl.*), which give to this hydra its distinctive colour.

Concerning these chlorophyll corpuscles there has been much discussion. By some they are regarded as plants (algæ) parasitic (or rather symbiotic, *i.e.* living together for mutual advantage) on the hydra. It is more probable, however, that they are true products of the animal cells. One of these is shown at E., as seen under the highest power, the parts shaded dark being in the cell itself green. Very commonly the green part forms (as seen in optical section) a ring surrounding a clear central part of the cell. There is no nucleus. As in the case of vegetable chlorophyll, the green colouring matter liberates oxygen in sunlight, as has been proved by keeping hydras under an inverted test-tube in sunlight and testing the bubble of gas collected.

Besides chlorophyll corpuscles, the endoderm cells may contain clear spaces (A., *vac.*), and sometimes irregular masses, which resemble food particles. If this be so, we have in these cells a process of *intracellular digestion*, solid food-particles being taken into the cell just as we find to be the case in such a single-cell organism as amœba. And in this connection it is interesting to note that these cells are capable of altering their form and pushing forth pseudopodial processes (A., *ps.*). If a hydra be teased up fresh, some of the endoderm cells may be seen in amœbiform motion. Occasionally thread-cells may be found in the endoderm. It may be that, as has been suggested, these have been taken in with the food and ingested.

The body-cavity, which lies within the endoderm cells, is subject to much variation of diameter. It extends from near the foot to the mouth, and sends prolongations into all the tentacles. It is not readily seen in green hydra (the outer clear space in Fig. 99 representing ectoderm and not the whole body-wall), but is more easily observed in the rarer reddish polyp, *Hydra fusca*.

Processes of Multiplication.—(1.) *Budding.*—When well fed the hydra multiplies by a process of budding. The bud appears first as a little green knob, then becomes pear-shaped, and ere long develops mouth and tentacles, the mouth appearing somewhat suddenly. It thus reaches a size which may fully equal half that of the parent, after which the connection between the parental body-cavity is obliterated by the ingrowth of ectoderm, and the daughter hydra separates to lead a free existence. Such a bud in an advanced stage is shown in Fig. 99, A. Sometimes there may be several buds in various stages of development, when the hydra has the appearance of a compound organism. This process is wholly asexual.

(2.) *Fission.*—It is said that hydra may also multiply by fission, a constriction appearing in the middle of the body, and separation of the organism into two parts occurring at this spot. In this case the separated portion acquires a new foot, and the basal portion acquires the circumoral tentacles. It is somewhat doubtful, however, whether this process occurs under natural conditions. But there is no doubt that if a hydra be artificially divided, each separated part develops into a new and perfect individual.

(3.) *Sexual Reproduction.*—In its sexual reproductive organs the hydra differs very markedly from any form we have as yet studied. It is hermaphrodite and self-fertilising; but the testes and ovaries are variable in number, and not definite in position. The testes, or spermaries, of which there may be as many as four or five in the same individual, arise as thickenings of the ectoderm from the active local growth of interstitial cells. The neuro-muscular cells here take the form of a thin external layer. Within the testes, which assume a mammiform appearance (Fig. 99, *ts.*), swarms of spermatozoa may be seen under a high power. Each has an oval head and vibratile tail (Fig. 79, G.), and is developed from one out of the many clear spherical bodies into which the spermatospores derived from the interstitial cells subdivide.

The ovary (Fig. 100, B., *ovy.*) is larger than the testis. There may be one or two in the same individual. It, too, originates in the ectoderm, and from the interstitial cells which undergo special differentiation over an area which embraces about half the circumference of the body, and is often, but not always, situated nearer the foot than the testes (Fig. 99, B.). As development of the ovary proceeds one central cell increases enormously at the expense of the others. This single ovum (Fig. 100, H.) is capable of amœboid movements, pushing forth pseudopodial processes at various parts of its surface. It contains a number of yolk-granules, and is provided with a well-marked germinal vesicle containing a germinal spot. In the green hydra chlorophyll corpuscles are developed within the ovum. Before impregnation by the spermatozoa, the germinal vesicle and its contained germinal spot are said to disappear. The ovum escapes by the bursting of the thin ovarian walls, but still remains attached to the shrunken remnant of the ovary

(Fig. 99, A., *ovm.*).   Impregnation now takes place by the
discharge of spermatozoa from one of the testes of the same
individual.   Both ovary and testis are ectodermal.

Development.—Cleavage would seem to be holoblastic and
regular, the cells protruding and retracting pseudopodial processes along the cleavage grooves.   A morula or mulberry mass
is thus formed ; and at the close of segmentation the cells are of
two kinds, (*a.*) prismatic cells, constituting an outer layer, and
(*b.*) polygonally flattened cells forming an inner mass
    On the outer surface of each external cell a thin chitinous
plate is, according to the most recent researches, formed, and
the plates from adjoining cells fuse together so as to give rise
to a continuous pellicle.   When this has assumed distinctness,
a second is formed in a similar way, and then a third, a fourth,
and so on, the various pellicles fusing together to form a thick
laminated encasing chitinous shell, *which is thus morphologically
one of the ectodermal tissues of the embryo.*   Between this cuticle
and the germ an elastic membrane is formed.
    After this a remarkable retrograde metamorphosis is said to
take place.   The cells undergo *histolysis* (cell dissolution), their
boundaries becoming indistinct and disappearing, and the
protoplasm of the cell is said to revert to a primitive unsegmented condition.   Within the protoplasm there is then formed
a small excentric cavity, the rudiment of the body-cavity, which
gradually increases in size until the germ assumes the condition
of a hollow sac, with homogeneous walls.   In this state it is set
free by the bursting of the external chitinous investment.   The
walls differentiate into a clear superficial zone (ectoderm) and a
darker inner zone (endoderm), and within the outer of these
layers neuro-muscular, and then interstitial cells, are differentiated.   The embryo elongates, and the mouth is suddenly
formed as a star-like cleft at one end.   Bud-like tentacular
processes make their appearance at about the same time.   The
endoderm cells are now differentiated, the inner elastic membrane
is thrown off, and the young hydra is set free.
    Several points seem to call for especial notice in this process

of development, and the product to which it gives rise. In the first place we may note the single ovum as the product of the ovary ; (2) its amœboid movements ; (3) the modification of part of the embryo to form a chitinous case for the rest ; (4) the curious histolysis after a certain stage of histogenesis has been reached ; (5) the formation of the primitive body-layers, and the contained digestive cavity, not by a process of invagination, but by delamination of the walls of a hollow sac-like embryo ; (6) the disruptive formation of the mouth ; (7) the absence in the perfected organism of any body-cavity. This distinguishes the hydra and its allies (cœlenterata) from all other metazoa (cœlomata), which have at some period a body-cavity.

It will be seen, also, that the development of hydra is con-tinuous. There is no alternation of generations. In this the hydra is peculiar among the group of hydroid polyps to which it belongs. It differs also from allied zoophytes in being solitary, and not a member of a colony, group, or stock (*hydrosome*). Such a stock consists, in general, of a stem and perhaps ramifying branches, from which the polyps of the colony (*hydranths*) are offshoots. The stem and branches are tubular, and the somatic cavity within them is in free communication with the digestive cavity of the hydranths. The generative organs (*gonophores*) arise in special regions, and arise as buds either from the polyp or from the tubular stem or branches. In either case it is at first a sac-like diverticulum of the somatic or gastric cavity. Within this diverticulum the generative products accumulate. The genital cells may, however, arise in parts other than those in which they are finally lodged, or even in some cases in the body-wall of the polyp before the gonophores have made their appearance. In such cases the generative cells subsequently migrate into the gonophores. The development of the oospores in some cases precedes that of the spermatospores ; and the ova and spermatozoa are, as a rule, finally lodged in different gono-phores.

So far we have regarded the gonophore as a simple bud or diverticulum attached to the polyp, or to the axial branches of the stock. But in the higher forms this develops into a *medusa*, which is detached and floats off to enjoy a free and separate existence as a provisional embryonic organism.

Such a medusa is figured in Fig. 101. Within the bell-shaped hollow umbrella or *nectocalyx* there hangs a tubular *manubrium* (*m.*). The cavity of the manubrium leads up into the umbrella, and there branches into four or more radiating canals (*r. c.*), which proceed outwards and lead into a circular canal (*c. c.*), running round the rim of the nectocalyx. The rim has a deli- cate shelf-like inward projection —the *velum* (*v.*)—which narrows the orifice of the bell. From the edge of the rim *tentacles* (*t.*) take their origin ; and on its margin there may be sense organs, called *marginal bodies* (*m. b.*), which may be eyespots or auditory vesicles. The sper- maries and ovaries are generally in different individuals, and are lodged either in the walls of the manubrium, or in those of the radiating canals (*g. o.*).

The nervous system consists of a double ring of nervous

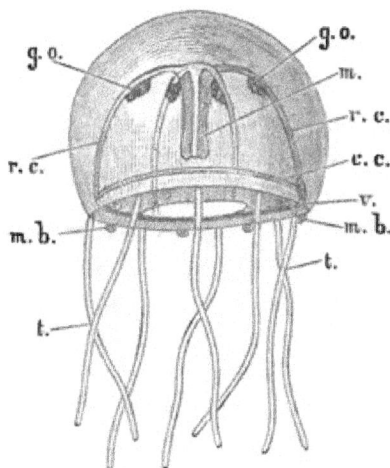

FIG. 101.—MEDUSA.

c. c. Circular canal. *g. o.* Genital organ. *m.* Manubrium. *m. b.* Marginal body (ocellus or otocyst). *r. c.* Radiating canal. *t.* Tentacles. *v.* Velum.

matter, containing both nerve-fibres and nerve-cells, which runs round the edge of the nectocalyx at the point of insertion of the velum. The upper nerve-ring lies entirely in the ectoderm, and is hardly differentiated from the epithelium. It sends out fibres to the enveloping epithelium, and these, or the epithelium cells, bear delicate sense-hairs. The lower nerve-ring, though still belonging to the ectoderm, is more completely differentiated from the epithelium, has larger fibres, and a greater number of ganglion cells. The two nerve-rings are separated by a thin membrane, through which, however, there pass strands of con- necting fibres. The entire under surface of the nectocalyx (*sub-umbrella*), and the surface of the manubrium, is covered with an epithelial layer, beneath which is a thin sheet of muscular tissue, not thicker than very thin paper. And between these two there is an exceedingly delicate plexus of nervous tissue, containing both nerve-fibres and ganglion cells. Beneath the epithelium of the outer surface of the nectocalyx there is neither muscular nor nervous layer. In the velum there are concentric muscular fibres, but probably no nervous tissue.

The nectocalyx is rhythmically contractile. Hence the medusa can swim lazily through the water. But if the margin containing the nerve-ring be cut off, immediate, total, and permanent paralysis of the umbrella results.

It should be noted that this medusa or free gonophore is in several respects more highly organised than the hydroid stock from which it takes its origin. It has (1) between the endoderm and ectoderm a thick mass of intermediate substance; it has (2) a well-defined nervous system ; and it has (3) sense organs.

The segmentation of the ovum is complete or holoblastic. A morula results, which passes into a single-layered, ciliated, *planula*. The segmentation cavity becomes the gastric cavity of the future polyp, but becomes lined with a second layer of cells derived from the first by delamination. The spherical or oval larva, now free, after a short locomotive existence, fixes itself by its aboral pole. At the opposite end a mouth appears ; tentacles bud out; fresh individuals are produced by a process of budding ; these, however, do not separate as in hydra viridis, but remain attached, and form part of the colonial stock.

Such, then, is the typical mode of reproduction in the hydrozoa, the sexual products ripen, and are fertilised in a separate medusoid gonophore. In some cases, however, the medusa never separates to lead a free existence; it is an *attached medusa*. In others, the attached gonophore is incomplete ; its essential parts are present, but in an unexpanded condition; it is a *disguised medusa*. In yet others development may go no further than the production of a hollow diverticulum of the body-wall of the parent; it is a *sporosac*. The free medusa in the course of its development passes in succession through the stages of sporosac, disguised medusa, and attached medusa. On the hypothesis of evolution these forms may be regarded either as (1) stages in the gradual evolution of a perfect medusa, or (2) stages in the degeneration of the perfect medusa, by arrested development owing to changed conditions. The latter view is the more probable. It is difficult to see how the disguised medusa and attached medusa, which have all the parts of the free medusa fitting it for independent existence, but never have an opportunity for employing them, could have arisen by any process of natural selection or otherwise; or how the possession of a swimming-bell that is never opened could be of any advantage to the individual or the species.

And now for the bearing of these facts on the system of generation in hydra. Here there is not even a sporosac. And it has been recently suggested that in hydra development has been arrested at a yet earlier stage, not only the medusa condition (free, attached, or disguised) having been suppressed, but even the sporosac condition also. This might very possibly be the result of a fresh-water habitat. For it has been shown that fresh-water forms rarely have free and active embryos, since they would tend to be swept out to sea, and this would result in the perishing of the race as a fresh-water type, or in the return of the race to the sea, its primitive home. That power of migration at some stage of its existence, which is so important for the well-being and extension of a marine species, is here abandoned under new conditions of existence.

If this view of the mode of generation of hydra be correct, this organism must be regarded as a *degenerate* type. Stellate nerve-cells represent the nervous system. The muscular system is not differentiated. But (on this view) Hydra is the descendant of organisms in which nervous system, sense organs, and muscular system were better developed during the medusoid stage. And where a form is descended from organisms more highly developed than itself it is said to be degenerate. For it sometimes happens that an organism is placed in conditions where the possession of organs which were of advantage to its ancestors in the life-struggle are no longer of advantage.

These useless organs are therefore lost, and the form—though admirably suited to the meaner conditions of its life—becomes degenerate through the meanness of these conditions.

On the other hand, the more generally, and until recently universally accepted view of the generative system of hydra is that its reproductive organs exhibit a primitive condition prior to the evolution of special sexual buds. In which case it has not degenerated from a higher form, but has simply never become so highly developed as the other hydrozoa.

CHAPTER XIX.

VORTICELLA AND PARAMŒCIUM.

VORTICELLA, the bell-animalcule, has a delicate clear transparent wine-glass or bell-shaped body, carried at the end of a longer or shorter elastic thread-like stalk or pedicle. Within the transparent outer sheath of this stalk is a central contractile thread or fibrilla (*c. f.*), disposed in an extended spiral form. In some of the largest forms, and under the highest powers, the sheath is seen to be continuous with the thin transparent outer wall of the body called the *cuticle*. The contractile thread is seen to be continuous with a delicate contractile myophan layer that passes up into the walls of the body and underlies the cuticle. These constitute the *ectosarc*. Beneath this, and perhaps hardly separable from it, is the cortical layer, which passes insensibly into the more fluid central substance of the body. These constitute the *endosarc*.

That which answers to the rim of the wine-glass is known as the *peristome* (Fig. 102, A., *ps.*). It is somewhat thickened and turned outwards, and is beset with cilia. Within the bell-shaped opening of the wine-glass is the *disc*, the edges of which are richly ciliated. The disc is, however, not large enough to fill the opening of the wine-glass, and is attached to one side. Thus there is left a crescent-shaped space, the *oral fossa* or *vestibule*. Into this a branch of the ciliary wreath descends. It passes obliquely downwards towards the centre of the bell, thus passing beneath the disc, and gradually narrows till it ends abruptly. An anal aperture is described as opening into the vestibule, but it is perhaps rather a weak spot, or *potential anus* (*p. a.*).

Within the body, close to the vestibule, is a single spherical *contractile vesicle* (*c. v.*). This may be seen slowly to enlarge until, when it has reached a certain size, it suddenly disappears. The diastole is slow, the systole rapid. When a little finely-divided carmine is mixed with the water in which a vorticella is under observation, the method of feeding may be watched. The fine particles are driven by the cilia into the vestibule, and accumulate at its blind end. After a while they suddenly burst through into the soft inner substance of the body, carrying with them an equal or even greater amount of water. Thus are formed *food-vacuoles*, which float freely for a time with a circulatory motion in the soft protoplasm of the body substance. Such digestible matter as the food may contain is extracted, and the indigestible refuse is passed out into the vestibule through the potential anus. It is probable that the contractile vesicle has the function of draining off the excess of water thus introduced within the body, in which case it must be regarded as excretory. At the same time it is not improbable that the water thus introduced within the body, and circulating through its substance, serves also to oxygenate the protoplasm, and thus ministers to a rudimentary respiratory function.

When a vorticella is disturbed, by a sudden shock given to the stage of the microscope for example, the long stem rapidly coils up like a cork-screw (Fig. 102, B.), and the animal is thus drawn back towards its point of attachment. At the same time the lips of the peristome close in and apply themselves to the disc, which is also retracted. The organism thus becomes pear-shaped. Soon, however, the stalk will uncoil and lengthen out again, the disc will be protruded, the peristome will open out, and the cilia will recommence their orderly vibrations. Nothing can be more beautiful than a colony of vorticellidæ alternately contracting and expanding, their stems rapidly coiling and uncoiling.

The cilia are developed on the ectosarc. Each is a delicate structureless process, which, when the vorticella is in the expanded condition, is in constant motion, alternately bending and straightening. Currents are produced in the water by their continuous motion, and these currents may be readily observed if a little powdered carmine be added to the water

beneath the cover glass. Perhaps the most striking and interesting point with regard to this ciliary action is the co-
ordination of the movements, and their apparently purposive

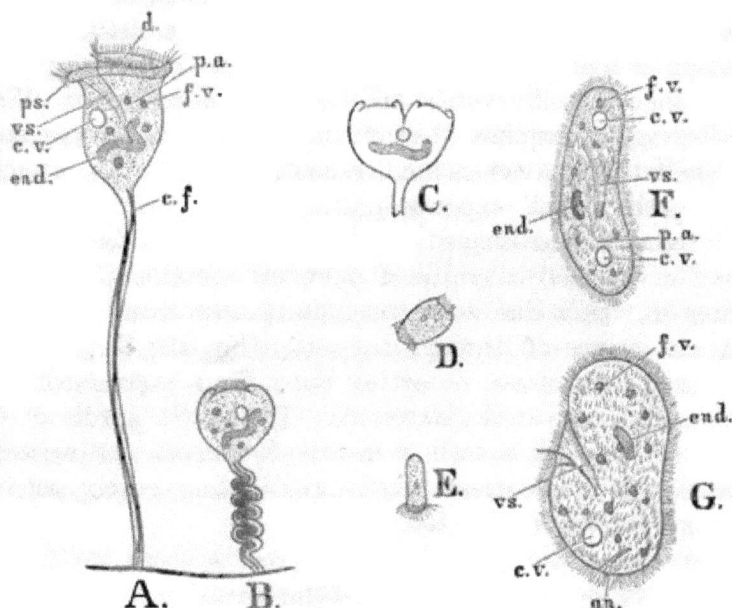

FIG. 102.—VORTICELLA, PARAMŒCIUM, NYCTOTHERUS.

A. Vorticella **extended.**   D. Free vorticella **after fission.**
B. Contracted.   E. Embryo vorticella.
C. Vorticella undergoing fission.   F. Paramœcium.
                          G. Nyctotherus.

an. Anal slit. *c. f.* Contractile fibre. *c. v.* Contractile vesicle. *d.* **Disc.**
end. Endoplast. *f. v.* Food-vacuole. *p. a.* **Potential anus.** *ps.* **Peristome.**
vs. Vestibule.

nature. This is accompanied in the metazoa by the development
of nerve-tissue; but nothing of the kind can be made out in
vorticella.

If a vorticella be stained with iodine or magenta, the **ectosarc**
is relatively uncoloured, but the cortical layer is stained. And
within **the cortical layer a** band-like structure, before hardly
distinguishable, **becomes** evident by its deeper stain. This is
the *endoplast* (*end.*). It is homologous with **the nucleus of a**
simple nucleated cell.

One mode of multiplication adopted by vorticella is that of *fission* (Fig. 102, C.). The animal assumes the pear-shaped condition of contraction, and becomes somewhat swollen. A notch then makes its appearance in such a way as to divide the vestibule into two halves. The notch becomes a cleft, and the cleavage is continued through the body in such a way as to divide the contractile vesicle and the band-like endoplast. When the cleavage is complete, there result two heads to the one stalk. But the fate of the two is not the same. One remains attached to the pedicle and expands into a complete vorticella. The other remains pear-shaped, and develops round the posterior region of the body a girdle of powerful vibratile cilia, by the lashing of which the zooid tears itself away from the parent stem and swims off through the water (Fig. 102, D.). After a short active existence it settles down in a convenient spot, adhering by its posterior extremity. The hinder girdle of cilia is lost or absorbed, a stalk is rapidly developed, the peristome expands, the adoral wreath of cilia exhibit their orderly activity, and the organism is a perfect vorticella.

In the case just described, the two zooids which result from fission are of equal size. But in some cases the fission is very unequal, so that a minute *microzooid* results. This swims off by means of its posterior girdle of cilia, but does not merely settle down and grow up into a vorticella. Its mission is different. It attaches itself to the base of the bell of another vorticella, and becomes completely absorbed into its larger mass. Such a process is called *conjugation*. Perhaps we may see in this process a foreshadowing of the union of spermatozoon and ovum in the metazoa. The result of the process would seem to be a rejuvenescence or access of vital power; for the normal mode of multiplication by binary division would seem to go on more rapidly after conjugation. The endoplasts of the two conjugating organisms seem to break up and disintegrate, a new endoplast being shortly afterwards reconstructed.

It is not improbable that, in some cases, conjugation is a preparatory process to *encystment*. But it would seem that encystment may take place without any such preparatory process.

The animalcule which is to be encysted separates from its stalk, contracts into a spheroidal body, around which a clear structureless envelope or *cyst* is formed. Within the cyst the endoplast and the contractile vesicle, permanently expanded, are visible. But ere long the contractile vesicle disappears, the band-like endoplast increases much in size, and becomes convoluted. It next becomes pinched in at intervals, so as to assume a beaded appearance. And at last each bead becomes separately pinched off from the rest. Thus the cyst appears full of bead-like spores. Then the cyst bursts, and these spores are discharged as minute oval or pear-shaped free-swimming animalcules, possessing an anterior crown of cilia, in the midst of which there is a potential mouth. These feed and multiply by fission, but ere long attach themselves by the anterior oral end (E.). The girdle of cilia is now lost; disc, peristome, and vestibule are developed, a gradually lengthening contractile pedicle is acquired, and the organism is once more a vorticella. This mode of reproduction (the occurrence of which is strongly vouched for) does not seem to be common among vorticellidæ. More frequently encystment is merely a temporary resting stage, during seasons of drought perhaps, or other uncongenial conditions, from which the vorticella emerges simply to resume its ordinary mode of life.

Vorticella may be found adherent to the roots of duckweed.

Paramœcium is found in pond-water or in vegetable infusions. From its peculiar form (Fig. 102, F.) it is often known as the slipper-animalcule. There is a thin superficial cuticle richly supplied with cilia. Since in paramœcium and its allies the cilia are more or less completely developed over the whole surface, they are known as the *Holotricha*, as opposed to the *Peritricha*, to which vorticella belongs, where there is a wreath of cilia, and other groups in which the cilia are differently arranged. Beneath the cuticle is a cortical layer, which is fibrillated in a direction perpendicular to the surface. If dilute acetic acid be added to the water in which paramœcium is under observation, the organism will become quiescent, and there will shoot out from the surface a number of stiff processes. These accessory structures are known as *trichocysts*. There is a bean-shaped endoplast (*end.*), associated with which, applied to one side, is a smaller endoplastule, homologous with the nucleolus. There are two contractile vesicles (*c. v.*), which, at the moment of **systole, are** seen under certain conditions to become starlike, as if there proceeded from them radiating canals, which thus for a moment became distended. The vestibule (*vs.*) is a ciliated funnel on the ventral surface. The process of feeding, giving rise to food-vacuoles (*f. v.*), is similar to that in vorticella. The indigestible residue is passed out by a potential anus (*p. a.*) or weak spot in the protoplasm of the ventral surface.

The movements of paramœcium through the water show a curious combination of backward or forward motion and rotation. Interesting observations have been made on the effect of a rise of temperature on these animalcules. It was found that the pulsation of the contractile vesicle grew more rapid until a maximum was reached of **30° C.** Beyond this point the pulsations grew less frequent, but the ciliary activity continued to increase up to 35° C., when the co-ordination of the movements of the cilia became imperfect, resulting in a peculiar aimless combination of direct and rotatory motions. At 40° C. the direct movements ceased, while rotation continued to 42°-45°, **when death** ensued.

Paramœcium multiplies by *fission*, which is transverse. *Conjugation* also takes place, but differs from that observed in vorticella by being transient **and not** permanent. Two paramœcia become locked together **by the** close application or union of **their** oral surfaces. In this condition they may remain for five **or six days.** During or subsequent to this process the endoplastule becomes fusiform and striated, and the endoplast turgid and enlarged. It would seem that the endoplast breaks up into a number of spheroidal bodies, which are ultimately cast out of the body, and not improbably constitute reproductive germs from which daughter paramœcia take their origin.

Many of the ciliated infusoria **are** parasites or messmates. In the rectum of the frog, for instance, examples of *Balantidium* and *Nyctotherus* (Fig. 102, G.) **are** almost sure to be found. The former **has** several contractile vesicles, the latter generally

but one (*c. v.*). The vestibule (*vs.*) in nyctotherus is very conspicuous, and on its lower edge there is a long stiff seta; the
anus also forms a distinct cleft (*an.*).

The fact to be especially noted in the forms which we have
studied in this Chapter is that they are unicellular organisms, or
protozoa, are never possessed of sexual organs, and are thus
very markedly distinguished from the multicellular organisms
or metazoa. Metazoa are typically produced by a segmentation
in the ovum, by which a number of cells result, which remain
so connected as to be constituents of a single individual.
Differentiation in the metazoa proceeds by the divergent development of different cells, a morphological differentiation
of structure accompanying a physiological differentiation of
function. In the higher forms the differentiated cells aggregate
into groups, and thus form tissues, which take their part
in the formation of more or less differentiated organs. In the
protozoa there is no segmentation of the ovum. The process of
fission does, indeed, in some respects resemble the first cleavage
of the ovum into two blastomeres, but it also serves to show the
fundamental distinction between the two groups of organisms.
For in the protozoa, after fission, each cell typically becomes a
distinct individual. There is thus no opportunity for that
physiological division of labour which is normal among the constituent cells of the metazoan.

Differentiation, therefore, where it goes on among the protozoa,
must take place within the single cell of which each individual
is framed. The forms we have studied show the nature of this
differentiation. At the best they remain lowly forms of life.
They have not learnt the secret of merging their individuality
for the common good, and being content to be the constituent
units, each with its own narrow field of work, of a nobler
organism.

23

CHAPTER XX.

THE AMŒBA.

The Amœba, or Proteus animalcule, consists of a minute irregular particle of jelly-like protoplasm (Fig. 103, A.). Individuals may generally be found in animal and vegetable infusions or in stagnant pond water. Each looks like a little patch of somewhat cloudy white of egg ; but, if it be carefully watched, it will be found to be constantly changing its form, and moving slowly across the stage of the microscope by pushing out blunt processes or *pseudopodia* (*ps.*). Careful observation discloses an external cortical layer, the *ectosarc* or *ectoplasm*, enclosing the inner substance, *endosarc* or *endoplasm*. The ectoplasm is clear, the endoplasm granular. When a pseudopodium is pushed out the ectoplasm precedes the endoplasm, which follows with a kind of rush. Within the granular endoplasm there is at least one, sometimes more than one, rounded endoplast (*n.*), with a smaller endoplastule within it. A rhythmically pulsating *contractile vesicle* (*c. v.*) exhibits the usual slow diastole and sudden systole. Oily granules, crystals, and food-vacuoles may also be seen to flow hither and thither within the endoplasm in vague circulation.

Food is taken in at any point of the body, and the indigestible residue is also extruded at any point. There is not even a potential mouth or anus. Sometimes a pseudopodial process seems to be wrapped over the morsel of food, in which case some of the ectoplasm becomes internal. It must be remembered, however, that the differentiation into these two layers is of the slightest.

The amœba multiplies by fission. After attaining a certain

354

size the endoplast divides into two, or sometimes the endoplastule is the first to divide, and then the endoplast divides also.
In Fig. 103, B., an amœba is figured, in which the division
of the endoplast has taken place, and the two endoplasts
(*n. n.*) are widely separated. Shortly afterwards the contractile

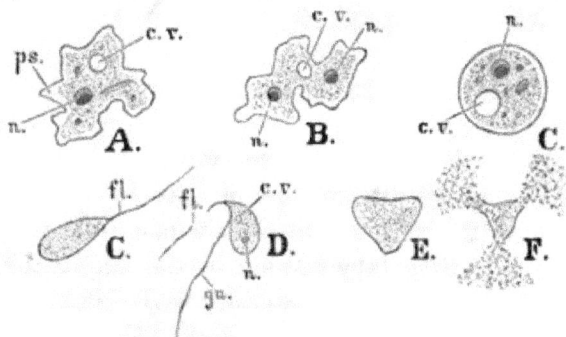

FIG. 103.—AMŒBA AND MONADS.

A. Amœba. B. Amœba undergoing fission. C. Amœba encysted. C.
Monad. D. Heteromita. E. Cyst resulting from fusion of two Heteromita.
F. Cyst bursting and giving rise to fluid full of germs.

c. v. Contractile vesicle. *fl.* Flagellum. *gu.* Gubernaculum. *n.* Endoplast.
p. s. Pseudopodium.

vesicle (*c. v.*) also divided, and a delicate line of division
seemed suddenly to make its appearance. Forty minutes
afterwards the two amœba were quite separate. In Fig. 103,
C., an amœba is shown in the encysted condition, the outer
layer being clear and tolerably distinct. In this case the
amœba subsequently resumed its usual irregular form; but
it is said that in some cases there emerge from the cyst crowds
of minute amœbæ. Conjugation is said sometimes to occur.

Let us note that we have in the amœba a simple cell that
has undergone very little differentiation. That cell, however,
performs several functions which characterise it as a living
organism. It responds to external stimuli, and is therefore sensitive in the sense in which we have before (p. xv.) used that
term. The responses take the form of changes of shape in the
cell. It receives and assimilates food; so that there must be a

rudimentary form of digestion. This process, and the energy involved in the constant change of form, are the indications of hidden metabolic processes which go on within the cell; and there can be no doubt that these metabolic changes involve processes of respiration and the excretion of waste products. Whether there is any part of the cell which merely forms a framework for the support of the rest it is difficult to say. Finally the simple cell is reproductive.

We have already seen that the metazoan differs from the protozoan in being a cell-aggregate. We may now note that as we advance towards the higher metazoa we find a more and more perfect division of labour among the cells—special cells taking upon themselves the various functions which, in the amœba, are performed in a rudimentary fashion by the single cell. Or we may say—to put the matter in another way—that in the higher metazoa there is such differentiation among the cells that special cells or groups of cells are enabled to cultivate, almost exclusively, some one or other of the protoplasmic faculties. Some (nerve) become eminently sensitive; others (muscle) eminently contractile; yet others (liver) eminently metabolic, secretory (salivary), or excretory. All or most are capable of reproducing by fission other cells like themselves; but to certain cells are intrusted the all-important function of continuing the race by reproducing complex individuals inheriting the peculiarities of the parent organism. So important are these cells that it has been suggested that all other tissues and organs exist that the ovary may produce an ovum which shall be duly fertilised.

Be this as it may, we see that the higher metazoan individual is a cell-aggregate in which there has gone on a large amount of differentiation and division of labour among the constituent cells. But here we must further note that accompanying the differentiation there is also integration. It would be of no use for the cells to become divergent in structure and function if, accompanying this essential process, there were not a process, equally essential, by which the cells are knit together into the tissues and organs of one individual body. For efficient advance in the life-scale integration and differentiation must go hand in hand.

And not only (1) do the parts of an organism get more different and yet at the same time more intimately connected, but the organism (2) tends to become more different from the environment, and yet, at the same time, requires to be, so to speak, more closely in touch with the conditions of the environment ; and yet further, (3) the organism not only becomes more different from other organisms, but at the same time becomes more dependent upon other organisms, so that the whole organic world becomes more and more a vast system of interdependent units.

These considerations help us to see what is the meaning of the terms higher and lower as applied to animals. The higher animals are those in which this joint process of differentiation and integration has been carried furthest. But it is well to avoid instituting comparisons of this nature between animals of widely different groups. Such a question as, Which is the higher animal—a bee or a bear ? is not very profitable. It is even well to limit comparison, so far as possible, to analogous cases, and to speak of higher and lower in respect to the performance of special functions. At the same time we may perhaps say generally that the more complex the conditions of life to which an organism is suited, and the more perfectly that organism is suited to these conditions of its life, other things equal, the more advanced it is.

A final question arises out of the fact that if we prepare an animal or vegetable infusion (e.g. hay or cod's head), and leave it to stand for a short time, we shall find it swarming with organisms—paramœcia, amœbæ, monads (Fig. 103, C. D.) and countless bacteria. Whence comes all this teeming life ? Two answers are or have been given to this question. The first (abiogenesis) is that the living forms take their origin by a regrouping of the molecules of the organic but lifeless matter in the solution, without the aid of any parental organisms. The second (biogenesis) is that the living forms are due to the presence in the solution of organic living germs, the products of parental organisms. Into the arguments brought forward by the advocates of these two answers we cannot enter here. The balance of scientific evidence would seem to be very largely in favour of biogenesis. For if a solution be sterilised by boiling, sufficiently prolonged to kill all germs, and then be preserved from contact with the dust- and germ-laden air by a plug of cotton wool, the molecules of the organic but lifeless matter in solution are utterly incapable of regrouping themselves into living forms, and the infusion remains barren of life. This is the negative evidence in favour of biogenesis.

Day by day positive evidence in favour of the same view increases in amount as the life-histories of these lowly forms of life are patiently worked out, and it is found that they " breed true." In Fig. 103, D., is figured a form of monad (*Heteromita*) found in an infusion of cod's head. It has two whip-like processes, a flagellum (*fl.*) and a gubernaculum (*gu.*), by which it can anchor itself. There is an endoplast (*n.*), and a contractile vesicle (*c. v.*). Such monads multiply by fission, but also by another process. Two individuals conjugate and fuse. A triangular cyst (E.) results. After a while the cyst bursts, and a homogeneous fluid is emitted (F.). The highest powers of the microscope fail to disclose in this emitted matter any germ of life, and there, at first sight, would seem to be an end of the matter. But wait and watch, and there appear in the field of the microscope, suddenly and as if by magic, countless minute points which prolonged watching shows to be growing. And when they have fully grown every distinct point is seen to be a monad. Thus we see that a solution in which the highest powers of the microscope disclose no trace of living matter, may nevertheless be swarming with germs. Let a drop of such solution evaporate into the air; there is no reason to suppose that the germs perish. On the contrary, there is much to lead us to believe that the germs have far greater powers of resisting high temperature, desiccation, and other adverse conditions, than the fully developed organism. We may thus see how the air comes to be laden with germs which, should they fall into an appropriate infusion, may give rise to the teeming life which we know to be so soon developed in it.

Biogenesis would seem, then, to be a law of life under the

conditions attainable in our biological laboratories. Let us be
careful, however, not to dogmatise on this subject. Whether
living forms ever spring from not-living matter, in ponds or in
the great ocean, we do not know. Nor do we know whether
abiogenesis has or has not taken place in times geologically
remote.

APPENDIX

CLASSIFICATION OF TYPES.

THE broadest division of the typical organisms considered in this volume is into *Protozoa*, or organisms which are unicellular, and *Metazoa*, or organisms which are multicellular, and in which there is physiological division of labour and morphological differentiation of structure among the cells. To the protozoa belong the amœba, destitute of a mouth, and possessed of the power of emitting pseudopodia (*Rhizopoda*), the vorticella and paramœcium (*Infusoria ciliata*), with ciliated bodies, and provided with a vestibule or oral fossa, and the monads (*Infusoria flagellata*), with one or more whip-like processes or flagella. The other organisms here considered belong to the metazoa.

The metazoa again are divided into *Cœlenterata*, to which the hydra belongs, in which there is no true body cavity and in which the symmetry is radial, and *Cœlomata*, in which a body cavity or cœlom is present.

The cœlomatous forms are roughly divided into the *Vertebrata*, in which an endoskeleton is developed, the axial portion of which consists of the skull and vertebral column, and the *Invertebrata*, in which an endoskeleton so constituted is absent. The Invertebrata form a very heterogeneous group, distinguished merely by negative characters. Those which we have considered fall into three divisions.

(1) The *Arthropods*, with bilateral symmetry, segmented bodies, jointed lateral appendages, and a ventral nerve-chain. To this group belong the crayfish, a *decapod crustacean*, breathing by means of branchiæ, in which the body is divided into cephalothorax and abdomen ; and the cockroach, an *orthopterous insect*, in which respiration is by tracheæ, and the body is divided into head, thorax, and abdomen.

(2) The *Vermes*, with bilateral symmetry, unsegmented (liverfluke) or uniformly segmented (earthworm) body, with lateral appendages which (when present) are not jointed, and with

paired excretory organs. **The** group is a large and diverse one, including the *Platyhelminthes* (liver-fluke and tapeworm), in which the body **is** flattened, and is either unsegmented (liver-fluke) **or** divided into a number of successive divisions, which **have** a tendency **to** separate individualisation (tapeworm), **without a** ventral nerve-chain, generally hermaphrodite, without **lateral** appendages, but provided with suckers (liver-fluke) or **suckers** and hooks (tape-worm); and the *Chœtopoda*, ringed **worms** (*Annelids*), possessed of a ventral nerve-chain and setose locomotive appendages, which may be numerous in each segment (*Polychœta*), or, as in the case of the earth-worm, few (*Oligochœta*).

(3) The *Mollusca*, bilaterally symmetrical (mussel) or somewhat asymmetrical (snail) animals, without lateral appendages, but possessed of a muscular locomotive 'foot,' unsegmented in the adult form, and with paired ganglia united by commissures which form an œsophageal nerve-ring, generally possessing a bivalve (mussel) or univalve (snail) shell, and a bilobed (mussel) or unilobed (snail) mantle. To this group belong the mussel, which has plate-like gills (*Lamellibranchiata*), no separate head, and in which the sexes are distinct; and the snail (*Gastropoda*), in which the head is distinct, and the mouth is provided with an odontophore (*Odontophora*), having a twisted univalve shell, and pulmonary respiration (*Pulmonata*).

The *Vertebrata* are bilaterally symmetrical animals, with a notochord, and in nearly all cases an internal skeleton, the vertebral column, and a dorsal nerve system generally protected by the arches of the vertebræ. There are never more than two pairs of limbs, and the jaws, which are never (as in the arthropods) modified appendages, move up and down. They are always possessed, at some period of life, of visceral arches and clefts. They are divided into three great groups.

1. The *Ichthyopsida*, including the *Pisces* (cod) and the *Amphibia* (frog). Respiration by gills always takes place during a part (frog), and sometimes the whole of life (cod). The exoskeleton is either absent or very slightly represented. The basioccipital region of the skull is generally unossified. The temperature of the blood varies with, and is not much above, that of the surrounding medium. The red blood corpuscles are **nucleated.** There are two systemic aortic arches. The urinary **organs are** persistent Wolffian bodies. There is no thoracic **diaphragm.** There is no true corpus callosum in the brain. The **hypoglossal** nerve does not perforate the brain-case. The amnion

is absent, and the allantois rudimentary. There are no mammary glands.

2. The *Sauropsida*, including the *Reptilia* and *Aves* (fowl and pigeon). There is no gill-respiration, but gill-arches are present in the embryo. There is almost always an epidermic exoskeleton in the form of scales or feathers. The occipital region of the skull is completely ossified, and there is one occipital condyle. The mandible consists of several bones, and articulates with the skull through the intervention of a quadrate bone. The ankle-joint is between the proximal and distal divisions of the tarsus. The temperature of the blood is near (*reptilia*) or considerably above (*aves*) the normal temperature of the surrounding medium. The red blood corpuscles are oval and nucleated. The left systemic aortic arch is often (*e.g.* in all birds) suppressed. The thoracic diaphragm is never complete or functional. The Wolffian bodies are replaced by true kidneys. There is no corpus callosum. The right oviduct and ovary may be (birds) rudimentary. There is an amnion and a well-developed allantois, the function of which is chiefly respiratory. There are no mammary glands.

3. The *Mammalia*, to which the rabbit belongs. There is no gill-respiration, but gill-arches are present in the foetus. The epidermic exoskeleton consists of hairs. The occipital region of the skull is completely ossified; and there are two occipital condyles. The mandible consists of a single bone on each side, and articulates directly with the squamosal of the skull. The quadrate becomes an auditory ossicle. The ankle-joint is between the tibia and the astragalus. The temperature of the blood is more or less constant, and considerably above the normal temperature of the surrounding medium. The red blood corpuscles are generally round and not nucleated. The right systemic aortic arch is suppressed. There is a complete and functional diaphragm. The Wolffian bodies are replaced by permanent kidneys. There is a corpus callosum in the brain. The foetus has an amnion and allantois; and in the higher forms a true placenta is formed. Mammary glands supply the young with nourishment.

GLOSSARY

Abiogenesis.—Also called spontaneous generation ; the doctrine that living beings can arise *de novo* in the midst of not-living matter. Used in contradistinction to *biogenesis* : the doctrine that every living cell, or group of cells, is a link in the chain of life.

Acinus.—The blind grape-shaped termination of a gland or hollow organ. The word *alveolus* is used with like signification.

Ætiology.—The doctrine of causes. Evolution and Natural Selection fall under this head.

Afferent Nerves.—Those which convey waves of metabolic change inwards to the central nerve-system ; in distinction from efferent nerves, which convey these waves outwards to an organ or gland.

Alecithal.—A term applied to the ovum, such as the rabbit's, in which there is little or no food-yolk, or in which the little there is, is distributed uniformly through the ovum. Used in contradistinction to *telolecithal* (fowl's ovum), in which the food-yolk is aggregated at one pole, and *centrolecithal* (crayfish ovum), in which it is aggregated at the centre.

Alternation of Generations.—The alternation of two forms of the same species. Under this general head come (1) *metagenesis*, when the one form is sexual and the other asexual (medusa) ; (2) *heterogamy*, when the second form (or forms) are perfectly or imperfectly sexual. The liver-fluke falls under this head if we regard the masses of cells from which the cercariæ arise as rudimentary ovaries. When the ova develop without the influence of spermatozoa, the term *parthenogenesis* (maiden-birth) is used.

Altrices.—Birds whose young are hatched in a callow, immature condition, as in the pigeon, as opposed to *præcoces*, in which they are active at hatching.

Alveolus.—*See* Acinus.

Ametabola.—Insects which undergo no metamorphosis, or in which the metamorphosis is incomplete (sometimes called *hemimetabola*), as opposed to *holometabola*, when the metamorphosis is complete.

Amniota.—Vertebrates in which the amnion, a fœtal membrane, is developed (reptiles, birds, and mammals), in distinction from *anamniota* (frog, cod), in which there is no amnion.

Amœboid.—A term applied to cells which change their form like the amœba.

Anabolism.—*See* Metabolism.

Analogous Organs.—Organs which perform the same function, but are not necessarily constructed on the same plan; used in contradistinction to homologous organs, when they are constructed on the same plan, but have not necessarily the same function. The wing of the insect and that of the bird are analogous, but not homologous; the wing of the bat and the paddle of the whale are homologous, but not analogous; the hind-limbs of the frog and the rabbit are both analogous and homologous.

Anastates.—*See* Metabolism.

Aponeurosis.—The strong sheath of connective tissue which invests a muscle. Those covering the muscles have been termed *aponeuroses of investment*, in contradistinction to the tendinous expansions or *aponeuroses of insertion.*

Appendicular.—Applied to that part of the vertebrate skeleton which is related to the appendages, as opposed to the *axial* skeleton.

Archenteron.—The digestive cavity of the embryo at the gastrula stage. After the formation of the mesoblast the term *mesenteron* is employed.

Arterial.—A term applied to blood which has been oxygenated in the lungs or gills, opposed to *venous* when the oxygen has been used up by the tissues. The pulmonary artery contains venous blood, the pulmonary vein arterial blood.

Articulation.—Union at a joint. The lower jaw is *articulated* to the skull.

Atrophy.—The wasting away of a part or organ.

Axial.—*See* Appendicular.

Biogenesis.—*See* Abiogenesis.

Blastocœle.—Also called *segmentation cavity;* the cavity contained within the *blastosphere* or hollow mass of cells (*blastomeres*) resulting from the segmentation of the ovum. Where segmentation gives rise, as in the fowl, to a cellular patch, it is termed the *blastoderm.* The hollow mass of cells produced by segmentation in the rabbit is termed the *blastodermic* **vesicle.** The *blastopore* **is** the orifice of the archenteron of the gastrula, produced by the invagination of the blastosphere.

Budding.—The asexual formation of new individuals, in hydra, by the formation of a pear-shaped outgrowth which develops into a new **hydra.** The process is also called *gemmation.*

Cæcum.—A blind pouch.

Canaliculi.—Minute tubes arising from the lacunæ of bone, or from the repeated branching of the hepatic ducts.

Capillaries.—The finest and most delicate blood-vessels, formed by the ultimate subdivision of the arteries, and giving rise to the most minute **factors** of the veins.

Capsule.—A little case in which some organ or part of an organ is enclosed.

Cells.—The nucleated elements of which the protozoa consist, and which in metazoa are built up into *tissues*. From the *undifferentiated cells* of the embryo arise such forms as the *amœboid* (white blood-corpuscles), the *columnar* (epithelium), the *flattened* or *squamous* (epithelium, endothelium), the *fibrous* (muscle, nerve), the *branched* (connective tissue), and the *ciliated* (tracheal epithelium).

Centrolecithal.—*See* Alecithal.

Ciliated.—A term applied to those cells or organisms which have delicate, motile, hair-like processes or *cilia*.

Cleavage.—A term applied to the splitting, division, or segmentation of the ovum into a number of cells or blastomeres; also to the splitting of the mesoblast into two layers.

Cœlomata.—Those organisms which have a separate body cavity, as opposed to the *cœlenterata*.

Columella.—(1) A bone connected with the organ of hearing in the frog and fowl; (2) the axis of the shell in the snail to which the *columellar muscles* are attached.

Commissure.—A nerve band joining two ganglia or two parts of the brain. The term is sometimes reserved for transverse bands, longitudinal connections being termed *connectives*.

Conjugation.—The union or fusion, temporary or permanent, of two protozoa. The phenomenon occurs in connection with the reproductive process.

Development.—Differential growth, involving a change in the organism as it passes from the ovum to the mature state.

Diœcious.—Having the sexes distinct, the individuals being male and female; in contradistinction to *monœcious* or *hermaphrodite*, when both sexes are combined in the same individual.

Differentiation.—The process by which cells or parts originally similar become different in structure. It is accompanied by a specialisation of function among the cells, sometimes spoken of as the physiological division of labour.

Distal.—Applied to the end of an appendage, bone, or limb which is furthest removed from the trunk: the other end being called *proximal*.

Distribution.—(1) The way in which organisms are distributed over the earth in space and in time; (2) the way in which nerves and blood-vessels branch out to supply organs. The tenth nerve (*e.g.*) is distributed to the larynx, heart, lungs, stomach, etc.

Diverticulum.—A blind tube opening out of another tube.

Ecdysis.—The throwing off of the outer skin or exoskeleton.

Embryo.—The organism in the stages which intervene between the segmentation of the ovum and birth or hatching. The term *fœtus* is used of mammals. **Embryonic** or fœtal appendages are provisional structures (such as the allantois and amnion) in connection with embryonic or fœtal life.

Encystment.—A phenomenon occurring among lowly organisms when they become motionless and are surrounded by a coat, case, or *cyst.*

Endoderm.—The inner layer of the hydra as opposed to ectoderm. *Endoplasm* or *endosarc*, the inner substance of a protozoon, as opposed to *ectoplasm* or *ectosarc*, its outer substance. *Endoplast*, a term applied to the nucleus of the protozoa.

Epibole.—The overgrowth of epiblast cells over the other cells of the segmented ovum, as in the case of the frog.

Evagination.—*See* Invagination.

Eversion.—A term applied to the opening out and expansion of the peristome of the vorticella.

Excretion.—The process by which waste products are got rid of or expelled from the body.

Facet.—(1) A smooth surface of bone on which another bone works, *e.g.* tubercular facet of a vertebra ; (2) the polygonal areas of the insect's or crustacean's eye.

Femur.—(1) The thigh of a vertebrate ; (2) the thigh-bone ; (3) the first joint in the leg of a cockroach.

Fertilisation.—The union of a spermatozoon with the ovum. The process is often called *impregnation.*

Fenestra ovalis and Rotunda.—Membranous "windows" in the periotic bone in connection with the organ of hearing. In the fowl the two often open externally into a *fenestral recess.*

Fission.—The splitting of a cell or a lowly organism into two more or less similar parts. The cleavage of the ovum is a special form of fission.

Fœtus. Fœtal membranes.—*See* Embryo.

Folding off of the embryo.—The process by which the embryo (*e.g.* fowl) is constricted from the yolk-sac.

Fontanelle.—A space occupied by membrane.

Foramen.—A hole in the skull or in a bone through which nerves or blood-vessels pass.

Fronto-parietal.—A bone of the frog's skull, also called *parieto-frontal.*

Ganglion.—A group of nerve-cells, *e.g.* Gasserian ganglion, sympathetic ganglia.

Gastrula.—The two-layered condition of the embryo, resulting from a pushing in or invagination of one portion of the blastosphere.

Germinal layers.—The cell-layers of the embryo, *hypoblast, epiblast,* and *mesoblast. Germinal vesicle,* the nuclear portion of the ovum containing nucleolar *germinal spots.* The *germinal disc* is a thickened part of the blastoderm in which the embryo is developed. The *germinal epithelium* is that part of the peritoneal lining from which the generative organs take their origin.

Gland.—An organ composed of one or more epithelial cells, the physiological function of which is secretion or excretion. The term is also applied to other organs, such as the lymphatic glands, thymus, pineal gland, hermaphrodite gland, etc.

Gonad.—A term applied to the reproductive organ, **male (***testis* or *spermary*)**, female (***ovary*) or **hermaphrodite (***ovotestis* **of snail).**

Hepatic.—Of or belonging to the liver. *Hepato-pancreas*, a term applied to the invertebrate digestive gland, as in the crayfish.

Hermaphrodite.—Combining male and female in the same organism (earthworm) or organ (ovotestis and hermaphrodite duct of snail).

Histology.—The microscopic study of tissues. *Histolysis* is the breaking up of a tissue during development. *Histogenesis* is the embryological development and origin of tissues. The term *Histozoa* has been employed as equivalent to *metazoa*.

Holoblastic.—A term applied to segmentation, where the **cleavage affects** the whole ovum ; opposed to *meroblastic*, **where the cleavage is only** partial and restricted.

Holometabola.—*See* Ametabola.

Homologous organs.—*See* **Analogous organs.**

Impregnation.—The fertilisation **of an** ovum by the entrance of a spermatozoon.

Ingestion.—The taking in of food particles. It may be *inter-cellular*, between **the** cells, or *intra-cellular* into the cells.

Intus-susception.—**The** incorporation of new material with the old during life and growth ; **as** opposed to accretion, the addition of layer upon layer.

Invagination.—The pushing **in** of a part of the surface to form a pouch ; the gastrula (*e.g.*) is produced by the invagination of the blastosphere to form a two-layered cup. *Evagination* is the converse process, where a pouch is pushed outwards, or where an invaginated pouch is turned inside out.

Irritability.—The property of **responding to a stimulus. Specialised in** nerve and muscle.

Karyokinesis.—A complex process undergone by the nucleus of many cells during division.

Katabolism and Katastate. *See* Metabolism.

Lacunar.—Having lacunæ or spaces.

Lamella.—A leaf-like plate, *e.g.* of bone. *Lamellar* structure is where there are a number of leaves.

24

Lamina.—A thin film or plate. *Lamina perpendicularis*, part of the ethmoid bone ; *Lamina spiralis* in the organ of hearing ; *Lamina terminalis*, the front wall of the brain. *Laminæ dorsales*, the ridges which bound the neural or medullary groove.

Larva.—The freed embryo of certain organisms, such as insects.

Ligament.—(1) A fibrous band of connective tissue by which the bones are connected at the joints or articular surfaces ; (2) a fibrous elastic band where the shell of the mussel is not calcified.

Liver.—A large secreting gland of the vertebrate. The term liver is often applied to the digestive gland of the invertebrate. Since this has often pancreatic properties, it is sometimes called the *hepato-pancreas*.

Lobe.—A part of an organ separately folded off. There are lobes of the brain, liver, lungs, and many other organs ; a *lobule* is a little lobe.

Lumen.—The central cavity of a fine tube, duct, or vessel.

Manubrium.—(1) A process of the sternum. (2) A process of the hammer bone or malleus of the ear. (3) The clapper of the bell in a medusa.

Maxilla.—(1) A jaw-bone in the vertebrate skull. (2) An appendage in the crayfish and the cockroach.

Medulla.—(1) The pith-like substance in the axis of a feather. (2) *Medulla oblongata*, the hinder part of the brain. (3) *Medullary folds, groove,* and *tube*, the first indications of the central nervous system, now often termed the neural folds, tube, etc. (4) The internal striated part of the kidney, as opposed to the outer dotted cortical part. (5) *Medullary cords*, strands of lymphatic tissue in a lymphatic gland.

Membrane.—A thin sheet of tissue, forming an investment or a distinct layer.

Meroblastic. *See* Holoblastic.

Mesenteron. *See* Archenteron.

Mesentery.—A fold of peritoneal membrane by which the alimentary canal is suspended in the body-cavity or cœlom.

Mesostates.—*See* Metabolism.

Metabolism.—A term applied to the chemical changes associated with living cells. Metabolism may result in the formation of more complex compounds (*anabolism*), or of less complex compounds (*katabolism*). Anabolism is associated with the storage of energy in the tissues ; katabolism with the activities of the organism. Intermediate compounds are called *mesostates ;* as terms of a katabolic process, *katastates ;* of an anabolic process, *anastates.*

Metameres.—Serially homologous divisions of an organism.

Metamorphosis.—The series of changes taking place during the free life of the organism by which it passes from the embryonic to the adult state. When the changes take place before birth or hatching the term *transformation* is used.

Metazoa.—Organisms composed of many cells, in which a physiological division of labour has taken place : **used** in contradistinction to *protozoa* where there is generally but **one** cell, or at most an aggregate of similar cell units. Sometimes called *histozoa*.

Morphology.—That branch of the science which deals with the form and structure of organs or organisms : whereas *physiology* deals with their functions ; and *histology* with their minute structure. Morphological structure is ascertained by *anatomical* **investigation**.

Morula.—A mulberry mass of cells **not enclosing a cavity, and resulting** from segmentation.

Neural—(1) Neural folds, groove, and tube (often called the medullary folds, tube, etc.), the first indications of the cerebro-spinal nervous system ; (2) Neural arch and spine, processes of the vertebræ.

Notochord.—A rod of cells derived from the hypoblast, and underlying the neural tube. It forms the primitive axial support of the body. Sometimes called the *chorda dorsalis*.

Nucleus.—The specially differentiated 'kernel' of a cell in which all vital changes seem to be initiated. Within the nucleus there may be one or more *nucleoli*.

Oral.—The oral end of an organism is that part which bears the mouth. The opposite end is called *aboral*.

Origin.—(1) Ætiological questions are spoken of as questions of origin (*e.g.* origin of species, origin of tissues). (2) The relatively fixed point to which a muscle is attached is termed its origin, in contradistinction to the relatively moveable part where it has its *insertion*.

Otoliths.—Concretions or foreign particles in an auditory sac.

Ovo-testis.—Also called *hermaphrodite gland*. An organ or gonad which produces **both ova** and spermatozoa.

Ovum.—The **cell produced by the** female, from which, generally after fertilisation by a spermatozoon, **a new** organism arises. Such reproduction is called *sexual:* where **the** organism is reproduced without ova (*e.g.* by budding or fission), the reproduction is called *asexual.* When the ovum develops without fertilisation the term *parthenogenesis* is used.

Palp.—A process supposed to be an organ of touch generally **in relation to** the mouth (*e.g.* labial palps of mussel, mandibular palp of **crayfish**).

Pancreas.—A vertebrate digestive gland. The invertebrate **digestive gland** (*e.g.* of crayfish) is sometimes called the *hepato-pancreas*.

Papilla.—A small projection (*e.g.* of the tongue).

Parietal.—(1) A bone of **the skull** which gives its name to one of the segments of **the brain-case**. (2) Of or belonging to the parietes or walls, *e.g.* parietal (or somatic) layer of the peritoneum, in contradistinction to the *visceral* (or splanchnic) layer.   *Parieto-frontal*, also called *fronto-parietal*, a bone in the skull of the frog.

Parthenogenesis.—*See* Ovum.

Periostracum.—Also called, as in the crayfish, *epiostracum*, the outer layer of the shell in the mussel.

Physiology.—*See* Morphology.

Placenta.—A vascular mammalian structure bringing the embryo into relation with the uterus of the mother. Villi on the allantoic chorion fit into depressions or crypts of the uterine wall, the conjoint structure forming the placenta. If the fœtal structures separate at birth from the maternal mucous membrane, the placenta is *non-deciduate ;* if the maternal mucous membrane is torn away with the fœtal structures and delivered as the *after-birth*, it is *deciduate*. According to the arrangement of the villi the placenta is discoidal (rabbit), dome-shaped, zonary, or diffuse.

Pleura.—(1) The membrane that covers the mammalian lungs.   (2) The lateral edges of the exoskeleton of the crayfish.

Portal.—A portal vein is one which instead of carrying the blood directly to the heart carries it to some organ or gland within which it breaks up into a capillary plexus. All vertebrates have a *hepatic-portal* system, with a capillary plexus in the liver. The frog has also a *renal portal* system, with a capillary plexus in the kidney.

Process.—A projecting portion proceeding from some cell, organ, or part of the body; *e.g.* transverse process of vertebra, pseudopodial processes of amœba.

Procœlous.—Of vertebræ, the centra of which are concave in front ; *amphi-cœlous* are concave at both ends ; *opisthocœlous* concave posteriorly.

Proctodæum.—That part of the alimentary canal which arises in development as a posterior epiblastic invagination, giving rise in the crayfish, *e.g.* to the *hind-gut*.

Præcoces.—*See* Altrices.

Protovertebræ.—A term sometimes used for *mesoblastic somites*.

Protozoa.—*See* Metazoa.   Also called *cytozoa*.

Protoplasm.—The elementary organic basis of living matter, sometimes termed *sarcode*.

Proximal.—*See* Distal.

Racemose.—A term applied to a gland in which the lobules contain a number of acini opening in clusters into the extremities of a branched duct.

Ramus.—A branch or division, (1) of a nerve, the spinal nerves, *e.g.* have dorsal and ventral rami ; (2) of the lower jaw, each lateral mandible being termed a ramus.

Rejuvenescence.—A term applied to the added energy of a protozoon after conjugation.   Also termed *adjuvenescence*.

Reproduction.—*See* Ovum.

Sac.—A hollow cavity, *e.g.* air sac, dart sac, scrotal sac.

Segmentation.—(1) The process by which an ovum is cleaved into a number of cells or blastomeres. (2) The division of the body into a number of divisions or segments. *Segmental organs,* a term applied to the excretory organs or *nephridia* of the earthworm. *Segmental tube,* the primitive excretory duct of the vertebrate.

Sella turcica.—A depression in the **basisphenoid bone of the skull, some-what resembling, in the human subject, a Turkish saddle.** In the **rabbit there is in this region a hole through the bone, the** *canalis craniopharyngeus.*

Septum.—A division or partition, *e.g. Septum nasi,* between the nasal chambers; *septum auricularum* between the auricles; *septum lucidum* between the hemispheres of the brain.

Serous membrane.—A membrane that secretes a thin **serous** fluid (*e.g.* peritoneum), in distinction from a *mucous membrane,* **which secretes a** thicker mucous fluid (*e.g.* that lining the nasal chamber).

Skeleton.—The supporting framework of the body, **consisting of** (*a*) connective tissue, (*b*) cartilage or gristle, (*c*) bone. In addition to this *endoskeleton* (or **instead** of it in invertebrates) there may be an *exoskeleton.*

Spermary.—A term used for the male gonad or testis.

Spinal cord.—The central nervous system **of the** vertebrate posterior to the brain. The *spinal nerves* are the nerves given off therefrom. The *spinal column* is the vertebral column which forms the backbone, and the arches of which protect the spinal cord. This is sometimes **called the** *spine.* The *neural spine* **is a** median dorsal process **of a** vertebra.

Splanchnic.—Equivalent to visceral. The terms splanchnic and somatic are the Greek-derived words **answering to** the Latin derivatives visceral and parietal.

Spleen.—A red organ attached to the mesentery. *Splenial,* a bone in the lower jaw.

Squame.—A plate-like portion **of the** antenna **of the** crayfish, also called the *scaphocerite. Squamosal,* a bone of the vertebrate skull.

Sternum.—(1) The vertebrate breast-bone. (2) **The ventral part of the** crayfish's, or insect's exoskeleton.

Stomodæum.—That part of the alimentary canal which arises in development as an anterior epiblastic invagination, giving rise in the crayfish, *e.g.* to the *fore-gut;* the *mid-gut* being derived from the mesenteron, and the *hind-gut* from the proctodæum.

Stroma.—The connective tissue framework of an organ such as the ovary.

Sympathetic nerve system.—A part of the general nerve system of the vertebrate which differentiates from the rest, and forms a chain of ganglia, connected with the cerebro-spinal nerves.

Systole.—The contraction of the ventricle or other part of the heart; the relaxation is called *diastole*.

Telolecithal. *See* Alecithal.

Tendon.—A fibrous band of connective tissue by which a muscle is attached to the endoskeleton.

Tergum.--The dorsal part of the exoskeleton of the crayfish and cockroach. Hence this dorsal region is often termed the tergal region, as opposed to the ventral or sternal region.

Testes.—(1) The male generative organs. (2) The posterior divisions of the corpora quadrigemina of the mammalian brain.

Thorax.—(1) The part of the cœlom in front of the diaphragm in the vertebrate. (2) The mid-region of the body in the crayfish and cockroach.

Trabecula.—(1) A band of tissue. (2) Special bands of tissue in the developing skull.

Trochanter.—(1) A process of the upper part of the femur. There are three such processes in the rabbit. (2) The second joint in the leg of the cockroach.

Typhlosole.—A longitudinal fold projecting inward into the intestine.

Umbilicus.—(1) The *inferior* and *superior umbilicus* are small holes in the bird's feathers ; (2) The *umbilical vesicle* and *cord* are fœtal structures in the rabbit.

Urinogenital System.—Sometimes called *urogenital ;* the reproductive and excretory organs, the ducts of which are closely related in the vertebrate.

Utriculus.—(1) A division of the vertebrate organ of hearing ; (2) A finger-like process of the mushroom gland of the cockroach.

Vascular.—Connected with the blood. The vascular system is the circulatory system. An organ is vascular when it is richly supplied with blood-vessels.

Venous.—The blood is venous when the corpuscles have given up their free oxygen to the tissues, and the plasma is laden with carbonic acid. The pulmonary *artery* contains venous blood.

Ventricle.—(1) The fleshy part of the heart, subdivided in the rabbit and the fowl ; (2) A cavity of the brain, *e.g.* 4th ventricle, lateral ventricle.

Vesicle.—(1) The *primitive vesicles of the brain* are the anterior enlargements of the neural tube ; (2) The *auditory vesicle* is the auditory sac ; (3) The *germinal vesicle* is the nuclear part of the ovum ; (4) The *umbilical vesicle* is a provisional fœtal organ in the rabbit ; (5) The *vesiculæ seminales* are accessory parts of the male reproductive organs.

Vestibule.—(1) Part of the vertebrate organ of hearing ; (2) That part of the female urinogenital organs of the rabbit which lies just within **the** vulva ; (3) That part of the urinogenital organs of the snail which lies just within the genital aperture ; (4) The ciliated funnel in vorticella and paramœcium.

Villus—A hair-like projection, multitudes of which give the small intestine of the rabbit a shaggy appearance. Villi are generally developed so as to give additional surface, absorptive (intestine) or vascular (placenta).

Yolk.—*Yelk* or *vitellus.* The egg of the fowl consists of (1) an external shell ; (2) An outer shell-membrane ; (3) An inner shell-membrane ; (4) The albumen or white of egg ; (5) The vitelline membrane surrounding (6) the *vitellus* or *yolk*, in one special area of which the *cicatricula or blastoderm*, the embryo, is developed. The great mass of the yolk is composed of *yellow yolk*, consisting of minute *yolk spheres* filled with numerous highly refractive granules. Besides the yellow yolk there are concentric layers and an internal flask-shaped mass of *white yolk*. The **term yolk is** therefore applied (1) to the vitelline mass in which the embryo is formed, (2) to the yolk spheres, which are to be regarded as stored food-stuff for the embryo during growth within the egg.

INDEX OF TYPES

INDEX.

Conjugation—Paramœcium, 352 ; Vorticella, 350.

Connective tissue, histology of, 67 ; development of, 119.

Contractile vesicle—Amœba, 354 ; Monad, 358 ; Paramœcium, 352; Vorticella, 347.

Coracoid, 47, 169.

Cornea, 129.

Cornicula laryngi—Rabbit, 39.

Coronal suture—Rabbit, 144.

Coronoid process—Rabbit, 148.

Corpus, or corpora. Adiposa—Cockroach, 266, Frog, 29. Bigemina ; *see* Optic lobes. Callosum—Rabbit, 218. Mamillare —Rabbit, 218. Quadrigemina — Rabbit, 218. Restiformia, 217. Striata, 219. Trepezoidea—Rabbit, 217.

Corti, fibres of—Rabbit, 135.

Coxa—Cockroach, 261.

Coxopodite—Crayfish, 236.

Crayfish, 231.

Cribriform plate—Rabbit, 142.

Cristæ acusticæ—Rabbit, 134.

Crop—Pigeon, 45.

Crura cerebri, 218.

Crustacea, 231.

Crystalline lens, 129.

Cuticle—Cockroach, 258 ; Crayfish, *see* Exoskeleton ; Earthworm, 282 ; Mussel, *see* Shell ; Snail, *see* Shell ; Vorticella, 347.

Cysticercus—Tapeworm, 334.

Cytozoa, xvi.

DACTYLOPODITE—Crayfish, **236.**

Dart sac—Snail, 301.

Deciduate placenta—Rabbit, **112.**

Dental capsule, 129.

Dental papilla, 128.

Dentary bone—Fowl, **154** ; Frog, **158.**

Dentine, 13, **71.**

Dermis, 82.

Development—Subdivisions of, xix ; Amphioxus, 114 ; Cercaria, 330 ; Cockroach, 275 ; Crayfish, 254 ; Earthworm, 289 ; Hydra, 341 ; Medusa, 344 ; Mussel, 305 ; Pond-snail, 319 ; Vertebrate, 94.

Dextrin, 181.

Diaphragm—Rabbit, 37.

Diaphysis, 121, 165.

Diastema, 148.

Digestion, 186.

Digestive gland—Crayfish, 245 ; Mussel, 309 ; Snail, 295.

Digits. *See* Manus and Pes.

Diplöe, 121.

Disc, Vorticella, 347.

Distribution, xx.

Ductus arteriosus, 211.

Duodenum—Frog, 24 ; Pigeon, 47 ; **Rabbit,** 38.

Dura mater, 44.

EAR. *See* Auditory organ.

Earthworm, 278.

Ecdysis—Crayfish, 231 ; **Frog, 5.**

Ectoderm—Hydra, 341.

Ectosarc—Amœba, 354.

Ectostracum—Crayfish, 238.

Egg membranes, 90.

Ejaculatory duct, 275.

Embryology, Vertebrate, 86.

Embryonic membranes, 108.

Enamel, 13, 71.

—— organ, 128.

Encystment—Amœba, 355 ; Vorticella, 350

Endoderm—Hydra, 341.

Endolymph, 133.

Endomysium, 72.

Endoneurium, 74.

Endophragmal system—Crayfish, 234.

Endopleurite—Crayfish, 234.

Endopodite—Crayfish, 236.

Endosarc—Amœba, 354.

Endosternite—Crayfish, 234.

Endostracum—Crayfish, 238.

Endothelium, 65.

Epiblast. *See* Germinal layers.

Epicranium—Cockroach, 258.

Epidermis, 83.

Epiglottis—Rabbit, 39.

Epimeron—Crayfish, 233.

Epiostracum—Crayfish, 238.

Epiotic—Fowl, 152 ; Frog, 157 ; Rabbit, 146.

Epiphysis, 121, 165.

Epipodite—Crayfish, 236.

Epithelium, 65.

Ethmoidal plane—Rabbit, 144.

Ethmoturbinal—Rabbit, 145.

Eustachian tube or recess—Fowl, 149; Frog, 8, 132, 161 ; Pigeon, 17 ; Rabbit, 14.

—— valve—Rabbit, 208, 213.

Exhalent siphon—Mussel, 306.

Exoccipital—Fowl, 149 ; Frog, 155 ; Rabbit, 141, 160.

Exopodite—Crayfish, 236.

Exoskeleton—Cod, 10 ; Cockroach, 258 ; Crayfish, 238 ; Frog, 7 ; Pigeon, 16 ; Rabbit, 12.

Eye—Cockroach, 272 ; Crayfish, 250 ; Liver-fluke, 327 ; Snail, 298 ; Vertebrate, 129.

FALCIFORM LIGAMENT—Pigeon, 47.

Fallopian tube—Rabbit, 43.

Falx cerebri—Rabbit, 220.

Fat body—Cockroach, 266 ; Frog, 29.

Fat cells, 68.

Feathers—Pigeon, 16, 18.

Femur, 174.

Fenestra ovalis—Fowl, 135, 152 ; Frog, 133, 157 ; Rabbit, 135, 146.

Fenestra rotunda—Fowl, 135 ; Rabbit, 135, 146.

Fertilisation. *See* Impregnation.

Fibrin, 59.

Fibula, 174.

Fibulare, 174.

Filoplume. *See* Feathers.

Fins—Cod, 9.

Fission—Amœba, 354 ; Hydra, 339 ; Paramœcium, 352 ; Vorticella, 350.

Fissure of Sylvius—Rabbit, 219.

Flagellum—Monad, 358 ; Snail, 301.

Floccular fossa—Rabbit, 143, 146.

Flocculus—Rabbit, 217.

Fœtal membranes, 108.

Folding-off of the embryo, 104.

Fontanelle—Frog, 158.

25

PRINTED BY T. AND A. CONSTABLE, PRINTERS TO HER MAJESTY.

AT THE EDINBURGH UNIVERSITY PRESS.